Handbook For Transistors

JOHN D. LENK

Consulting Technical Writer

PRENTICE-HALL, INC., *Englewood Cliffs, New Jersey*

Library of Congress Cataloging in Publication Data

LENK, JOHN D date
Handbook for transistors.

1. Transistors. I. Title.
TK7871.9.L3638 621.3815'28 75-2294
ISBN 0-13-382259-1

To *Irene*, *Karen*, *Mark*, *Lambie*, and *Heather*

10 9 8 7 6 5 4 3 2 1

Printed in the United States of America

PRENTICE-HALL INTERNATIONAL, INC., *London*
PRENTICE-HALL OF AUSTRALIA, PTY. LTD., *Sydney*
PRENTICE-HALL OF CANADA, LTD., *Toronto*
PRENTICE-HALL OF INDIA PRIVATE LIMITED, *New Delhi*
PRENTICE-HALL OF JAPAN, INC., *Tokyo*
PRENTICE-HALL OF SOUTHEAST ASIA (PTE.) LTD., *Singapore*

is called *doping*. Each hole will behave in a manner similar to that of an electron in an *N*-type semiconductor except that the hole will move from the positive to the negative terminal of a power source applied to the semiconductor. Such a semiconductor is called a *P* type since electric current will flow due to positive charges (holes). In this case, the aluminum, gallium, or indium atoms are called *acceptors*.

It is not practical to differentiate between an electron current and a current of holes; however, both types of conduction make rectification in crystal diodes and amplification in transistors possible.

1-2. THE BASIC CRYSTAL JUNCTION (OR DIODE)

When a free electron meets a moving hole in a semiconductor material, the electron occupies the free space and a positive or negative charge no longer exists; that is, the charge is neutralized. When a *P*-type and an *N*-type crystal are joined to make a single semiconductor, as shown in Fig. 1-1, current will flow in one direction only.

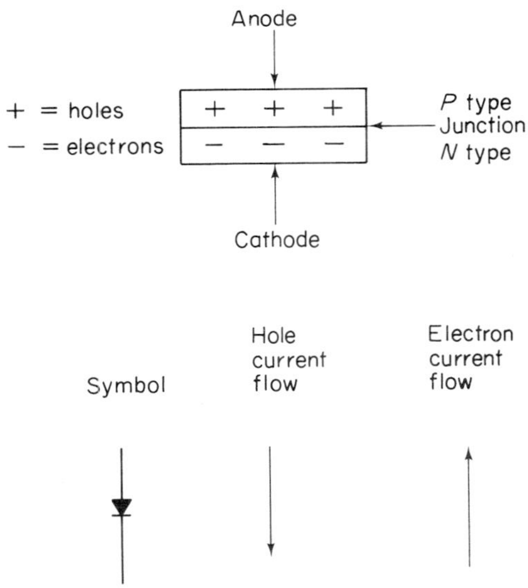

FIGURE 1-1 Joining *P*-type and *N*-type crystals to form the basic crystal junction.

As an example, when a power source is connected to the semiconductor as shown in Fig. 1-2, the semiconductor is said to be *forward biased*. The holes will be repelled toward the junction by the positively charged battery terminal; whereas the electrons are pushed toward the junction by the battery's negative terminal.

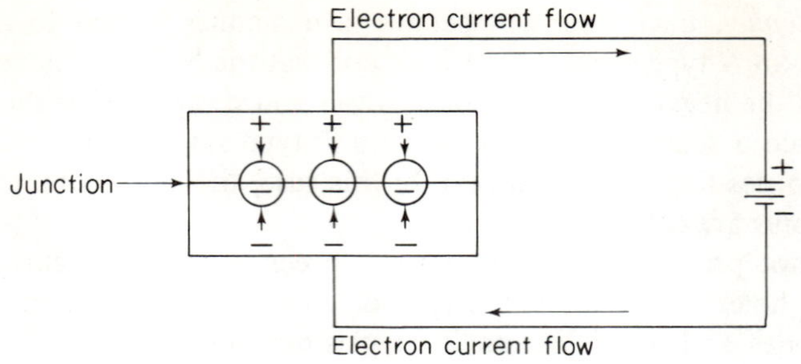

FIGURE 1-2 Forward-bias connection in basic junction (or diode).

At the junction, the electrons combine with the holes. Electrons enter the semiconductor at the *N* terminal to replace the electrons that have combined with the holes. Likewise, electrons leave the *P* terminal by attraction of the positive voltage and create new holes. This movement of electrons, from the negative voltage source through the junction and from the positive terminal of the semiconductor to the positive voltage source, creates a current flow. Thus, current will flow in a semiconductor when the semiconductor is forward biased.

When the polarity of the power source is reversed, as shown in Fig. 1-3, the semiconductor is said to be *reverse biased.* The holes are moved away from the junction by the negative voltage; whereas the electrons are drawn away from the junction by the positive voltage. Thus, there is little or no combining of electrons and holes at the junction, and no current will flow.

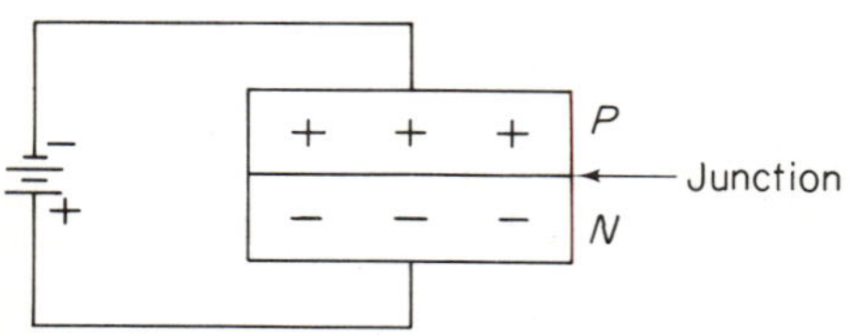

FIGURE 1-3 Reverse-bias connection in basic junction (diode).

In practical terms, there will always be a few electrons and holes near the junction, allowing a very small current to pass. This small current is known as *leakage current* and is usually in the order of a few microamperes (or possibly picoamperes).

When *P*-type and *N*-type regions are formed in the same crystal, the semiconductor is known as a *diode* or *rectifier*. The boundary between the two regions is termed a *junction*. The *P*-region terminal is called the *anode*; whereas the *N*-region terminal is called the *cathode*.

Usually, when such semiconductors are used with signals, the semiconductors are called *diodes* or *signal diodes*. When the device is used for conversion of alternating current (ac) to direct current (dc), the semiconductor is called a *rectifier*.

1-3. THE BASIC TWO-JUNCTION OR BIPOLAR TRANSISTOR

Like a diode, a transistor can be used to prevent (or limit) the flow of current in one direction. The prime use for a transistor, however, is to *control the amount* of current in a circuit. This is done by adding a second junction to the basic diode junction, discussed in Sec. 1-2. For this reason, such transistors are called *two-junction transistors* or, possibly, *bipolar transistors.* The author prefers the former term but recognizes that the latter term is also in common use.

There are two possible arrangements for the two junctions in transistors: *NPN*, where a positive semiconductor material (holes) is placed between two negative semiconductor materials (electrons), and *PNP*, where a negative material (electrons) is placed between two positive materials (holes).

With either junction arrangement, the basic two-junction transistor will have *three* elements. These elements, shown in Fig. 1-4 as an *NPN* arrangement, are the *emitter*, which emits electrons; the *collector*, which collects electrons; and a *base*, which controls the flow of electrons by controlling the charge concentration at the two junctions on either side of the base.

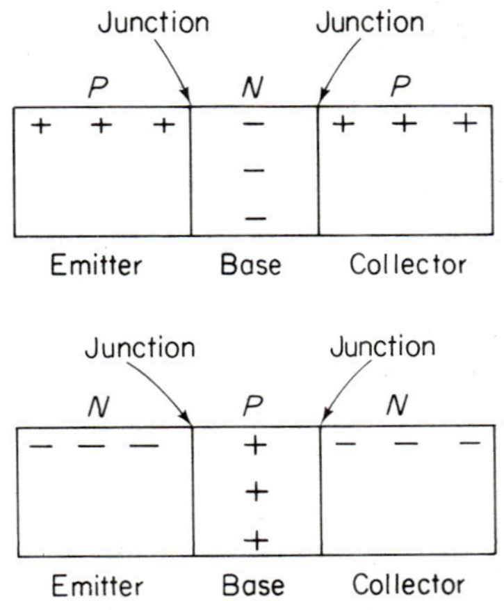

FIGURE 1-4 *NPN* and *PNP* two-junction transistor arrangements.

Figure 1-5 shows how a two-junction transistor operates in its basic circuit. As shown, the emitter-base junction will pass current easily because the junction is forward biased. The collector-base junction will not pass current (except for a small leakage current) since the junction is reverse biased. (The term *back bias* is often used in place of reverse bias.)

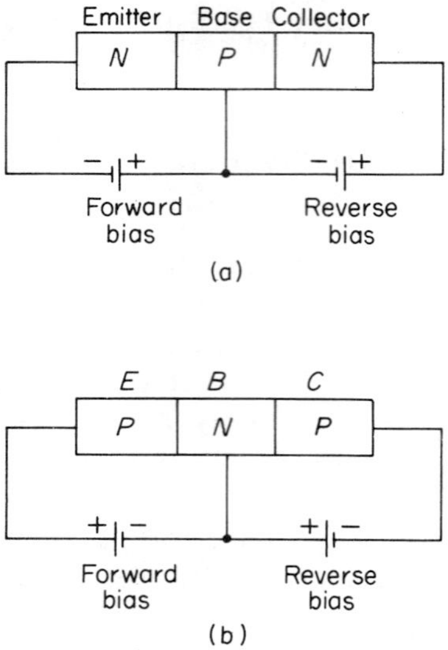

FIGURE 1-5 Basic two-junction transistor bias circuits.

It should be noted that the polarities of bias voltages for an *NPN* transistor differ from those of a *PNP* transistor; however, the net results are the same. For example, as shown in Fig. 1-5(a), the emitter is negative with respect to the base (*NPN*) to produce a forward bias. In Fig. 1-5(b), the emitter is positive with respect to base (*PNP*) to produce the same forward bias. Similarly, the collector has a reverse bias for both *NPN* and *PNP*, even though the polarities are reversed. Also, it should be noted that for normal operation an *NPN* has its base biased *positively* with respect to its emitter. Conversely, a *PNP* base is negative with respect to its emitter.

1-3.1 Basic two-junction transistor operation

During normal operation of a two-junction transistor, current will flow between emitter and base and between emitter and collector but not between collector and base. Most of the current flows between emitter and collector because of the large voltage difference existing between these elements (the sum of the emitter-base voltage and the collector-base voltage). This produces a large number of charge carriers (positive holes in a *PNP* or negative electrons in an *NPN*) that diffuse through this base region when passing from emitter to collector (or vice versa). Few of these charge carriers combine with the charges (positive in *NPN*, negative in *PNP*) in the base.

More charge carriers will be pulled out of the emitter and made available for the collector if the base-emitter current is increased. This can be accomplished by making the base more negative in a *PNP* transistor or by making the base more positive in an *NPN* transistor.

If the base-emitter voltage is decreased, less charge carriers will be pulled from the emitter and less emitter-collector current will flow.

Since very little voltage (approximately 0.2 V for germanium and 0.5 V for silicon) is required to produce a large current flow in the emitter, input power to a transistor is low. Most of the emitter current flows in the collector circuit, where the voltage is made much larger. As a result, a relatively large amount of power can be controlled in an external load (connected in series with the collector circuit) by a small amount of power in the emitter circuit. The power gain of a transistor (the ratio of power output to power input) can be 40,000 or higher in some applications.

1-3.2 Basic NPN *transistor operation*

In an *NPN* transistor, the base-emitter forward bias causes electrons from the emitter to move into the base, as shown in Fig. 1-6(a). Electrons flow through the thin base into the collector region, with only about 10% of the electrons combining with the base holes. The remaining 90% of the electrons are attracted to the positive collector voltage. In most two-junction transistors, the base is kept as thin as possible to maintain the base current at a minimum. This is done by adding a minimum of impurities to the base material. In effect, base material is only slightly *P* or slightly *N*, in comparison to emitter or collector materials that are heavily doped.

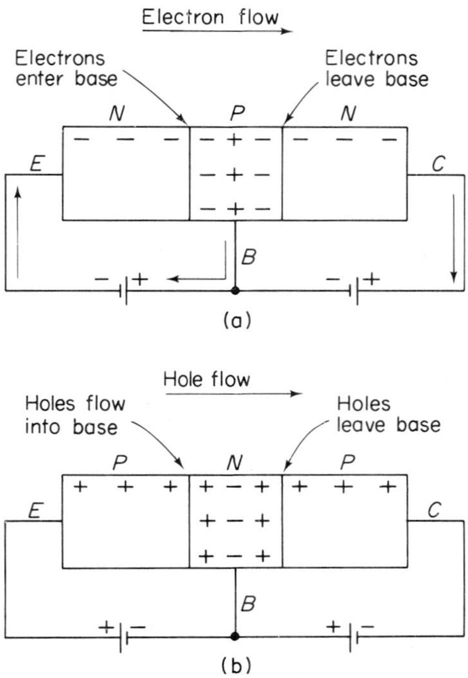

FIGURE 1-6 Basic *NPN* and *PNP* two-junction transistor operation.

1-3.3 Basic PNP *transistor operation*

In a *PNP* transistor, the forward bias between the emitter and base causes holes to flow into the base, as shown in Fig. 1-6(b). The base is relatively thin (and of high purity) so that most of the holes pass through into the collector region. The holes are then attracted to the negative collector voltage, where they are neutralized by electrons from the power source.

1-3.4 Basic two-junction transistor symbols and reference designations

Transistors are indicated by appropriate symbols in schematic diagrams. In rare cases, transistors will be shown by block representations. Figure 1-7 shows the basic two-junction transistor symbols and the corresponding block representations for both *NPN* and *PNP* transistors.

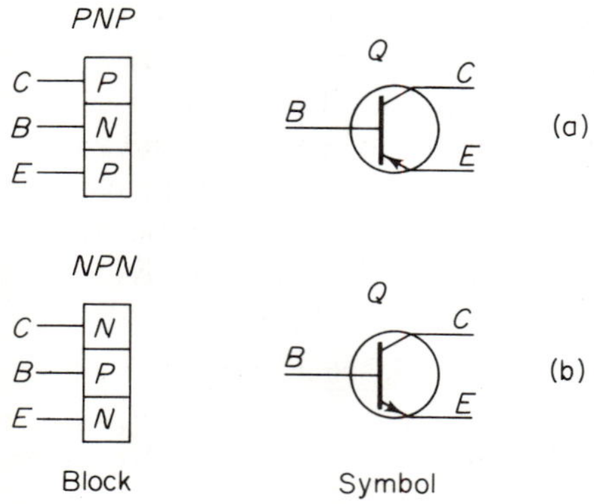

FIGURE 1-7 Basic two-junction transistor symbols and reference designations.

The only difference between an *NPN* and a *PNP* transistor symbol is the arrowhead on the emitter. The arrowhead points away from the base element for an *NPN* and toward the base for a *PNP*.

Usually, the letters *C* (collector), *E* (emitter), and *B* (base) shown in Fig. 1-7 are omitted from schematic diagrams since the elements can be identified from the symbol.

The appropriate letter portion of a transistor reference designation is *Q*. Transistors should be identified as Q_1, Q_2, Q_{101}, Q_{301}, etc.; however, transistors can also be identified by the letter *T* or even *V*, although these letter designations are obsolete.

Two-junction transistors are covered in much greater detail in Chapter 2.

1-4. FIELD EFFECT TRANSISTORS

The basic field effect transistor (or FET) consists of *P* and *N* junctions similar to that of two-junction transistors; however, the operation of an FET is quite different.

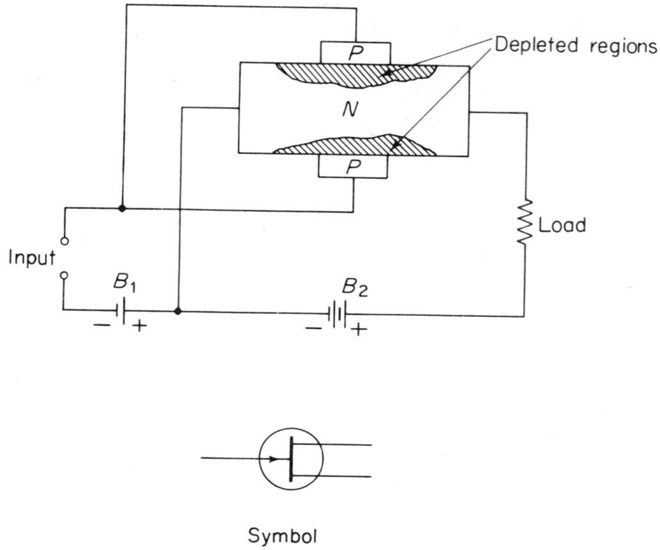

FIGURE 1-8 Basic depletion-type field effect transistor (FET).

As shown in Fig. 1-8, the primary current flow is produced by battery B_2 and flows through the *N* section and the load resistor *R*. In an *N*-type material, the current flow consists primarily of electrons.

Battery B_1 applies a reverse bias to both *P* regions, with respect to the *N* region. The negative battery terminal forms a closed circuit with the *P* regions through the signal-applying circuit at the input. A sine wave signal applied to the input will increase or decrease the reverse-bias voltage applied by B_1.

When reverse bias is applied to the FET, the bias causes current carrier to move away from the *PN* junctions, leaving depletion regions. The greater the encroachment of the depletion region on the *N* region, the less carriers will be present in the *N* region for current conduction. The input signal can add to, or subtract from, the reverse bias and thus move the depletion regions more or less into the *N* region. As a result, the input signal can control the amount of current in the *N* region.

It is possible to control a large amount of current flow with a small input signal. The large current through the *N* region also passes through the load resistor and produces a large voltage drop across the resistor. Thus, a small input voltage variation will produce a large load (or output) voltage variation.

In addition to the *depletion* FET just described, there is also an *enhancement* FET, where the bias enhances or increases the current carriers. Likewise, some FETs operate in both the *depletion and enhancement* modes. Field effect transistors are covered in much greater detail in Chapter 3.

1-5. UNIJUNCTION TRANSISTORS

Unlike the two-junction transistor or the FET, the unijunction transistor (or UJT) is a *negative resistance* device. Under proper conditions, the output current of a UJT can increase, even though the input signal decreases.

The UJT is formed from an *N*-type silicon material to which another element is bonded, forming a *PN* junction, as shown in Fig. 1-9. A UJT has one emitter lead, two base leads, but no collector lead.

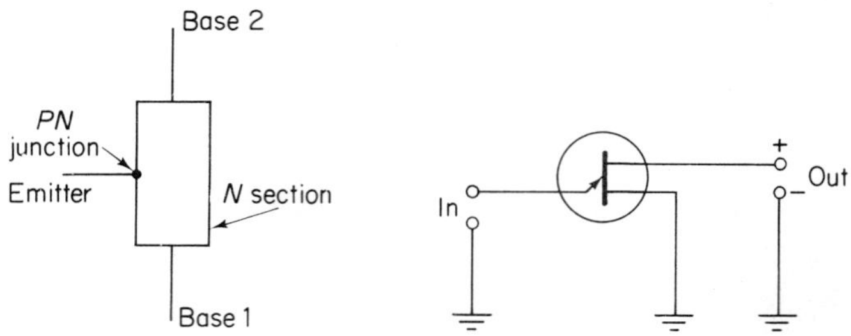

FIGURE 1-9 Basic unijunction transistor (UJT).

When the voltage applied to the emitter is in reverse-bias form (or zero voltage), the base material acts as a conventional resistor. Because the emitter taps this "resistance," a voltage difference exists between the emitter and ground.

When a forward bias is applied to the emitter, the usual transistor junction characteristics are formed and the resistance between the emitter and base 1 decreases, causing a current increase. The emitter voltage can then be decreased, but with a continued increase in current flow. In effect, once the UJT is "turned on," it will not "turn off" until the circuit is broken or the input voltage is completely removed.

Unijunction transistors are covered in much greater detail in Chapter 4.

1-6. BASIC PHYSICAL CHARACTERISTICS OF TRANSISTORS

All types of transistors (two-junction, FET, UJT) are housed within cases and provided with terminal leads for connection to external circuits. Cases are required to protect the semiconductor material. Case leads are required since it is not practical to connect the external circuits directly to elements (collector, base, emitter, etc.) on the semiconductor material (referred to as the pellet, wafer, block, or chip, depending on style and manufacturer).

Figure 1-10 shows the outline, mounting dimensions, and terminal

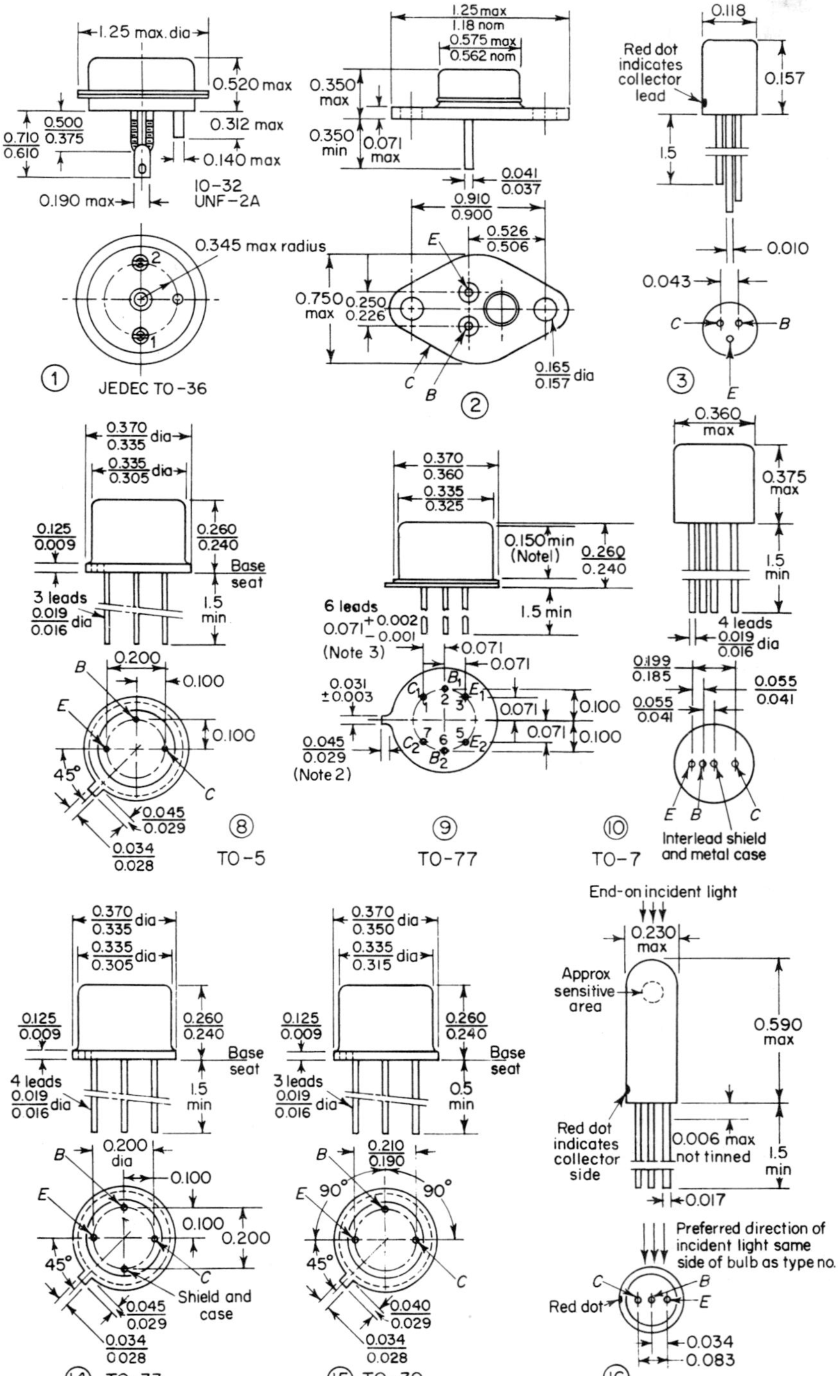

FIGURE 1-10 Outlines, dimensions, and terminal lead arrangements for some typical transistor cases. (Lenk, *Practical Semiconductor Databook for Electronic Engineers and Technicians*, Prentice-Hall.)

lead arrangements for some typical transistor cases. Many other case styles are available, but those shown in Fig. 1-10 represent a cross section of cases in common use.

The cases are generally metal or plastic (or possibly ceramic in isolated instances). Power transistors use metal cases that act as a *heat sink* to dissipate and carry away heat from the semiconductor material. As discussed in Sec. 1-7, power transistors are often used with external heat sinks attached to the metal cases. In other applications, the metal cases are mounted directly on metal chassis to help dissipate the heat.

Note that the transistor cases may have any number of terminal leads; however, terminal leads are usually arranged in groups of two, three, four, or possibly six where more than one transistor is contained within one case.

The TO-5 case style is typical for small, low-power transistors (less than 1-W power dissipation). The collector, emitter, and base are provided with one terminal lead each but are electrically isolated from the metal case.

The TO-33 is similar to the TO-5, except that four leads are available. The fourth lead provides for electrical connection to the case and a shield inside the case. A typical application for this arrangment is where the transistor must be mounted on a nonconducting surface (such as a plastic component board) and a good ground is required for the case (to minize pickup of stray signals, etc.).

The case styles labeled 1 and 2 in Fig. 1-10 are typical of large, high-power transistors. Here, only two terminal leads are provided. The case and its mounting provisions are connected directly to the third transistor element. For example, in case style 2, the collector is connected directly to the metal case. As discussed in Chapter 6, power transistors used in *RF* work often have the collector or emitter connected directly to the case.

In case style 1, the case is held onto the mounting surface by a nut and lockwasher threaded on the stud at the bottom center. Case style 2 is attached by means of hardware passing through the two holes at either end of the case mounting flange.

A power supply problem is created when transistors are designed with one element connected directly to the case or mounting. Typically, chassis or mounting surfaces are connected to the negative side of the power supply. *NPN* transistors are typically operated with the collector at a voltage more positive than the emitter. Thus, if the collector of an *NPN* is mounted on a negative chassis, the transistor will not operate (or will be destroyed). For this reason, *PNP* is generally used where the collector is connected to the case. Likewise, *NPN* is used where the emitter is connected to the case.

Transistor datasheets usually include drawings that show outline dimensions and terminal lead arrangements. The datasheets include any special information such as connections between transistor elements and case.

1-7. TRANSISTOR TEMPERATURE RELATIONSHIPS

There are two basic temperature-related problems in working with transistors of all types.

First, datasheets specify transistor parameters (such as gain and frequency range) at a given temperature. Many of these parameters will change with temperature. Since transistors rarely operate at the exact temperature shown on the datasheet, it is important to know the parameters at the actual operating temperature. To compound this problem, changes in parameters can affect transistor temperature. For example, an increase in current gain or power dissipation will usually result in a temperature increase. For these reasons, Chapters 2, 3, and 4 contain information on the relationship between temperature and operating parameters of the transistors.

Second, in addition to knowing the effects of temperature on parameters, it is important to know how *heat sinks* or transistor mounting can be used to offset the effects of temperature. For example, if a transistor is used with a heat sink or is mounted on a metal chassis that acts as a heat sink, an increase in temperature (from any cause) can be dissipated into the surrounding air.

The following paragraphs describe methods for determining the proper dissipation characteristics of transistor cases and heat sinks.

1-7.1 Effects of temperature on power dissipation

Silicon transistors are less temperature-sensitive than germanium transistors. Similarly, silicon transistors can be operated at higher temperatures than germanium transistors. As a guideline, *do not exceed 100°C for germanium transistors or 200°C for silicon transistors.* These guidelines apply no matter what the actual power dissipation may be in a given transistor.

The power dissipation capabilities of a transistor must be carefully considered when using any type of transistor. Of course, in small-signal circuits, the power dissipation is usually less than 1 W and heat sinks are not needed. In such circuits, the only concern is that the rated power dissipation (as shown on the datasheet) not be exceeded. As a guideline, *do not exceed 90% of the maximum power dissipation, as shown on the datasheet, for small-signal circuits.*

Transistor manufacturers specify maximum power dissipation in a variety of ways on their datasheets. Some manufacturers provide *safe operating area curves* for temperature and/or power dissipation of transistors. Other manufacturers specify *maximum power dissipation*, in relation to a given ambient temperature or a given case temperature. Still others specify a *maximum junction temperature* or a *maximum case temperature.*

1-7.2 Thermal resistance

Transistors designed for power applications usually have some form of *thermal resistance* specified to indicate the power dissipation capability. Thermal resistance can be defined as *the increase in temperature of the transistor junction* (*with respect to some reference*) *divided by the power dissipated,* or °C/W. For example, if the temperature increases 30°C for each watt dissipated, the thermal resistance is 30°C/W.

In power transistors, thermal resistance is normally measured from the transistor semiconductor material junction to the case. This results in the term θJC. On those transistors where the case is bolted directly to the mounting surface with an integral threaded bolt or stud, the terms θMB (thermal resistance to mounting base) or θMF (thermal resistance to mounting flange) are used. These terms take into account only the thermal paths from junction to case or mount. For power transistors in which the semiconductor material is mounted directly on a header or pedestal, the total internal thermal resistance from junction to case (or mount) varies from about 50°C/W to less than 1°C/W.

Datasheets for power transistors usually specify thermal resistance at a given temperature. For each increase in temperature from this specified value, there is a change in temperature-dependent characteristics of the transistor. Since there is a change in temperature with changes in power dissipation, the junction-to-ambient temperature will also change, resulting in a characteristic change. Thus, the transistor characteristics can change with ambient temperature changes and with change produced by variations in power dissipation.

1-7.3 Thermal runaway

The main problem in operating a two-junction transistor at or near its maximum power dissipation limits is a condition known as *thermal runaway*. When current passes through a transistor junction, heat is generated in the semiconductor material. Unless all this heat is dissipated by the case (an impossibility), the junction temperature increases. This, in turn, causes more current to flow through the junction, even though the voltage, circuit values, etc., remain the same. In turn, this causes the junction temperature to increase even further, with a corresponding increase in current flow. If the heat is not dissipated by some means, the two-junction transistor will burn out and be destroyed.

Thermal runaway is not a problem with FETs since FETs usually have a negative thermal coefficient. That is, current flow decreases with increases in temperature. Likewise, UJT characteristics are not drastically affected by temperature.

1-7.4 Operating transistors without heat sinks

If a transistor is not mounted on a heat sink, the thermal resistance from case-to-ambient air, or θCA, is so large in relation to that from junction to case (or mount) that the total thermal resistance from junction to ambient air, or θJA, is primarily the result of the θCA term.

Figure 1-11 lists values of case-to-ambient air thermal resistances for a number of popular transistor cases. It will be seen that the large, heavy-duty cases such as TO-3 will have a small temperature increase (for a given wattage) in comparison to such cases as the TO-5. That is, the heavy-duty cases will dissipate the heat into the ambient air.

Case	θCA (C°/W)
TO-3	30
TO-5	150
TO-8	75
TO-18	300
TO-36	25
TO-39	150
TO-46	300
TO-60	70
TO-66	60

FIGURE 1-11 Case-to-ambient thermal resistance for typical transistor cases.

The information in Fig. 1-11 can be used to approximate the maximum power dissipation of transistors (without heat sinks) when such information is not shown on the datasheet. Assume that a germanium transistor with a TO-5 case is to be used and that it is desired to find the absolute maximum power dissipation (without a heat sink).

The θCA factor for a TO-5 is 150, as shown in Fig. 1-11. As discussed in Sec. 1-7.1, germanium transistors should not be operated above 100°C. Assuming a 25°C ambient temperature, the transistor temperature must not be allowed to increase more than 75°C maximum.

$$\text{Maximum power dissipation} = \frac{\text{maximum allowable temperature increase}}{\theta CA} = \frac{75}{150} = 0.5 \text{ W}$$

Thus, the TO-5 case could dissipate 0.5 W and permit only a 75°C increase in temperature. As discussed in Chapter 2, however, current gain for a two-junction transistor will double when the temperature is raised from 25 to 100°C. Assuming that the voltage remains constant, the maximum dissipation allowable would be 0.25 W. This would be an absolute maximum figure, assuming a germanium transistor, a TO-5 case, and an ambient temperature of 25°C. For practical design purposes, the 0.25-W figure would be safe if the case were mounted on a metal chassis, which would act as a heat sink.

1-7.5 Operating transistors with heat sinks

After about 1 W (or less) it becomes impractical to increase the size of the case to make the case-to-ambient thermal resistance term comparable to the junction-to-case term. For this reason, most power transistors are designed for use with an external heat sink. Sometimes, the chassis or mounting area serves as the heat sink. In other cases, a heat sink is attached to the case. Either way, the primary purpose of the heat sink is to increase the effective heat-dissipation area of the case and to provide a low heat-resistance path from case to ambient.

To design (or select) a heat sink for a given application properly, the thermal resistance of both the transistor and heat sink must be known. For this reason, power transistor datasheets will specify the θJA that must be combined with the heat sink thermal resistance to find the total power dissipation capability. (Note that some power transistor datasheets specify a *maximum case temperature* rather than θJA. Maximum case temperature can be combined with heat sink thermal resistance to find maximum power dissipation.)

Heat sinks are generally used with two-junction transistors. This is because most FETs and UJTs are low-power devices, operating with less than 1-W power dissipation. The following information, however, applies to heat sinks used with any type of transistor.

Heat sink capabilities. The thermal resistance of a heat sink can be calculated if the following factors are known: material, mounting provisions, exact dimension, shape, thickness, surface finish, and color. Even if all of these factors are known, the thermal resistance calculations are approximate.

As a rough guideline,

$$\text{Heat sink thermal resistance (in °C/W)} = \sqrt{\frac{1500}{\text{area}}}$$

where the area (total area exposed to the air) is in square inches, the material is 0.125-in. thick aluminum, and the shape is a flat disc.

From a practical standpoint, it is better to accept the manufacturer's specifications for a heat sink. The heat sink thermal resistance actually consists of two series elements: the thermal resistance from the case to the heat sink that results from conduction (θCS) and the thermal resistance from the heat sink to the ambient air caused by convection and radiation (θSA).

Heat sink ratings. Commercial fin-type heat sinks are available for various transistor case sizes and shapes. Such heat sinks are especially useful when the transistors are mounted in Teflon sockets, which provide no thermal conduction to the chassis or printed-circuit board.

Commercial heat sinks are rated in terms of thermal resistance, usually in terms of °C/W. When heat sinks involve the use of washers (between the sink and transistor), the °C/W factor usually includes the thermal resistance between the transistor case and sink, or θCA. Without a washer, only the sink-to-ambient, or θSA, thermal resistance factor is given. Either way, the thermal resistance factor represents temperature increase (in degrees Celsius) divided by wattage dissipated.

For example, if the heat sink temperature rises from 25 to 100°C (a 75°C increase) when 25 W are dissipated, the thermal resistance is $\frac{75}{25} = 3$. This could be listed on the datasheet as 3 θSA or simply as 3°C/W.

All other factors being equal, the heat sink with the lowest thermal resistance (°C/W) is best; that is, a heat sink with 1°C/W is better than a 3°C/W heat sink. Of course, the heat sink must fit the transistor case and the space around the transistor. Except for these factors, selection of a suitable heat sink should be no particular problem.

Practical heat sink considerations. To operate a transistor at its full power capabilities, there should be no temperature difference between the case and ambient air. This will occur only when the thermal resistance of the heat sink is zero and the only thermal resistance is that between the junction and case. It is not practical to manufacture a heat sink with zero resistance. The greater the ratio θJC/θCA, however, the nearer the maximum power limit (set by θJC) can be approached.

When transistors are to be mounted on heat sinks, some form of electrical insulation must be provided between the case and the heat sink (unless a grounded-collector or grounded-emitter circuit is used). Because good electrical insulators usually are also good thermal insulators, it is difficult to provide electrical insulation without introducing some thermal resistance between case and heat sink. The best materials for this application are mica, beryllium oxide (Beryllia), and anodized aluminum. A comparison of the properties of these three materials (for case-to-heat sink insulation of a TO-3 case) is as follows:

Material	Thickness (in.)	θCS (°C/W)	Capacitance (pF)
Beryllia	0.063	0.25	15
Anodized aluminum	0.016	0.35	110
Mica	0.002	0.4	90

Note that a "capacitance" column appears in the table. Any isolation between collector (or emitter) and the chassis (as produced by the washer between the case and heat sink) will also result in capacitance between the two metals. This capacitance can be a problem and must be considered in *RF* circuit design; however, the problem of collector- (or emitter-)to-chassis capacitance is usually noticed only at frequencies above 100 MHz.

Small, general-purpose transistors such as those with a TO-5 case are not generally used with fin-type heat sinks. For these transistors, a beryllium oxide washer can be used to provide insulation between the case and a metal chassis or printed-circuit board. The larger, power-type transistors such as those with a TO-3 case may use fin-type heat sinks, or they can also be mounted on a metal chassis. For these transistors, a beryllium cup should also be used between the transistor and heat sink. Figure 1-12 shows both types of mounting.

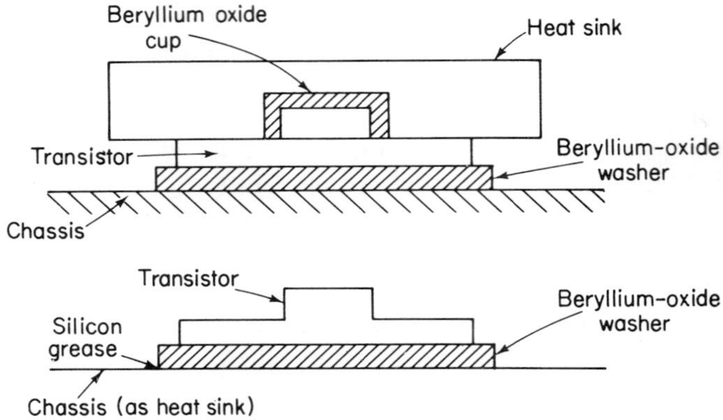

FIGURE 1-12 Typical mounting arrangements for transistor heat sinks. (Lenk, *Handbook of Simplified Solid-State Circuit Design*, Prentice-Hall.)

1-7.6 Calculating power dissipation under steady-state operating conditions

For design purposes, the no-signal dc voltage and current can be used to calculate power dissipation when a transistor is operated under steady-state conditions. The collector voltage and current are used for two-junction transistors. The drain voltage and current are used for FETs. There are other currents that result in power dissipation (collector-base leakage, emitter-base current, etc.). These can be ignored, however, and the power dissipation (in watts) can be considered as the dc collector (or drain) voltage times the collector current.

Once the power dissipation has been calculated, the *maximum power dissipation capability* must be found. Under steady-state conditions, the maximum dissipation capability is dependent on three factors: the sum of the series thermal resistances from the transistor junction to ambient air, the maximum junction temperature, and the ambient temperature.

The following are some examples of how power dissipation can be calculated.

Some power transistor datasheets specify a maximum case temperature. Assume that it is desired to find the maximum power dissipation for such transistors under the following conditions: ambient temperature 25°C, heat sink thermal resistance 3°C/W, and a maximum case temperature of 130°C.

Subtract the ambient temperature from the maximum permitted case temperature:

$$130°\text{C} - 25°\text{C} = 105°\text{C}$$

Divide the case temperature by the heat sink thermal resistance:

$$\tfrac{105}{3} = 35 \text{ W maximum power}$$

Note that the case temperature of any transistor can be calculated by

$$\text{Maximum case temperature}$$
$$= \text{maximum power} \times (\theta\text{SA}) + \text{ambient temperature}$$

Other power transistor datasheets specify a maximum junction temperature and a junction-to-case thermal resistance (θJC). Assume that it is desired to find the maximum power dissipation for such transistors under the following conditions: ambient temperature 25°C, heat sink thermal resistance 3°C/W, junction-to-case thermal resistance 2°C/W, and a maximum junction temperature of 200°C (typical for a silicon power transistor).

Note that the thermal resistance of the heat sink includes any thermal resistance produced by the washer between the transistor case and heat sink. (If this factor is not known, add 0.5°C/W thermal resistance for any washer between case and heat sink. This thermal resistance is high and allows a safe tolerance.)

Find the total junction-to-ambient thermal resistance θJA:

$$\theta\text{JA} = \theta\text{JC} + \theta\text{SA} \quad \text{or} \quad 5 = 2 + 3$$

Find the maximum permitted power dissipation:

$$\text{Maximum power}$$
$$= \frac{\text{maximum junction temperature} - \text{ambient temperature}}{\theta\text{JA}}$$
$$= \frac{200 - 25}{5} = 35 \quad \text{(maximum)}$$

If the transistor is to be used without a heat sink but under the same conditions and with a TO-3 case, the maximum power is calculated as follows:

Find the total junction-to-ambient thermal resistance θJA.

$$\theta JA = \theta JC + \theta CA \quad \text{or} \quad 32 = 2 + 30$$

The value of 30 for a TO-3 case is obtained from Fig. 1-11.

Next, find the maximum permitted power dissipation:

$$\frac{200 - 25}{32} = 5 \text{ W} \quad \text{(approx)}$$

1-7.7 Calculating power dissipation under pulse operating conditions

When transistors are operated by pulses, the maximum permitted power dissipation is much greater than with steady-state operation. The following are some examples of power dissipation calculations for both nonrepetitive and repetitive pulses.

For a single, nonrepetitive pulse, the transient thermal resistance must be calculated. Usually, the transient thermal resistance is shown on the safe operating area curves of the datasheet in the form of a *power multiplier* (given for a specific case temperature and a given pulse width).

For example, a 2N3055 transistor has power multipliers of 2.1 for 100-ms pulses, 3.0 for 1-ms pulses, and on up to 7.7 for 30-μs pulses.

The steady-state (or direct-current) maximum power is multiplied by these factors to find the maximum permitted single pulse power. Assuming that the maximum permitted steady-state power is 100 W for a given set of conditions, then the single-pulse maximum power would be 300 W if a 1-ms pulse is used.

Usually, the datasheets specify the power multipliers for a given case temperature. If the case temperature is increased, the factor must be derated. The temperature derating factor is found by

$$\text{Derating factor} = \frac{\text{case temperature} - \text{ambient temperature}}{\text{max junction temperature} - \text{ambient temperature}}$$

For example, assume that a transistor has a maximum junction temperature of 200°C, that the case temperature under these conditions is 130°C, that the ambient temperature is 25°C, and that the maximum steady-state power is 35 W. Further, assume that the transistor datasheet specifies a power multiplier of 3.0 for 1-ms pulses, with a case temperature of 25°C. The derating factor is

$$\frac{130°\text{C} - 25°\text{C}}{200°\text{C} - 25°\text{C}} = 0.6$$

To find the maximum single pulse power,

$$\begin{aligned} &\text{Maximum power} \\ &= \text{multiplier} \times (1 - \text{derating factor}) \times \text{steady-state power} \\ &= 3 \times (1 - 0.6) \times 35 = 42 \text{ W} \end{aligned}$$

These calculations for single-pulse operation are based on the assumption that the heat sink capacity is large enough to prevent the heat sink temperature from rising between pulses.

For repetitive pulses, both the case and heat sink temperatures will rise. This temperature increase must be taken into account when determining the maximum power dissipation.

The maximum permitted power dissipation for a transistor operated with repetitive pulses is calculated by

$$\text{Maximum permitted power} = \frac{\text{multiplier} \times (\text{max junction temperature} - \text{ambient temperature}}{\theta\text{JC} + (\text{multiplier} \times \text{duty cycle }\% \times \theta\text{JA}}$$

Assume that it is desired to find the maximum permitted power of the same transistor described for steady-state operation but now operated by 1-ms pulses repeated at 100-Hz intervals. The conditions are power multiplier, 3.0; θJC 2°C/W; θJA 5°C/W (including heat sink); maximum junction temperature, 200°C; ambient temperature, 25°C; and a duty cycle of 10% (1 ms on and 9 ms off).

$$\text{Maximum permitted power} = \frac{3(200 - 25)}{2 + (3 \times 0.1 \times 5)} = 150\text{ W}$$

The case temperature of a transistor operated by repetitive pulses is

$$\text{Case temperature} = \text{peak pulse power} \times \text{duty cycle }\% \times \theta\text{JA} + \text{ambient temperature}$$

Peak-pulse power is obtained by multiplying the collector (or drain) voltage by collector (or drain) current. This assumes that the transistor is operated in a grounded-emitter or grounded-base configuration and that the transistor is switched full-on and full-off by the pulses.

For example, assume an ambient temperature of 25°C, a duty cycle of 10%, a total θJA of 5°C/W, and peak pulses of 120 W (say a collector voltage of 60 V and a collector current of 2 A). The case temperature is

$$\text{Case temperature (or } T_C) = (120 \times 0.1 \times 5) + 25 = 85°\text{C}$$

1-8. TRANSISTOR PARAMETER DATA

In Chapters 2, 3, and 4, transistor characteristics and test procedures are described from the practical viewpoint. The information in this section supplements the practical procedures by providing a summary of transistor operating characteristics (or parameters) from the theoretical approach. While a technician is not usually called upon to solve design problems with the aid of transistor parameters, a senior laboratory technician must be familiar with the nomenclature of all commonly used transistor parameters in order to interpret datasheets properly.

Transistor parameters are vaguely similar to vacuum-tube parameters, in that they both show the relationship of output to input under various conditions; however, transistor parameters are far more complex. The major reason for this is that transistor input–output factors depend on one another, even when output terminals are open- or short-circuited to ac signals. In effect, transistor input and output circuits are *interconnected resistance networks*.

Current and voltages at transistor inputs and outputs are different under open- or short-circuited conditions because there is interaction between circuit elements under both conditions, and feedback of energy from output to input occurs when signals are passing through. This feedback produces what is known as *transfer circuit constants* under both open- and short-circuited conditions of the input and output terminals. These parameters are called forward and reverse transfer resistance, forward and reverse conductance or admittance, forward and reverse amplification factor, etc. As a result, two entire sets of open- and short-circuit voltage, current, and impedance relationships exist (one for input and one for output).

To further confuse matters, each manufacturer has his own set of letters and subscripts to describe the open- and short-circuit parameters. In many cases, the parameters are mixed on the datasheets; that is, a single datasheet will contain combinations of parameters taken from more than one parameter system.

To reduce confusion in interpreting transistor datasheets, the technician must

1. Have an understanding of the commonly used parameters.
2. Have a means of converting from one parameter to another.
3. Have a means of determining parameters at different frequencies, temperatures, etc., when only one condition is specified on the datasheet.

Since the effects of temperature, frequency, etc., are different for different types of transistors (two-junction, FET, UJT), these parameter relationships are discussed in the corresponding chapter (Chapters 2, 3, and 4, respectively). In this section, we shall concentrate on understanding the commonly used parameters and show how one parameter can be converted to another.

1-8.1 Understanding commonly used transistor parameters

There are a number of systems used to define and measure transistor parameters. The most commonly used include *r* or resistance, *h* or hybrid, *y* or admittance, *z* or impedance, *s* or scattering, and *g* or transconductance.

We are mostly concerned with the h parameters (for two-junction transistors) and y parameters (for FETs).

***r* parameters.** The r or *resistance* parameters are not in common use today. This is because r parameters do not take into account any gain between input and output; however, r parameters do provide a foundation for understanding other parameter systems.

A two-junction transistor can be considered as a three-terminal resistor network as shown in Fig. 1-13. A T network is formed by emitter resistance r_e, collector resistance r_c, and base resistance r_b. Since one of the terminals (the base terminal in this case) is common to both input and output, a more practical representation is the four-terminal networks of Fig. 1-14.

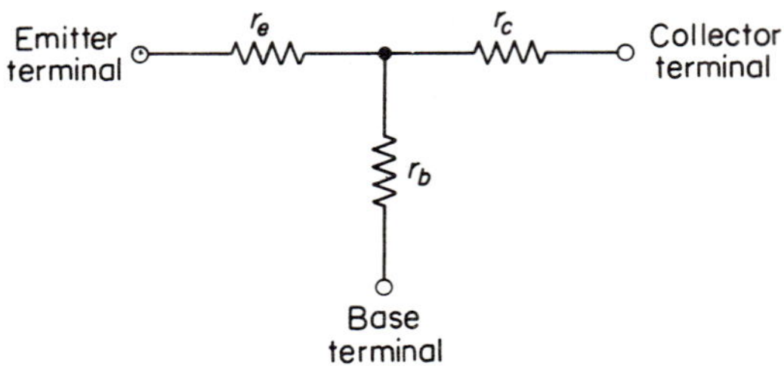

FIGURE 1-13 Basic three-terminal resistance network equivalent of a transistor. (Lenk, *Practical Semiconductor Databook for Electronic Engineers and Technicians*, Prentice-Hall.)

The networks of Fig. 1-13 are passive, and the resistance values represented are determined by dc measurement. The networks of Fig. 1-14 are active in that they show a generator (G) in the output to indicate the amplifying function of a transistor. The networks of Fig. 1-14 are a close approximation of true transistor characteristics at low frequencies and at dc but not at high frequencies.

As frequency increases, the intercapacitance of transistors produces impedance. Also, the external circuits in which the transistors are used affect the network. For this reason, it is more convenient to consider transistor networks as "black boxes" similar to that shown in Fig. 1-15. To analyze the black box, the voltages and currents at input and/or output are measured or signals are applied to input/output, and the signal effect is evaluated.

No matter what parameter system is used, I_1 indicates input current, V_1 indicates input voltage, I_2 indicates output current, and V_2 indicates output voltage.

In considering r parameters, the concern is with the effects of resistance (or voltage divided by current) since $r = E/I$.

For example, r_{11} indicates input resistance, obtained by using values of V_1 and I_1, with the output open-circuited. The first 1 of the subscript indicates V_1 voltage and the second 1 of the subscript indicates the I_1 current.

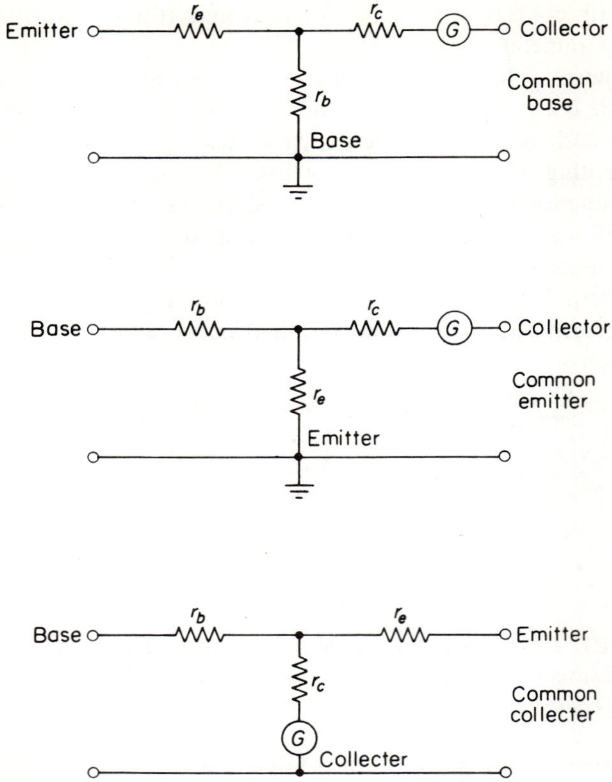

FIGURE 1-14 Equivalent transistor network for dc and low-frequency operation. (Lenk, *Practical Semiconductor Databook for Electronic Engineers and Technicians*, Prentice-Hall.)

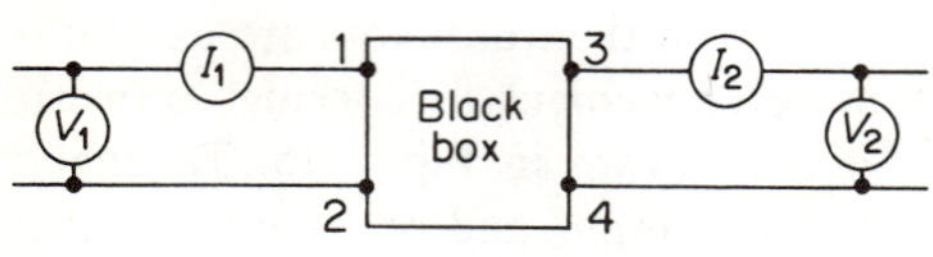

FIGURE 1-15 Equivalent black-box representation of transistor parameters. (Lenk, *Practical Semiconductor Databook for Electronic Engineers and Technicians*, Prentice-Hall.)

r_{21} indicates forward transfer resistance, obtained using values of V_2 and I_1. Here, a test voltage V_1 is applied to the input. The current I_1 and the output voltage V_2 are measured.

r_{12} indicates reverse transfer resistance, or V_1/I_2.

r_{22} indicates output resistance, or V_2/I_2.

Note that the double-number subscript implies that the first variable is always divided by the second.

***h* parameters.** The h or hybrid system of parameters is the most popular method to indicate two-junction transistor characteristics (at least this is true on datasheets). Unlike r parameters (which are measured under open-circuit conditions), the h parameters are based on combinations of constant current and constant voltage, in both open- and short-circuit conditions.

Using the black box of Fig. 1-15, the four basic h parameters are

$h_{11} = V_1/I_1$ (input resistance, output terminals 3 and 4 shorted, $V_2 = 0$)
$h_{12} = V_1/V_2$ (reverse voltage ratio, input terminals 1 and 2 open, $I_1 = 0$)
$h_{21} = I_2/I_1$ (forward current ratio, output terminals 3 and 4 shorted, $V_2 = 0$)
$h_{22} = I_2/V_2$ (output conduction, input terminals 1 and 2 open, $I_2 = 0$)

The use of number subscripts for h parameters has been replaced (in many practical applications) by letter subscripts.

The first subscript indicates the characteristic: i for input, o for output, f for forward transfer, and r for reverse transfer.

The second subscript indicates the circuit configuration: b for common base, c for common collector, and e for common emitter.

Using this system, h_{11} or input resistance can be indicated by h_{ib} for common-base input, h_{ic} for common collector, and h_{ie} for common emitter.

h_{12} or reverse transfer voltage ratio can be indicated by h_{rb}, h_{rc}, and h_{re} for common base, collector, and emitter, respectively.

h_{22} or output conductance can be indicated by h_{ob}, h_{oc}, and h_{oe} for common base, collector and emitter, respectively.

h_{21} or forward current transfer ratio (usually the most important parameter) can be indicated by h_{fb}, h_{fc}, and h_{fe} for common base, collector, and emitter, respectively.

When h parameters were first used, the common-base configuration was the most popular. Today, the common-emitter configuration is used to the greatest extent. For this reason, practically all two-junction transistor datasheets list h_{fe}. Some datasheets mix the configurations. For example, a typical two-junction transistor datasheet lists h_{fe}, h_{ib}, h_{rb}, and h_{ob}.

Practical hybrid parameter applications. Figure 1-16 shows the basic grounded-emitter circuit used for most h parameters, together with the hybrid equivalent circuit.

Figure 1-16 also shows the equations for the most important transistor circuit characteristics: A_i or signal current gain, R_i or input resistance, A_e or A_v signal voltage gain, and A_p or signal power gain. Note that these equations use the letter subscripts rather than the number subscripts. This matches most two-junction transistor datasheets.

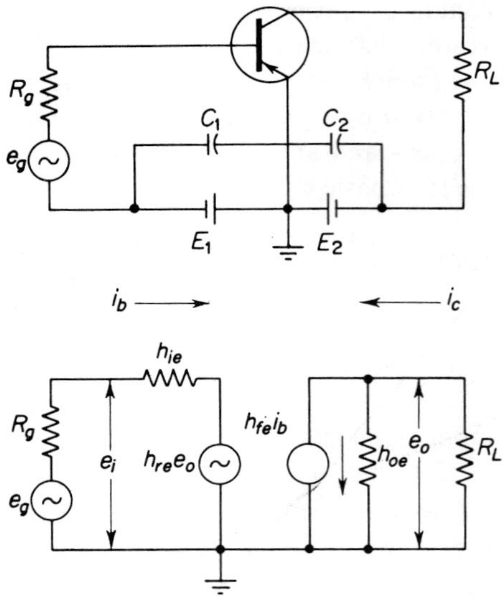

h_{oe} = Output admittance or conductance
h_{ie} = Emitter input resistance
h_{re} = Reverse voltage transfer ratio
h_{fe} = Forward current ratio

$A_i = h_{fe}/1 + h_{oe}R_L$ = Signal current gain

$$R_i = h_{ie} - \frac{h_{fe}h_{re}R_L}{1 + h_{oe}R_L} = \text{Input resistance}$$

$$A_e = \frac{e_o}{e_i} = \frac{1}{h_{re} - \dfrac{h_{ie}}{R_L}\,\dfrac{1 + h_{oe}R_L}{h_{fe}}} = \text{Voltage amplification}$$

$A_p = A_e A_i$

FIGURE 1-16 Grounded emitter and equivalent hybrid circuit. (Lenk, *Practical Semiconductor Databook for Electronic Engineers and Technicians*, Prentice-Hall.)

The equations of Fig. 1-16 differ somewhat from those of Sec. 1-8.2; however, the net results are the same. The equations of Fig. 1-16 are based on a common-emitter circuit. The same equations apply to common-base or common-collector circuits; however, the second subscript *e* must be changed to *b* or *c*, as applicable.

The following is an example of how *h* parameters can be applied to a practical problem. Assume that it is desired to find signal current gain for a two-junction transistor with an h_{fe} of 55 and an h_{oe} of 0.1 μmho, using a 10,000-Ω load resistor.

Using the equation of Fig. 1-16, A_i is

$$A_i = \frac{55}{1 + (0.1 \times 10^{-6}) \times 10{,}000} = 54.9$$

Using the equation of Sec. 1-8.2, A_i is

$$A_i = \frac{55 \times 0.0001}{(0.1 \times 10^{-6}) + 0.0001} = 54.9$$

***y* parameters.** In general, y parameters are used with FETs. This is particularly true when FETs are used in *RF* applications. For this reason, y parameters are discussed in Chapters 3 and 6.

***z* and *g* parameters.** Although used to some extent in circuit design, z and g parameters are not of great importance to technicians and are not discussed in detail. Since it may be necessary for a technician to convert from one of these parameters to the more popular h- or y-parameter systems, however, Sec. 1-8.2 contains the conversion data.

***s* parameters.** The s or *scattering*-parameter system is used primarily for design of high-frequency circuits. Since the s-parameter system is rarely used on existing transistor datasheets, it will not be discussed here. However, Fig. 1-17 shows the signal flow diagram that forms the basis for s-parameters. Likewise, conversion data for the s-parameter and h-parameter systems are included in Sec. 1-8.2.

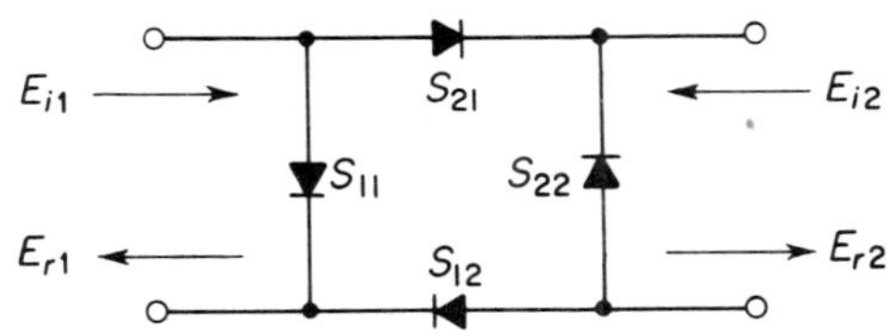

FIGURE 1-17 Scattering parameters for linear two-port networks. (Lenk, *Practical Semiconductor Databook for Electronic Engineers and Technicians*, Prentice-Hall.)

1-8.2 Transistor parameter conversion

Conversions among parameter types for y, z, h, and g parameters.

h to y:

$$y_{11} = \frac{1}{h_{11}}, \qquad y_{12} = -\frac{h_{12}}{h_{11}}, \qquad y_{21} = \frac{h_{21}}{h_{11}}, \qquad y_{22} = \frac{\Delta h}{h_{11}}$$

where $\Delta h = h_{11}h_{22} - h_{12}h_{21}$.

y to h:

$$h_{11} = \frac{1}{y_{11}}, \qquad h_{12} = -\frac{y_{12}}{y_{11}}, \qquad h_{21} = \frac{y_{21}}{y_{11}}, \qquad h_{22} = \frac{\Delta y}{y_{11}}$$

where $\Delta y = y_{11}y_{22} - y_{12}y_{21}$.

h to z:

$$z_{11} = \frac{\Delta h}{h_{22}}, \qquad z_{12} = \frac{h_{12}}{h_{22}}, \qquad z_{21} = -\frac{h_{21}}{h_{22}}, \qquad z_{22} = \frac{1}{h_{22}}$$

z to h:

$$h_{11} = \frac{\Delta z}{z_{22}}, \qquad h_{12} = \frac{z_{12}}{z_{22}}, \qquad h_{21} = -\frac{z_{21}}{z_{22}}, \qquad h_{22} = \frac{1}{z_{22}}$$

where $\Delta z = z_{11}z_{22} - z_{12}z_{22}$.

h to g:

$$g_{11} = \frac{h_{22}}{\Delta h}, \quad g_{12} = -\frac{h_{12}}{\Delta h}, \quad g_{21} = -\frac{h_{21}}{\Delta h}, \quad g_{22} = \frac{h_{11}}{\Delta h}$$

g to h:

$$h_{11} = \frac{g_{22}}{\Delta g}, \quad h_{12} = -\frac{g_{12}}{\Delta g}, \quad h_{21} = -\frac{g_{21}}{\Delta g}, \quad h_{22} = \frac{g_{11}}{\Delta g}$$

where $\Delta g = g_{21}g_{22} - g_{12}g_{21}$.

z to y:

$$y_{11} = \frac{z_{22}}{\Delta z}, \quad y_{12} = -\frac{z_{12}}{\Delta z}, \quad y_{21} = -\frac{z_{21}}{\Delta z}, \quad y_{22} = \frac{z_{11}}{\Delta z}$$

y to z:

$$z_{11} = \frac{y_{22}}{\Delta y}, \quad z_{12} = -\frac{y_{12}}{\Delta y}, \quad z_{21} = -\frac{y_{21}}{\Delta y}, \quad z_{22} = \frac{y_{11}}{\Delta y}$$

z to g:

$$g_{11} = \frac{1}{z_{11}}, \quad g_{12} = -\frac{z_{12}}{z_{11}}, \quad g_{21} = \frac{z_{21}}{z_{11}}, \quad g_{22} = \frac{\Delta z}{z_{11}}$$

g to z:

$$z_{11} = \frac{1}{g_{11}}, \quad z_{12} = -\frac{g_{12}}{g_{11}}, \quad z_{21} = \frac{g_{21}}{g_{11}}, \quad z_{22} = \frac{\Delta g}{g_{11}}$$

g to y:

$$y_{11} = \frac{\Delta g}{g_{22}}, \quad y_{12} = \frac{g_{12}}{g_{22}}, \quad y_{21} = -\frac{g_{21}}{g_{22}}, \quad y_{22} = \frac{1}{g_{22}}$$

y to g:

$$g_{11} = \frac{\Delta y}{y_{22}}, \quad g_{12} = \frac{y_{12}}{y_{22}}, \quad g_{21} = -\frac{y_{21}}{y_{22}}, \quad g_{22} = \frac{1}{y_{22}}$$

Conversions among common-emitter, common-base, and common-collector parameters of the same type for y and h parameters.

Common-emitter y parameters in terms of common-base and common-collector y parameters.

$$y_{11e} = y_{11b} + y_{12b} + y_{21b} + y_{22b} = y_{11c}$$
$$y_{12e} = -(y_{12b} + y_{22b}) = -(y_{11c} + y_{12c})$$
$$y_{21e} = -(y_{21b} + y_{22b}) = -(y_{11c} + y_{21c})$$
$$y_{22e} = y_{22b} = y_{11c} + y_{12c} + y_{21c} + y_{22c}$$

Common-base y parameters in terms of common-emitter and common-collector y parameters.

$$y_{11b} = y_{11e} + y_{12e} + y_{21e} + y_{22e} = y_{22c}$$
$$y_{12b} = -(y_{12e} + y_{22e}) = -(y_{21c} + y_{22c})$$

$$y_{21b} = -(y_{21e} + y_{22e}) = -(y_{12c} + y_{22c})$$

$$y_{22b} = y_{22e} = y_{11c} + y_{12c} + y_{21c} + y_{22c}$$

Common-collector y parameters in terms of common-emitter and common-base y parameters.

$$y_{11c} = y_{11e} = y_{11b} + y_{12b} + y_{21b} + y_{22b}$$

$$y_{12c} = -(y_{11e} + (y_{12e}) = -(y_{11b} + y_{21b})$$

$$y_{21c} = -(y_{11e} + y_{21e}) = -(y_{11b} + y_{12b})$$

$$y_{22c} = y_{11e} + y_{12e} + y_{21e} + y_{22e} = y_{11b}$$

Common-emitter h parameters in terms of common-base and common-collector h parameters.

$$h_{11e} = \frac{h_{11b}}{(1 + h_{21b})(1 - h_{21b})} + h_{22b}h_{11b} \approx \frac{h_{11b}}{1 + h_{21b}} = h_{11e}$$

$$h_{12e} = h_{11b}h_{22b} - \frac{h_{12b}(1 + h_{21b})}{(1 + h_{21b})(1 - h_{12b})} + h_{22b}h_{11b} = 1 - h_{12c}$$

$$h_{12e} \approx \frac{(h_{11b}h_{22b})}{(1 + h_{21b})} - h_{12b}$$

$$h_{21e} = -h_{21b}(1 - h_{12b}) - \frac{h_{22b}h_{11b}}{(1 + h_{21b})(1 - h_{12b})} + h_{22b}h_{11b} = -(1 + h_{21c})$$

$$h_{21e} \approx -\frac{h_{21b}}{1 + h_{21b}}$$

$$h_{22e} = \frac{h_{22b}}{(1 + h_{21b})(1 - h_{12b})} + h_{22b}h_{11b} \approx \frac{h_{22b}}{1 + h_{21b}} = h_{22c}$$

Common-base h parameters in terms of common-emitter and common-collector h parameters.

$$h_{11b} = \frac{h_{11e}}{(1 + h_{21e})(1 - h_{12e})} + h_{11e}h_{22e} \approx \frac{h_{11e}}{1 + h_{21e}}$$

$$h_{11b} = \frac{h_{11c}}{h_{11c}h_{22c}} - h_{21c}h_{12c} \approx -\frac{h_{11c}}{h_{21c}}$$

$$h_{12b} = h_{11e}h_{22e} - \frac{h_{12e}(1 + h_{21e})}{(1 + h_{21e})(1 - h_{12e})} + h_{11e}h_{22e}$$

$$h_{12b} \approx \frac{h_{11e}h_{22e}}{1 + h_{21e}} - h_{12e}$$

$$h_{12b} = h_{21c}(1 - h_{12c}) + \frac{h_{11c}h_{22c}}{h_{11c}h_{22c} - h_{21c}h_{12c}}$$

$$h_{12b} \approx (h_{12c} - 1) - \frac{h_{11c}h_{22c}}{h_{21c}}$$

$$h_{21b} = -h_{21e}(1 - h_{12e}) - \frac{h_{11e}h_{22e}}{(1 + h_{21e})(1 - h_{12e})} + h_{11e}h_{22e}$$

$$h_{21b} \approx -\frac{h_{21e}}{1 + h_{21e}}$$

$$h_{21b} = h_{12c}(1 + h_{21c}) - \frac{h_{11c}h_{22c}}{h_{11c}h_{22c} - h_{21c}h_{12c}} \approx -\frac{1 + h_{21c}}{h_{21c}}$$

$$h_{22b} = \frac{h_{22e}}{(1 + h_{21e})(1 - h_{12e})} + h_{11e}h_{22e} \approx \frac{h_{22e}}{1 + h_{21e}}$$

$$h_{22b} = \frac{h_{22c}}{h_{11c}h_{22c}} - h_{21c}h_{12c} \approx \frac{h_{22c}}{h_{21c}}$$

Common-collector h parameters in terms of common-base and common-emitter h parameters.

$$h_{11c} = \frac{h_{11b}}{(1 + h_{21b})(1 - h_{12b})} + h_{22b}h_{11b} \approx \frac{h_{11b}}{1 + h_{21b}} = h_{11e}$$

$$h_{12c} = 1 + \frac{h_{21b}}{(1 + h_{21b})(1 - h_{12b})} + h_{22b}h_{11b} \approx 1 = 1 - h_{12e}$$

$$h_{21c} = h_{12b} - \frac{1}{(1 + h_{21b})(1 - h_{12b})} + h_{22b}h_{11b} \approx -\frac{1}{1 + h_{21b}}$$

$$= -(1 + h_{21e})$$

$$h_{22c} = \frac{h_{22b}}{(1 + h_{21b})(1 - h_{12b})} + h_{22b}h_{11b} \approx \frac{h_{22b}}{1 + h_{21b}} = h_{22e}$$

Expressions for voltage gain, current gain, input impedance, and output impedance in terms of y, z, h, and g parameters.

Voltage gain:

$$A_v \text{ or } A_e = \frac{z_{21}Z_L}{\Delta z} + z_{11}Z_L$$

$$= -\frac{y_{21}}{y_{22}} + Y_L$$

$$= -\frac{h_{21}Z_L}{h_{11}} + \Delta h Z_L$$

$$= \frac{g_{21}Z_L}{g_{22}} + Z_L$$

where Z_L = load impedance,
Y_L = load admittance
$\Delta h = h_{11}h_{22} - h_{12}h_{21}$.

Current gain:

$$A_i = -\frac{z_{21}}{z_{22}} + Z_L$$

$$= -\frac{y_{21}Y_L}{\Delta y} + y_{11}Y_L$$

$$= \frac{h_{21}Y_L}{h_{22}} + Y_L$$

$$= -\frac{g_{21}}{\Delta g} + g_{11}Z_L$$

where $\Delta y = y_{11}y_{22} - y_{12}y_{21}$

$\Delta g = g_{11}g_{22} - g_{12}g_{21}$.

Input impedance:

$$Z_{\text{in}} = \Delta z + \frac{z_{11}Z_L}{z_{22}} + Z_L$$

$$= y_{22} + \frac{Y_L}{\Delta y} + y_{11}Y_L$$

$$= \Delta h + \frac{h_{11}Y_L}{h_{22}} + Y_L$$

$$= g_{22} + \frac{Z_L}{\Delta g} + g_{11}Z_L$$

where $\Delta z = z_{11}z_{22} - z_{12}z_{21}$.

Output impedance:

$$Z_{\text{out}} = \Delta z + \frac{Z_{22}Z_s}{z_{11}} + Z_s$$

$$= y_{11} + \frac{Y_s}{\Delta y} + y_{22}Y_s$$

$$= h_{11} + \frac{Z_s}{\Delta h} + h_{22}Z_s$$

$$= \Delta g + \frac{g_{22}Y_s}{g_{11}} + Y_s$$

where Y_s = source admittance

Z_s = source impedance.

Conversion of s (scattering) and h parameters.

$$h_{11} = 1 - S_{11} + S_{22} - \frac{\Delta S}{1} - S_{11} - S_{22} + \Delta S$$

$$h_{12} = \frac{2S_{12}}{1} - S_{11} + S_{22} - \Delta S$$

$$h_{21} = -\frac{2S_{21}}{1} - S_{11} - S_{22} - \Delta S$$

$$h_{22} = 1 - S_{11} - S_{12} + \frac{\Delta S}{1} - S_{11} + S_{22} - \Delta S$$

$$S_{11} = \Delta h + h_{11} - h_{22} - \frac{1}{\Delta h} + h_{11} + h_{22} + 1$$

$$S_{12} = \frac{2h_{12}}{\Delta h} + h_{11} + h_{22} + 1$$

$$S_{21} = -\frac{2h_{21}}{\Delta h} + h_{11} + h_{22} + 1$$

$$S_{22} = -\Delta h + h_{11} - h_{22} + \frac{1}{\Delta h} + h_{11} + h_{22} + 1$$

where $\Delta h = h_{11}h_{22} - h_{12}h_{21}$
$\Delta S = S_{11}S_{22} - S_{12}S_{21}$.

Hybrid h parameter relationships using *letter* subscripts.

Number Subscript	Common Emitter	Common Base	Common Collector
h_{11}	h_{ie}	$h_{ib} \approx h_{ie}/1 + h_{fe}$	$h_{ic} = h_{ie}$
h_{12}	h_{re}	$h_{rb} \approx (h_{ie}h_{oe}/1 + h_{fe}) - h_{re}$	$h_{rc} = 1 - h_{re}$
h_{21}	h_{fe}	$h_{fb} \approx -h_{fe}/1 + h_{fe}$	$h_{fc} = -(1 + h_{fe})$
h_{22}	h_{oe}	$h_{ob} \approx h_{oe}/1 + h_{fe}$	$h_{oc} = h_{oe}$

2

Two-Junction (Bipolar) Transistors

This chapter discusses basic two-junction transistor circuits from a practical standpoint. The discussion covers such subjects as transistor types, the three basic transistor circuits, how transistors operate in circuits, how signals are reproduced and amplified by transistors, how gain is produced in a two-junction transistor amplifier circuit, how the necessary bias voltages are applied to transistors, interpreting two-junction transistor datasheets, and the effects of frequency and temperature on two-junction transistor characteristics. The chapter concludes with a discussion of practical two-junction transistor testing methods.

2-1. BASIC TWO-JUNCTION TRANSISTOR TYPES

There are many types of two-junction transistors in use today and new types are being developed constantly. Two-junction transistor types can be classified by the method of construction. This is convenient since many of the operating characteristics of a transistor are directly related to the construction.

The following paragraphs describe the basic construction for the most commonly used two-junction transistors. Figure 2-1 shows the outer construction for two-junction transistors in general use, as well as some special-purpose transistors.

2-1.1 Outer construction of two-junction transistors

As shown in Fig. 2-1, two-junction transistors come in various sizes and shapes. As discussed in Chapter 1, the larger transistors are bolted to the

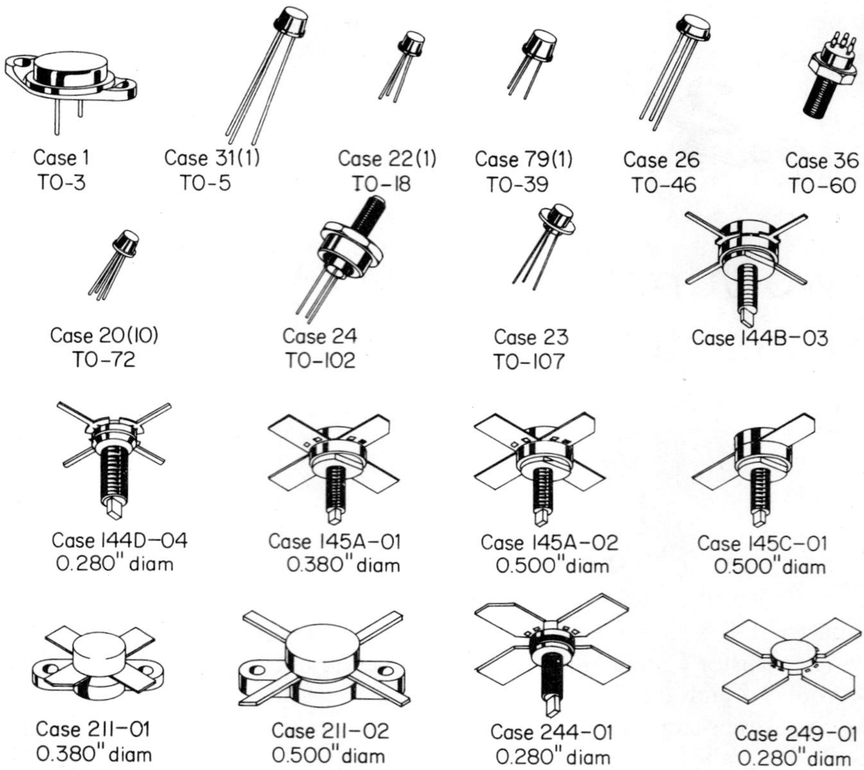

FIGURE 2-1 Case styles for typical two-junction transistors (courtesy Motorola).

chassis so that the flanges help dissipate the heat generated by high-power transistors. There are two methods in general use for attaching power transistors to the chassis. In one case, the attaching bolts pass directly through the transistor flanges (which act as heat sinks). In another configuration, a threaded terminal is provided to permit the transistor to be bolted to the chassis.

The smaller-type transistors are usually wired directly into the circuits, with the wire leads providing the only mechanical support. In some cases, smaller transistors are plugged into miniature sockets.

When transistors are mounted on insulated boards rather than on the metal chassis, it is often necessary to provide a separate heat sink for the transistor, as discussed in Chapter 1. Heat is dissipated into the surrounding air (rather than into the chassis) by the fins of the heat sinks.

2-1.2 Inner construction of two-junction transistors

The three regions (*NPN* or *PNP*) of a typical junction transistor are made from the same crystal. Several methods are used to produce two-junction transistors. The *grown-junction* method has been used for some time and is still in common use. When an *NPN* transistor is to be produced by the grown-junction method, a pellet of *P*-forming impurity (such as aluminum, gallium, or indium) is added to molten germanium into which a single crystal is submerged and pulled out. The *P* impurity melts and sticks to the crystal. Then *N*-type material is added by resubmerging the crystal in the molten germanium. One problem with the grown-junction method is that the thickness of the layers is difficult to control. Also, the base consists of a very thin layer, making it difficult to attach a terminal lead.

These problems are overcome in the *alloyed-junction* transistor, which is used for most high-power applications. The alloyed-junction transistor is produced by melting an impurity pellet to each side of a base. The impurity pellets are of the opposite-type material (*P* impurity for an *N* base and vice versa). The base is a thin wafer of semiconductor material to which is attached a contact or terminal ring rather than a terminal lead. A *PNP* alloyed-junction transistor is shown in Fig. 2-2.

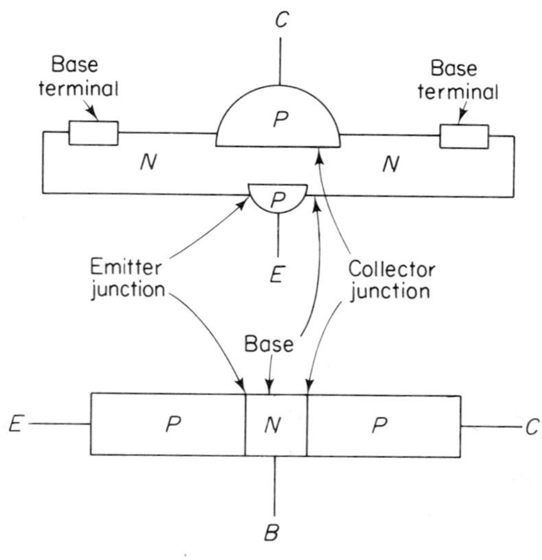

FIGURE 2-2 Typical *PNP* alloyed-junction transistor.

Special problems are created when transistors are to be operated at higher frequencies. The operating frequency of a transistor is limited by the thickness of the base material and by the area of the emitter and collector regions. The thicker the *N* base, the longer the transit time for electrons moving between the emitter and collector. When the base thickness is increased to a point where the time consumed in travel between emitter and collector does not permit a change in the direction of current flow, the upper operating frequency limit of the transistor is reached.

Larger emitter and collector regions increase the capacity between the two regions. Capacitors tend to store electrons. If the capacitance is high, the storage time is long and the transistors will not operate efficiently at high frequencies.

Base thickness can be reduced just so far in practical applications. Likewise, a reduction of emitter and collector size produces a corresponding reduction in power handling capabilities of the transistor.

There are several ways to eliminate or minimize these problems. The *surface-barrier* transistor provides satisfactory operation up to about 70 MHz. Surface-barrier transistors are formed by an electrochemical process. As shown in Fig. 2-3, a *PNP* surface-barrier transistor is formed by placing an *N*-type material between two fine streams of an electrolytic solution and applying a dc voltage that causes the solution to etch away the semiconductor material until the desired thickness is reached. The polarity of the dc voltage is then reversed. This causes the metallic solution to deposit small metal dots on each side of the thinned *N* region. The *N*-type material becomes the base, and the dots become the collector and emitter.

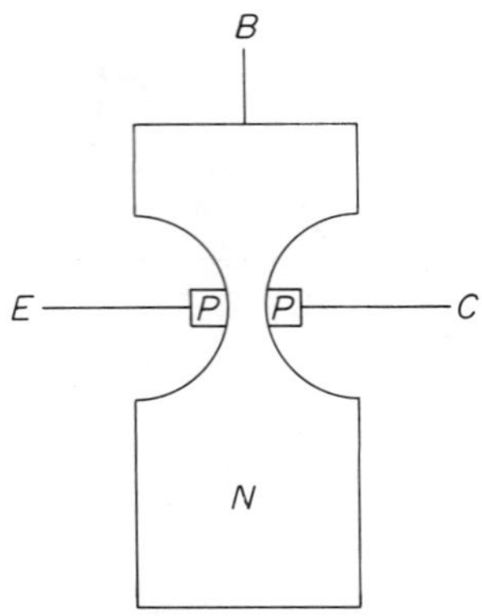

FIGURE 2-3 Typical *PNP* surface-barrier transistor.

The emitter-collector voltage of a surface-barrier transistor is about 6 V maximum so power capabilities are relatively low.

Where both high frequency and high power are required, the *mesa* (Spanish word for *table* or *hill*) transistor, or some form of *planar* transistor, is used effectively. These transistors permit operation at very high frequencies (about 100 MHz for silicon and 1000 MHz for germanium) with corresponding high power.

As shown in Fig. 2-4(a), the collector of a *mesa* transistor is formed by a layer of *P*-type semiconductor material diffused to a thin layer of *N*-type impurity which becomes the base. An emitter is added by alloying. The table-like structure is produced by etching away the excess material. Since the base area is made small by this etching, the capacity between the transistor

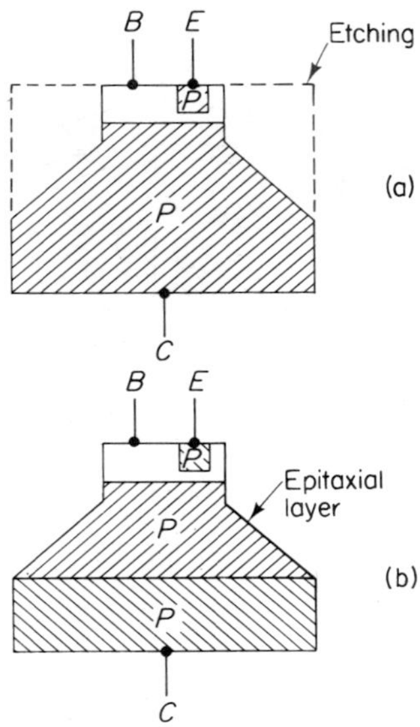

FIGURE 2-4 Typical *mesa* transistor.

elements remains low, permitting high-frequency operation; however, the large collector permits use of high voltages and corresponding high power.

An improved version of the *mesa* transistor is shown in Fig. 2-4(b) where a high-resistance layer of *P*-type material is added before the *N*-type base is diffused. This high-resistance layer (termed an *epitaxial* layer) allows higher collector voltages to be used.

As shown in Fig. 2-5(a), the collector of a *planar* transistor is formed by a wafer of *N*-type silicon, which has been passivated (coated) with an oxide layer. A circular trench is etched out of the oxide and filled with a *P*-type base. A disc-shaped area of the oxide is etched at the center and filled with an *N*-type emitter that is diffused to the base. The oxide layer protects the emitter-base and collector-base junctions from shorts or contamination.

An improved version of the planar transistor is shown in Fig. 2-5(b) where a high-resistance epitaxial layer is added to permit higher collector voltages.

High-frequency operation can also be obtained with a *tetrode* transistor (Fig. 2-6). The tetrode operates like a two-junction transistor, except that there is an additional terminal on the base. A voltage (termed *transverse* voltage) is applied between the two base terminals. This voltage causes the transistor action to occur at one side or the other of the transistor.

In the example of Fig. 2-6, the base is made negative by the transverse voltage. This restricts the flow of electrons between emitter and collector. At the bottom of the base, the transverse voltage is near zero, permitting normal electron flow. Thus, electron flow is heaviest at the bottom. This has the effect of reducing the junction area, with a corresponding reduction of capacitance. The lower capacitance permits higher-frequency operation.

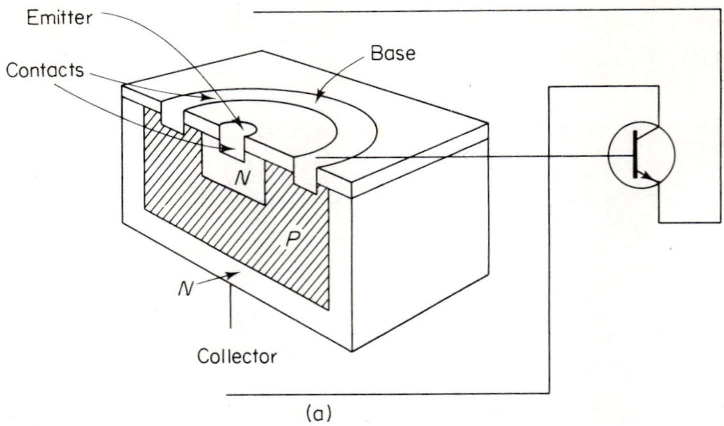

FIGURE 2-5 Typical planar transistor.

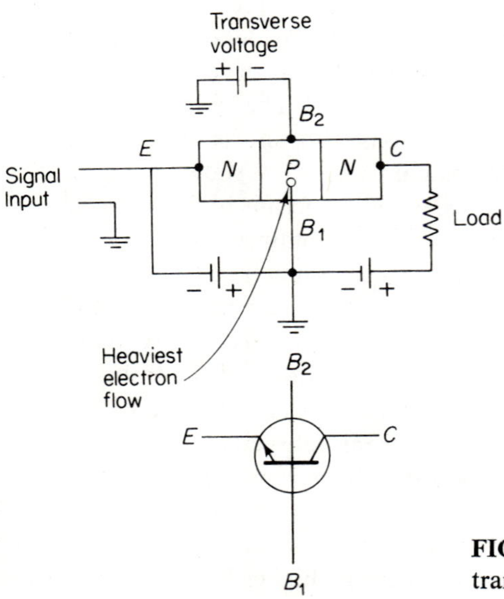

FIGURE 2-6 Typical tetrode transistor.

2-2. BASIC TWO-JUNCTION TRANSISTOR CIRCUIT CONNECTIONS

There are three basic two-junction transistor circuits:

1. Common base (or grounded base).
2. Common emitter (or grounded emitter).
3. Common collector (or grounded collector).

Each of the three basic circuits is discussed fully in later sections of this chapter. Figure 2-7 summarizes the three basic methods of connection. As shown, the major difference in the three basic circuits depends on which terminals are used for input and for output.

Each configuration in Fig. 2-7 is identified by the transistor element that is *common to both the input and output.* Although it is frequently called the *grounded* element, the common element is not necessarily connected to ground. Figure 2-7 also shows the power supply connections for both *NPN*- and *PNP*-type transistors.

In the common-base circuit, the input signal is applied between the base and emitter and the output signal appears between base and collector. Although there is no current gain between input and output of a common-base circuit, it is possible to obtain considerable *power gain.* This is due to the difference in resistance between input and output circuits.

In the common-emitter circuit, the input signal is applied between the base and emitter and the output signal appears between emitter and collector. This provides a very low input impedance and a very high output impedance. Note that the output signal is *out of phase* with the input.

Common-emitter circuits are most often used since there is current gain, resistance gain, and power gain, as discussed in Sec. 2-5.

In the common-collector circuit, the input signal is applied to the base and the output signal appears at the emitter. This provides a very low output impedance and a very high input impedance. Normally, this condition would produce no gain. Because the emitter-collector (output) circuit conducts more current than the base circuit, however, there is some current gain. The common-collector circuit is also known as an *emitter follower.* (The output signal is taken from the emitter and follows the input as to phase.)

Figure 2-7 also shows the similarities and differences in characteristics for the three basic circuits.

The actual input and output impedances of the three circuits will vary with transistor types used (*PNP* or *NPN*) and to some extent with individual units of the same type.

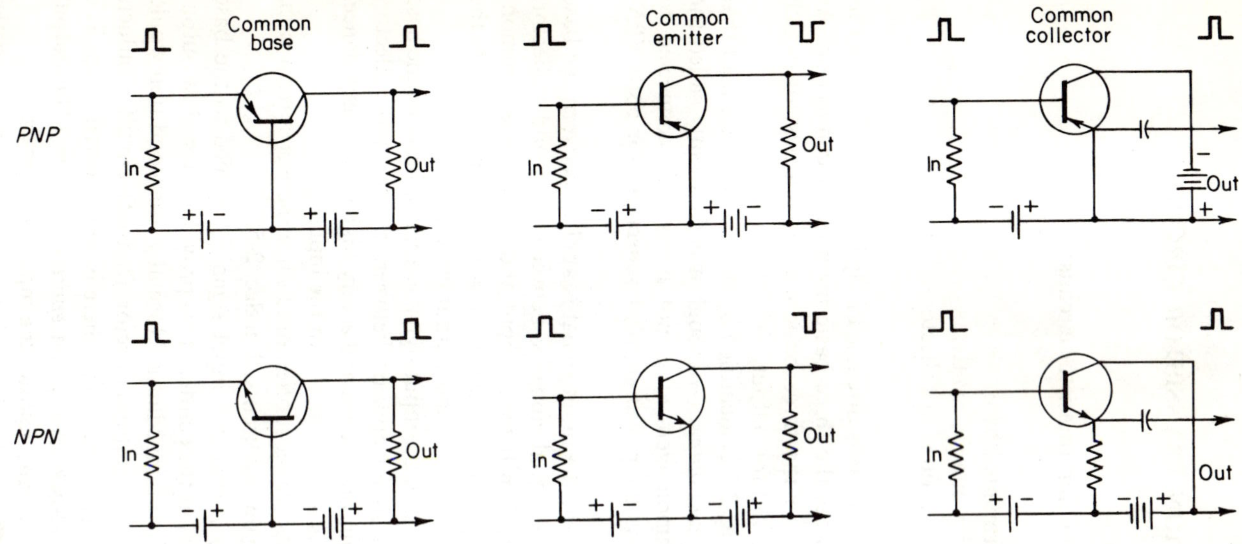

Characteristic	Common base	Common emitter	Common collector
Input resistance	30–150 Ω	500–1500 Ω	20–500 kΩ
Output resistance	300–500 kΩ	30–50 kΩ	50–1000 Ω
Voltage gain	500–1500	300–1000	Less than 1
Current gain	Less than 1	25–50	25–50
Power gain	20–30dB	25–40dB	10–20dB

FIGURE 2-7 Summary of two-junction transistor connections. (Lenk, *Practical Semiconductor Databook for Electronic Engineers and Technicians*, Prentice-Hall.)

2-3. THE BASIC TWO-JUNCTION AMPLIFIER

Operation of two-junction transistors can best be understood by reviewing how such transistors amplify signals.

Figure 2-8 shows a *PNP* transistor connected in a basic common-emitter circuit. The base-emitter circuit is forward biased; whereas the base-collector circuit is reverse biased. An input signal is applied across resistor R_1; the output is taken from across resistor R_2.

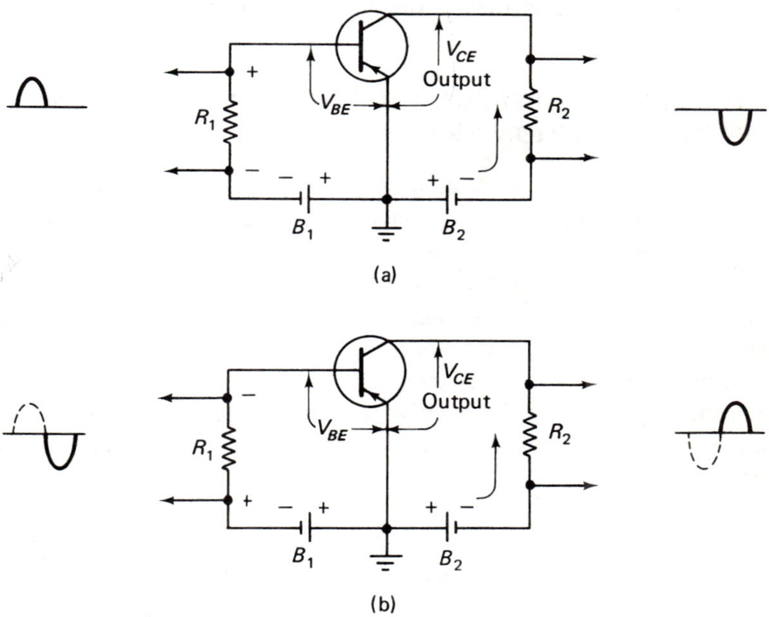

FIGURE 2-8 Operation of two-junction transistors in amplifier circuits. (Lenk, *Handbook of Modern Solid-State Amplifiers*, Prentice-Hall.)

Under no-signal (quiescent) conditions, current flows in the input circuit, causing a steady value of current to flow in the output circuit. When one alternation of an input signal is applied, as shown in Fig. 2-8(a), a voltage is developed across input resistor R_1. This voltage, positive at the base end of R_1, subtracts from the bias voltage provided by B_1, causing the base-to-emitter voltage V_{BE} to become less negative.

Under these conditions, the voltage from the collector to emitter will increase, the collector becoming more negative than it was during the no-signal condition. As the signal voltage increases in a positive direction, the output voltage (taken from collector to ground) increases in a negative direction. This action illustrates signal *phase reversal*, a characteristic of the common-emitter stage. Amplification occurs because the collector current is many times greater than the base-emitter current. As shown in Fig. 2-8(b), the opposite conditions occur on the next alternation of the input signal.

The total voltage V_{BE} is a combination of the input signal and B_1; the total voltage V_{CE} is a combination of output signal and B_2. In practical circuits, the input is measured from base to ground, with the output measured from collector to ground.

If an *NPN* transistor is placed in the same circuit, operation will be identical except that polarities of the voltages are reversed.

2-4. RULES FOR LABELING, BIASING, AND POLARITIES IN TWO-JUNCTION TRANSISTORS

The following general rules can be helpful in a practical analysis of how two-junction transistors operate. The rules apply primarily to a class A amplifier but also remain true for many other circuits.

In *PNP* transistors, holes flow from the emitter to the collector. Thus, the collector must be negative, *relative to the emitter*. In the external circuit, electrons flow from emitter to collector.

In the *NPN* transistor, electrons flow from the emitter to the collector. Hence, the collector must be positive, *relative to the emitter*. In the external circuit, aided by the bias voltage, electrons flow from the collector to the emitter.

The *middle letter* in *PNP* or *NPN* always applies to the *base*.

The *first two letters* in *PNP* or *NPN* refer to the *relative bias polarities of the emitter* with respect to either the base or collector. For example, the letters *PN* (in *PNP*) indicate that the emitter is positive with respect to both the base and collector. The letters *NP* (in *NPN*) indicate that the emitter is negative with respect to both the base and collector.

The dc electron current flow is always against the direction of the arrow in the emitter.

If electron flow is into the emitter, electron flow will be out from the collector.

If the electron flow is out from the emitter, electron flow will be into the collector.

The collector-base junction is *always reverse biased*.

The emitter-base junction is generally forward biased.

A base input voltage that opposes or decreases the forward bias also decreases the emitter and collector currents. For example, a negative input to the base of an *NPN*, or a positive input to a *PNP* base, decreases both currents.

A base input voltage that aids or increases the forward bias also increases the emitter and collector currents (positive to *NPN*, negative to *PNP*).

2-5. COMMON-EMITTER AMPLIFIER

The common-emitter circuit shown in Fig. 2-9 is the most widely used two-junction transistor amplifier configuration. Figure 2-9 shows the power supply connections for both *NPN* and *PNP* transistors in theoretical circuits. In most practical circuits, a single power supply is used.

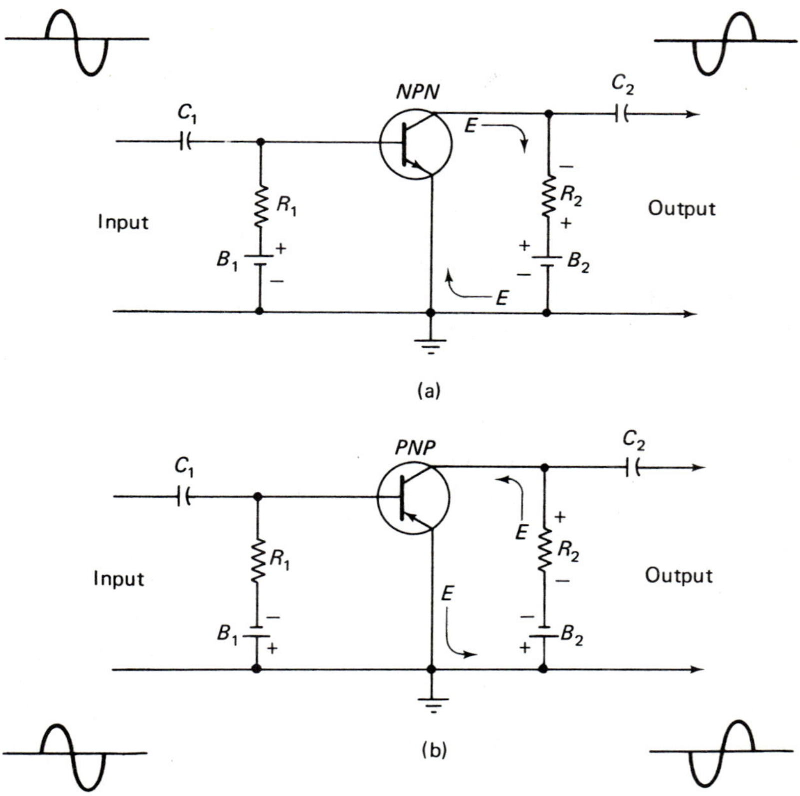

Typical Characteristics
Input resistance 500–1500 Ω
Output resistance 30–50 kΩ
Voltage gain 300–1000
Current gain 20–50
Power gain 25–40 dB

FIGURE 2-9 Basic common-emitter amplifier stage. (Lenk, *Handbook of Modern Solid-State Amplifiers*, Prentice-Hall.)

In the circuits of Fig. 2-9, the emitter is common to both the input and output circuits. Although the emitter is frequently called the grounded element, the emitter is not necessarily connected to ground. Battery B_1 forward biases the emitter-base junction. Battery B_2 reverse biases the collector-base junction. In practice, the voltage of B_2 is generally larger than that of B_1 and both voltages are taken from the same source.

The input signal is applied between the base and emitter; the output signal appears between the emitter and collector. This provides a very low input impedance and a very high output impedance. The output signal is out of phase with the input.

Common-emitter circuits are the most often used since there is *current gain* and *voltage gain*, thus resulting in a large *power gain*. For example, if the output impedance is 10 times higher than the input impedance, there will be a power gain of 10, even if the base-emitter (input) and emitter-collector (output) currents are equal (no current gain). Since it is possible to control a large output current with a small input current in a practical common-emitter circuit, there is also current gain.

Capacitor C_1 is used to block out any dc components in the input signal. The input signal voltage is developed across resistor R_1, which also serves to provide a closed circuit for current flow in the emitter-base circuit. This is essential since a two-junction transistor is a *current-operated* device.

The flow of collector current produces a voltage drop, with the indicated polarity, across R_2. The ac component of this voltage is the output signal of the amplifier. Capacitor C_2 is the coupling capacitor, which blocks out the steady dc component and passes the ac component.

The phase relationship between the input and output signals is shown by the sine waves appearing in Fig. 2-9. If the input voltage increases the forward bias of the base-emitter junction, the total emitter current is increased, producing an increase in the collector current in like degree. This increase in collector current, in turn, produces a larger drop across R_2. In the *NPN* circuit [Fig. 2-9(a)] the input voltage must be positive if it is to aid the forward bias of the emitter-base junction. In the *NPN* amplifier [Fig. 2-9(b)] a similar action takes place but with reverse polarities. In all cases where the common emitter is used, as the input voltage rises, the output voltage falls and vice versa. Hence there is a *phase reversal* between input and output signals (that is, input and output signals are 180 deg out of phase).

Practical common-emitter circuit. In a practical circuit, both junctions are biased from a single power source. There are several ways of supplying the bias, as discussed in Sec. 2-9. A typical method of supplying bias to a common-emitter amplifier stage from a single supply is shown in Fig. 2-10. The circuit is a common-emitter amplifier using an *NPN* transistor. V_{CC} is the bias for both junctions.

Since the emitter-base junction must be forward biased and the collector-base junction reverse biased, in the *NPN* transistor this means that

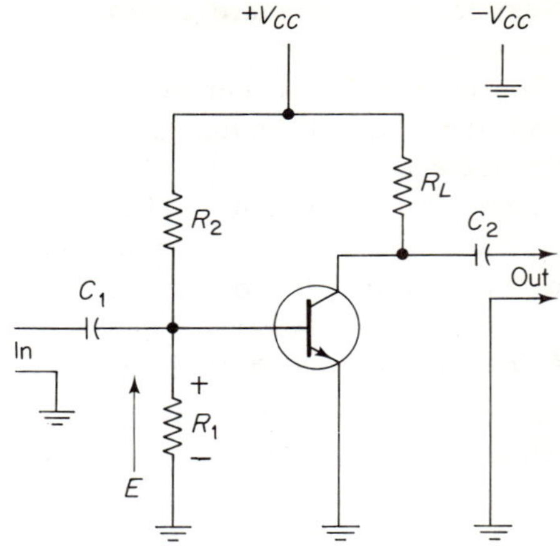

FIGURE 2-10 Practical common-emitter amplifier operated from a single power supply.

the base must be negative relative to the collector and positive relative to the emitter. Thus, the collector is at the most positive potential of the circuit and the emitter is at the most negative potential.

Note that when we say the base must be less positive than the collector and more positive than the emitter, we are talking in relative terms. For example, if the base voltage is +2 V and the emitter voltage is +1.5 V, the emitter is more negative than the base, even though both voltages are positive. Similarly, the emitter is more negative than the base if the emitter voltage is −2 V and the base voltage is −1.5 V.

Since the collector is connected to the positive side of V_{CC}, it is at the most positive potential. Similarly, since the emitter is connected to the negative side of V_{CC}, it is at the most negative potential. Resistors R_1 and R_2 form a voltage divider across V_{CC}. The flow of current through this divider produces a voltage drop across R_1 with the indicated polarity. Note that this voltage opposes that of V_{CC}. Hence, the voltage applied to the base is less positive than the collector voltage, although more positive than the emitter voltage.

If a *PNP* transistor is used in the circuit of Fig. 2-10, the action is the same except that the polarities of all voltages and the direction of current flow are reversed.

Because the emitter-base junction in the common-emitter configuration is forward biased, the input resistance of this circuit is low. Because the collector-base junction is reverse biased, the output resistance is high. Typically, the input resistance of the common-emitter amplifier stage is between 500 and 1500 Ω and the output resistance is between 30 and 50 kΩ. As discussed in later sections, however, it is possible to set input and output resistances of an amplifier stage by proper selection of resistors (typically, R_1 and R_L of Fig. 2-10).

Normally, the current gain, voltage gain, and power gain of common-emitter amplifiers are 25 to 50, 300 to 1000, and 25 to 40 dB, respectively.

To summarize the common-emitter amplifier, the input signal is applied between the base and emitter and the output signal appears between the emitter and collector. This provides a moderately low input impedance and a very high output impedance. The output signal is out of phase with the input. The common-emitter amplifier produces the highest power gain of all three configurations. Its voltage and current gains are fairly high.

2-5.1 Gain in two-junction transistor amplifiers

Many terms are used to express the gain of common-emitter amplifiers as well as other amplifier configurations. For example, there are the terms *alpha* and *beta*, in addition to *current gain*, *voltage gain*, *resistance* (or *impedance*) *gain*, *power gain*, *voltage amplifier*, and *power amplifier*. The terms are interrelated and are often interchanged (properly and improperly). To minimize this confusion, the following is a summary of how these terms are used throughout this handbook.

Alpha and *beta* are applied to two-junction transistors connected in the *common-base* and *common-emitter* configurations, respectively. Both terms are a measure of the *current gain* of the transistor. Alpha is always less than 1 (typically 0.9 to 0.99). Beta is more than 1 and can be as high as several hundred (or more). The relationships between alpha and beta are

$$\text{Alpha} = \frac{\text{beta}}{\text{beta} + 1} \qquad \text{Beta} = \frac{\text{alpha}}{\text{alpha} - 1}$$

The terms *alpha* and *beta* do not necessarily represent the current gain of the stage in which the transistors are used. Instead, the current gain of the stage *cannot be greater* than the alpha or beta of the transistor.

Current gain can be applied to the transistor or stage and is a measure of change in current at the output for a given change in current at the input. Current gain is equal to the change in output current divided by change in input current. For example, when a 1-mA change in input current produces a 10-mA change in output current, the current gain is 10. Since alpha (common base) is always less than 1, there is no current gain in common-base amplifier circuits. (Instead, there is a slight loss.)

Resistance gain is simply the ratio of output resistance (or impedance) divided by input resistance (or impedance). For example, if the input resistance is 1 kΩ and the output resistance is 15 kΩ, the resistance gain is 15. The input and output resistances (or impedances) are dependent on circuit values as well as transistor characteristics. As discussed throughout this handbook, many amplifier characteristics are directly dependent on the relationship between transistor characteristics and values of circuit components.

Resistance gain, by itself, produces no usable gain for the amplifier stage;

however, the resistance gain has a direct effect on the stage voltage and power gains. For example, using the previous values (current gain of 10, resistance gain of 15), it is possible to have a voltage gain of 150. Assume that the input resistance is 1 kΩ and a 1-mV signal is applied at the input. This results in an input current change of 1 μA. With a current gain of 10, the output current is 10 μA. This 10-μA current passes through a 15-kΩ output resistance to produce a voltage change of 150 mV. Thus, the 1-mV input signal produces a 150-mV output signal (a voltage gain of 150).

Voltage gain is generally applied to the stage and is equal to the difference in output voltage divided by a difference in input voltage.

Power gain is also generally applied to the stage and is equal to the difference in output power divided by the difference in input power.

Except in a common-base amplifier, power gain is always higher than voltage gain since power is based on the square of voltage (power $= E^2/R$) as well as the square of current (I^2R). Using the same values, the input power of the stage is 1×10^{-9}, the output power is 1.5×10^{-6}, and the power gain is 1500.

As in the case of current gain, both voltage gain and power gain are dependent on transistor characteristics and circuit values.

The function of a *voltage amplifier* is to receive an input signal consisting of a small voltage of definite waveform and to produce an output signal consisting of a voltage with the same waveform but much larger in amplitude. For example, as a radio wave cuts across the antenna of a radio receiver, the wave induces in that antenna a fluctuating voltage, usually on the order of microvolts. The voltage amplifier of the receiver amplifies this voltage to produce a similarly fluctuating voltage that is large enough to operate a *power amplifier* that, in turn, operates the power-consuming loudspeaker.

Those transistors designed for voltage amplification usually have high betas with little current-carrying capability. On the other hand, power-amplifier transistors have large current-carrying capacity but relatively low betas. Typically, there is at least one (and usually two) voltage-amplifier stage ahead of the power-amplifier stage. This permits a low voltage input signal (say from an antenna, tape head, or industrial transducer) to operate a power-consuming device (radio or stereo loudspeaker or industrial servo-motor).

2-6. COMMON-BASE AMPLIFIER

The common-base circuit is shown in Fig. 2-11, where both junctions are biased from a single power source. In this configuration, the base is common to both the input and output circuits. Although the base is frequently called the grounded element, the base is not necessarily connected to ground.

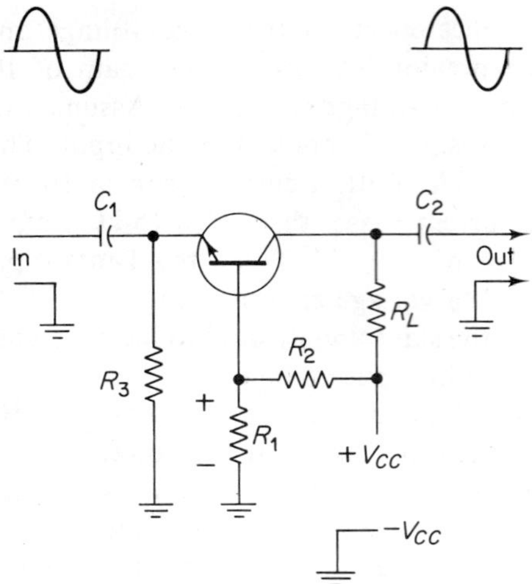

FIGURE 2-11 Practical common-base amplifier operated from a single power supply.

Capacitor C_1 is used to block out any dc in the input signal. The input signal is voltage developed across R_3, which also serves to provide a closed circuit for current flow in the emitter-base circuit. The flow of collector current produces a voltage drop across R_L. The ac component of this voltage is the output signal of the amplifier. Capacitor C_2 is the coupling capacitor that blocks out the steady dc component and passes the ac component. The phase relationship between the input and output signals is shown by the sine waves.

The collector is connected to the positive side of source V_{CC}, placing it at the most positive potential. The emitter is at the most negative potential. Resistors R_1 and R_2 form a voltage divider across V_{CC}. Current flows through this divider, causing a voltage drop across R_1 with the indicated polarity. Since the polarity of this voltage drop is opposed to the polarity of V_{CC}, the potential of the base is less positive than the potential of the collector, although more positive than the potential of the emitter.

The forward-biased emitter-base junction produces low impedance; whereas the reverse-biased collector-base junction produces high impedance. Typically, the input resistance of the common-base amplifier stage is between 30 and 150 Ω and the output resistance is between 300 and 500 kΩ. It is possible, however, to set input and output resistance of an amplifier stage by proper selection of resistors (typically, R_3 and R_L of Fig. 2-11).

Typically, the current gain, voltage gain, and power gain of common-base amplifiers are 0.9 to 0.99, 500 to 1500, and 20 to 30 dB, respectively.

To summarize the common-base amplifier, the input signal is applied between the base and emitter and the output signal appears between the base and collector. This provides an extremely low input impedance and a very high output impedance. The output signal is in phase with the input. The common-base amplifier produces high voltage gains and modest power gains, even though there is no current gain. This is possible because of the *resistance gain.*

As an example, assume that the input resistance is 100 Ω, the output resistance is 15 kΩ, the current gain is 0.9 (less than 1), and the input signal is 1 mV. With 1 mV across 100 Ω, there is a 10-μA current change. With a current "gain" of 0.9, the output current is 9 μA. This 9-μA current passes through a 15-kΩ output resistance to produce a voltage change of 135 mV. Thus, a 1-mV input signal produces a 135-mV output signal (a voltage gain of 135).

2-7. COMMON-COLLECTOR (OR EMITTER-FOLLOWER) AMPLIFIER

The common-collector circuit is shown in Fig. 2-12, where both junctions are biased from a single power source. The common-collector circuit is also known as an *emitter-follower* circuit since the output is taken from the emitter resistance and the output follows the input (in phase relationship).

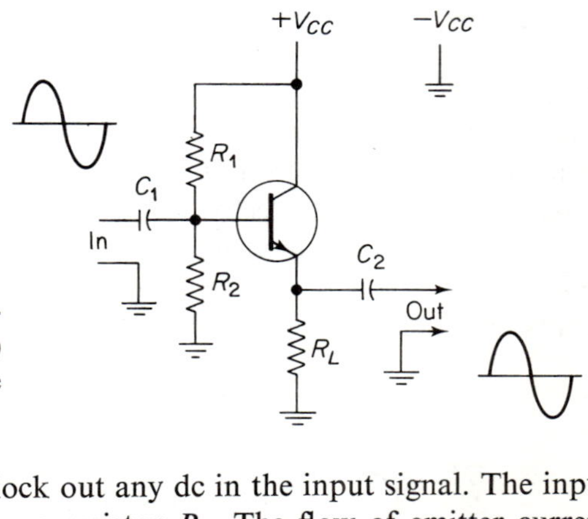

FIGURE 2-12 Practical common-collector (emitter-follower) amplifier operated from a single power supply.

Capacitor C_1 is used to block out any dc in the input signal. The input signal voltage is developed across resistor R_2. The flow of emitter current produces a voltage drop across R_L. The ac component of this voltage is the output signal. Capacitor C_2 is the coupling capacitor that blocks out the steady dc and passes the ac. The phase relationship between the input and output signals is shown by the sine waves.

Source V_{CC} makes the collector positive relative to the base, thus reverse biasing the collector-base junction. Resistors R_1 and R_2 form a voltage divider across V_{CC}, which forward biases the emitter-base junction.

Typically, the input resistance of the common-collector stage is between 20 and 50 kΩ and the output resistance is between 50 and 1000 Ω. It is possible, however, to set input resistances of the stage by proper selection of resistors (typically R_2 and R_L of Fig. 2-12). The current gain, voltage gain, and power gain of common-collector stages are 25 to 50, less than 1, and 10 to 20 dB, respectively.

To summarize the common-collector stage, the input signal is applied to the base and the output signal appears at the emitter. This provides extremely high input impedance and low output impedance. The output signal is in phase with the input. The common-collector stage produces modest current and power gains, even though there is no voltage gain. In general, the common-collector circuit gain (and hence the power gain) is limited by the current gain (beta) of the transistor.

2-8. USING LOAD LINES TO EXAMINE TWO-JUNCTION TRANSISTOR CHARACTERISTICS

Load lines are useful tools in interpreting two-junction transistor characteristics. Many datasheets show two-junction transistor characteristics by means of curves that are reproductions of displays obtained with an oscilloscope-type curve tracer (Sec. 2-13). The collector current versus voltage curves shown in Fig. 2-13 are typical.

Such curves are obtained by applying a series of stepped base currents and then sweeping the collector voltage over a given range. Several curves are made in rapid succession at different base currents. Gain can be found by noting the difference in collector current for a given change in base current, while holding a fixed collector voltage, as discussed in Sec. 2-13.

It is also possible to examine the combined effect of base current and a collector load resistor of given value by superimposing a load line on the curves. For example, assume that a collector load of 1000 Ω is used to the

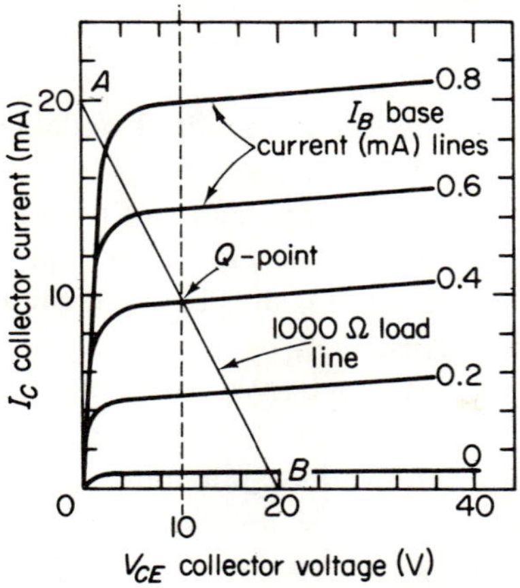

FIGURE 2-13 Typical load line drawn on characteristic curve of common-emitter transistor. (Lenk, *Handbook of Simplified Solid-State Circuit Design*, Prentice-Hall.)

transistor displayed in Fig. 2-13. When the collector current reaches 20 mA (I_B or base current at approximately 0.8 mA), the collector voltage drops to zero. Similarly, when the collector current drops to zero, collector voltage rises to 20 V (base current at zero).

If a load line is connected between these two extreme points, marked *A* and *B* on Fig. 2-13, the instantaneous collector voltage and collector current can be obtained for any base current along the line. This permits an operating point to be selected by graphic means. For example, assume that it is desired to operate the collector at 10 V with no signal applied. Draw a line up from the 10-V point, and note that it intercepts the load line at about 0.4-mA base current (and about 10-mA collector current). The transistor circuit bias resistors are then chosen to provide a 0.4-mA base current.

There are some schools of transistor design that use the load line method to select final component values for bias networks. This method is convenient and does provide some fairly accurate approximations; however, there is a drawback. The datasheet curves are "typical" for a transistor of a given type and represent an average gain (or in some cases the minimum gain). Also, transistor gain is temperature and frequency dependent. Thus, the selection of load values, bias values, and operating point on the basis of static gain curves is subject to error. If the transistor's beta shifts, the operating point must shift, requiring different bias and load values.

As discussed in Sec. 2-13, the problem of variable gain can be overcome by means of *feedback to stabilize the gain.* The feedback method bias design makes the circuit characteristics dependent on *relationship of circuit values* rather than on transistor gain characteristics.

2-9. TWO-JUNCTION TRANSISTOR BIAS NETWORKS

Practically all two-junction transistor circuits require some form of bias. As a minimum, the collector-base junction must be reverse biased. That is, current should not flow between collector and base. Any collector-base current that does flow is a result of leakage or breakdown. Breakdown must be avoided by proper design. Leakage, usually listed as I_{CBO} on datasheets (Sec. 2-10), is an undesirable (but almost always present) condition that must be reckoned with in practical use.

Under no-signal conditions, the emitter-base junction can be forward biased, reverse biased, or zero biased (no bias); however, emitter-base current must flow under some conditions of operation. For example, some current flows all the time in class A and class AB amplifiers. In class B and C amplifiers, current flows only in the presence of an operating signal. With any class of operation, the emitter-base circuit must be biased so that current can flow under some conditions (generally in the presence of an input signal).

The desired bias is accomplished by applying voltages to the corresponding transistor elements through bias networks, usually composed of resistors. The following paragraphs describe several basic bias networks. These circuits (or variations of them) represent most of the bias methods used with two-junction transistors.

Purposes of bias networks. The bias networks serve more than one purpose. Typically, the bias network resistors (1) set the operating point, (2) stabilize the circuit at the operating point, and (3) set the approximate input–output impedances of the circuit.

The basic purpose of the bias network is to establish collector-base-emitter voltage and current relationships *at the operating point of the circuit.* (The operating point is also known as the *quiescent point*, *Q point*, *no-signal point*, *idle point*, or *static point.*) Since transistors rarely operate at this *Q* point, the basic bias networks are generally used as a reference or starting point for design. The actual circuit configuration and (especially) the bias network values are generally selected on the basis of dynamic circuit conditions (desired output voltage swing, expected input signal level, etc.).

Once the desired operating point is established, the next function of the bias network is to stabilize the circuit at this point. Although there are many bias networks, each with its own advantages and disadvantages, one major factor must be considered for any network. The bias network must maintain the desired current relationships in the presence of temperature and power supply changes and possible transistor replacement. In some cases, frequency changes and changes caused by component aging must also be offset by the bias network.

Effects of improper bias stabilization. Two undesirable conditions can result when adequate bias stabilization is not provided. First, any changes in temperature, power supply voltage, and (possibly) frequency produce changes in collector-emitter and/or base-emitter junction current. For example, an increase in temperature or power supply voltage will increase currents. In turn, this will *shift the operating point of the circuit*. Such a shift can produce distortion and a change in frequency response as well as other undesired effects. In any event, the circuit is no longer at the optimum operating point for which it is designed.

The other undesirable effect of inadequate bias stabilization is *thermal runaway*, described in Sec. 1-7.3. Adequate bias stabilization will prevent any drastic change in junction currents, despite changes in temperature, voltage, and so on. Thus, proper bias stabilization will maintain the amplifier circuit at the desired operating point (within practical limits) and will prevent thermal runaway.

Circuit impedances. The resistors used in bias networks also have the function of setting input and output impedances of the circuit (particularly in the case of two-junction amplifiers). From a theoretical standpoint, the input–output impedances of a circuit are set by a wide range of factors (transistor beta, transistor input–output capacitance, etc.). For practical purposes, however, the input–output impedances of a typical circuit (operating at frequencies up to about 100 kHz) are set by the bias network resistors. For example, the output impedance of a common-base or common-emitter amplifier is approximately equal to the collector resistor (between collector and power supply).

2-9.1 Basic two-junction bias design considerations

The first step in bias design is to determine the characteristics of both the circuit and the transistor to be used. For example, will be circuit be used as an amplifier, oscillator, or switch? What class of operation (A, B, AB, or C) will be required? How much gain (if any) is wanted? What power supply voltages are available? What transistor is to be used? Must input and/or output impedances be set at some arbitrary value?

Once these preliminary factors have been decided, the design data should be recorded. This requires reference to the transistor datasheet (Sec. 2-10) as well as some arbitrary decisions.

For example, note the following:

If the power supply voltage is set arbitrarily by an available source, make sure that this voltage is below the collector-emitter breakdown voltage (BV_{CEO}) discussed in Sec. 2-10 and 2-13.

The maximum collector-current rating must never be exceeded in any bias circuit design. Always consider the possibility that the emitter-base junction may be at the maximum forward bias (under extreme conditions) and will result in maximum collector-current flow.

Never exceed the maximum power dissipation of the transistor. Keep in mind that the power dissipation capability decreases as temperature increases. *When the exact operating temperature is not known, use from one-half to one-fourth of the 25°C rating*, found on the datasheet.

When considering the effect of collector leakage (I_{CBO}), keep in mind that leakage increases as temperature increases. *When the exact operating temperature is not known, use 10 times the 25°C rating.*

Operating load current and resistance. Once the basic design characteristics have been tabulated, the next step is to calculate the operating load current and resistance. In some circuits, an arbitrary load resistance must be used (to achieve a given circuit output impedance, for example). In other circuits, an arbitrary load current is required (to obtain a given power output). In many amplifier circuits, both load current and load resistance are chosen solely to obtain a given operating point.

No matter what basis is used, two major factors determine operating load current for a two-junction transistor: *leakage current* (I_{CBO}) *and maximum rated collector current.*

Obviously, the load current cannot exceed the maximum rated current for the transistor (maximum I_C). Likewise, the load current must not be less than the leakage current (I_{CBO}) or no current will flow.

If the circuit is to be operated from a battery, an operating load current near the low end should be selected to minimize battery drain. However, *the load current should be no less than 10 times the leakage current.*

At the high end, the maximum load current should not exceed the maximum power dissipation, divided by the collector voltage at the operating point.

As an example, assume that the collector is to be at 10 V under no-signal conditions (from a 20-V source) and that the maximum power dissipation of the transistor is 50 mW. Under these conditions, the maximum load current should be 5 mA (50 mW/10 V = 5 mA).

The operating load current should be midway between these two extremes. For example, assume that the maximum load current is 5 mA (as previously determined) and that the I_{CBO} is rated at 0.1 mA.

Multiply the I_{CBO} by 10 to find a minimum load current of 1 mA. Subtract the minimum 1 mA from the maximum 5 mA and divide by 2 to find an average or operating load current of 2 mA ($5 - 1 = 4$; $\frac{4}{2} = 2$).

Once the operating load current has been established, the next step is to calculate the load resistance value.

When operating at frequencies up to about 100 kHz, the load resistance value can be calculated on a direct-current basis. At higher frequencies, it may be necessary to select a load resistance on the basis of impedance. This is discussed in Chapters 5 and 6.

When the operating load current is flowing at the selected no-signal point, the collector load resistance should drop the collector voltage to one-half the supply voltage. This is typical for all class A circuits. For example, a source of 20 V is dropped to 10 V through the load resistance. Using the previous example, the desired 10 V is divided by 2 mA (the operating load current) to produce a load resistance of 5000 Ω.

Once the desired (or arbitrarily selected) load current and load resistance have been determined, the final step in the basic bias circuit design is to select the resistances for the emitter and base that will produce the correct operating point (no-signal collector current). Although there are many bias networks, each with its own advantages and disadvantages, one major factor must be considered for any network. The basic bias network must *maintain* the desired current within practical limits in the presence of temperature and frequency changes. This is often referred to as *bias stability* or *bias stabilization.*

2-9.2 Two-junction transistor bias stabilization techniques

There are several methods for providing bias stabilization of two-junction transistors. All of these methods involve a form of *negative feedback* or *inverse feedback*; that is, any change in transistor currents produces a corresponding voltage or current change that tends to offset the initial change.

There are two basic methods for producing inverse or negative feedback: *inverse-voltage feedback* and *inverse-current* feedback. Current feedback is more commonly used than voltage feedback since two-junction transistors are primarily current-operated devices rather than voltage-operated devices (such as vacuum tubes and FETs).

Figure 2-14 shows a typical inverse-current bias network using an *NPN* transistor. This system is commonly called *emitter feedback*. Other bias

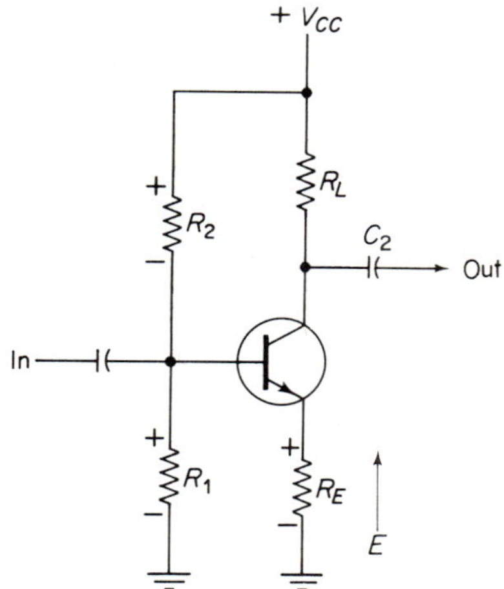

FIGURE 2-14 Typical bias network with inverse-current feedback (emitter feedback).

networks using the same principle are discussed in the following sections. However, the use of an emitter feedback resistance in any bias circuit can be summed up as follows:

Base current (and, consequently, collector current) depends on the *differential in voltage* between base and emitter. If the differential voltage is lowered, less base current (and, consequently, less collector current) will flow. The opposite is true when the differential is increased. All current flowing through the collector (ignoring collector-base leakage, I_{CBO}) also flows through the emitter resistor. The voltage drop across the emitter resistor is thus dependent (in part) on the collector current.

Should the collector current increase (for any reason), emitter current and the voltage drop across the emitter resistor will also increase. This increase tends to decrease the differential between base and emitter, thus lowering the base current. In turn, the lower base current tends to decrease the collector current and offset the initial collector-current increase.

For example, assume that the drop across R_1 is 1 V and that the drop across R_E is 0.5 V. This results in a base-emitter differential of 0.5 V. Now assume that the collector current increases, raising the current through R_E so that there is a drop of 0.8 V across R_E. This results in a differential of 0.2 V, which reduces both base and collector currents and tends to offset the initial increase in collector current.

Note that any form of negative or inverse feedback in an amplifier circuit tends to *oppose all changes*, even those produced by the signal being amplified. Thus, inverse or negative feedback tends to reduce and stabilize gain as well as undesired change. This principle of stabilizing gain by means of feedback is used in many types of amplifiers, as discussed in Chapters 5 and 6.

2-9.3 Some representative two-junction transistor bias networks

Figures 2-15 through 2-21 illustrate typical bias schemes used with two-junction transistors. All the bias networks use inverse-current feedback (emitter feedback) of some form. Note that the bias network resistors, in addition to providing the required voltage relationships at emitter, base, and collector, also set the approximate *input and output* impedance of the circuit, as shown by the equations on the illustrations. Likewise, when the circuits are used as amplifiers, approximate *voltage and current gains* are set by *resistance ratios*. The equations in the illustrations hold true for operation up to about 100 kHz (and higher in many instances). As operating frequencies increase beyond about 100 kHz, other factors enter into gain and impedance relationships. These factors are discussed in Chapters 5 and 6.

Maximum gain and minimum stability. In the circuit of Fig. 2-15, the value of R_B is selected to provide a given base current at the Q point. For an *NPN* silicon transistor, it can be assumed that the base voltage will be about

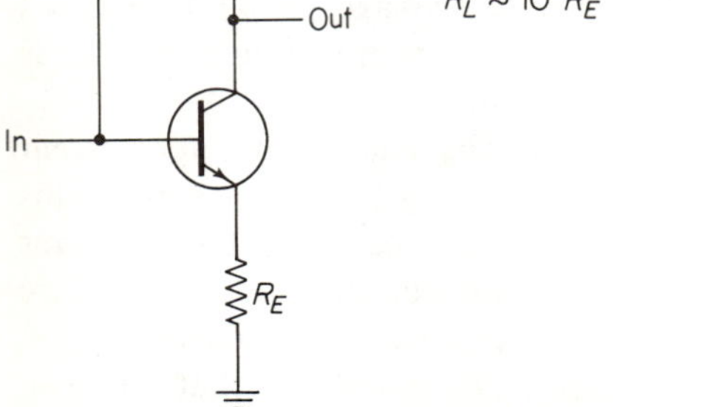

FIGURE 2-15 Bias network with maximum gain and minimum stability.

0.5 to 1 V more positive than the emitter. (The base of *PNP* silicon transistors will be 0.5 to 1 V more negative than the emitter.) If germanium transistors are used, the voltage difference will be about 0.2 to 0.5 V.

Some transistor datasheets show the collector current for a given base current or a given base-emitter voltage. When such information is available on datasheets, it is shown in the form of curves such as on Fig. 2-22. Note that the curves show a collector current of 2.4 mA with a base current of 0.2 mA and with a base-emitter differential of 0.5 V. These curves, usually called *transfer characteristic curves*, make it possible to calculate a more accurate value of R_B. In practice, however, the calculated value of R_B is only a trial value. A variable R_B is used when the circuit is in experimental form. R_B is adjusted until the transistor is at the desired operating point. Then R_B is measured and a fixed resistance of corresponding value is used in the final circuit.

The emitter voltage is dependent on the drop across R_E. As shown by the equations, the value of R_E is dependent on a tradeoff between stability and gain. An increase in the value of R_E, in relation to R_L, increases stability but decreases gain. Typically, the value of R_E is less than 1 kΩ and is generally not greater than 0.2 times R_L. An R_E that is 0.1 times R_L can be considered as typical.

The output impedance is approximately equal to R_L; whereas the input impedance is approximately equal to R_E times beta of the transistor. Since input impedance is dependent on transistor beta, the input impedance is subject to wide variation from transistor to transistor and with changes in frequency. Generally, this is considered a disadvantage.

The circuit voltage gain is approximately equal to the ratio of R_L/R_E times beta; whereas the current gain is approximately equal to beta. Again, this produces a wide variation in gain. The circuit of Fig. 2-15 offers the greatest possible voltage gain but the least stability of all the basic bias circuits described here.

Maximum gain and improved stability. In the circuit of Fig. 2-16, the value and characteristics are essentially the same as for the circuit of Fig. 2-15; however, the stability of the Fig. 2-16 circuit is increased by connecting base resistance R_B to the collector rather than to the power supply. If collector current increases, the drop across R_L also increases lowering the voltage at the collector. This lowers the base voltage and current, thus reducing the collector current. The feedback effect is combined with that produced by the emitter resistor R_E to offset any variation in collector current. Because of the increased stability produced by the feedback, however, gain for the Fig. 2-16 circuit is slightly less than for the Fig. 2-15 circuit.

Maximum stability. In the circuit of Fig. 2-17, the output impedance is approximately equal to R_L; whereas the input impedance is approximately equal to R_B. In theory, the input impedance is equal to R_B in parallel with R_E

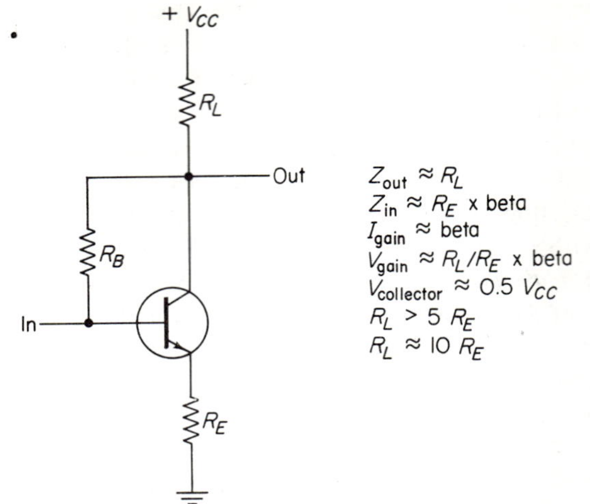

FIGURE 2-16 Bias network with maximum gain and improved stability.

times (beta + 1). Unless beta is very low, however, the R_E times (beta + 1) factor will be much greater than R_B. Thus, the value of R_B (or slightly less) can be considered as the circuit input impedance.

The circuit of Fig. 2-17 offers more stability than the other bias networks but with a tradeoff of lower gain and lower input impedance.

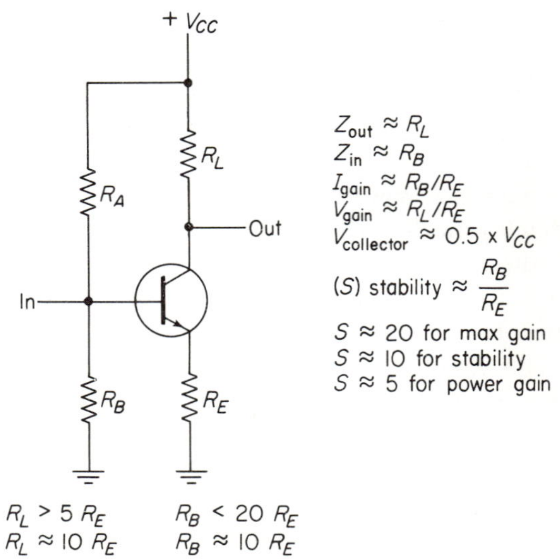

FIGURE 2-17 Bias network with improved stability. Circuit charactertistics are dependent upon circuit values.

The value of R_L is determined by the desired collector voltage and current or by an arbitrary need for a given output impedance. The value of R_E is dependent on a tradeoff between stability and gain. An increase in the value of R_E, in relation to R_L, increases stability and decreases voltage gain. An increase in the value of R_E, in relation to R_B, increases stability and decreases current gain. The value of R_E is typically less than 1 kΩ and should not be greater than 0.2 times R_L. An R_E that is 0.1 times R_L can be considered as typical.

The value of R_B is dependent on tradeoffs among the value of R_E, current gain, stability, and the desired input impedance. As a basic rule, R_B is about 10 times R_E. A higher value of R_B will increase current gain and decrease stability. Input impedance of the circuit will be approximately equal to R_B (actually slightly less). Thus, if the input impedance is of special importance in the circuit, the value of R_B is often selected on that basis. This may require a specific value of R_E to maintain stability and current gain relationships. Of course, any change in R_E results in a change of the voltage gain (assuming that R_L remains the same). The value of R_A is selected to provide a given base voltage at the Q point.

In addition to stability, the major advantage of the Fig. 2-17 circuit is that the input and output impedances, as well as voltage and current gain, are not dependent on transistor beta. Instead, circuit characteristics are dependent on circuit values.

Diode temperature stabilization. In the circuit of Fig. 2-18, the basic characteristics are the same as for the Fig. 2-17 circuit except that temperature stability is increased by connecting diode D between the base and R_B. Diode D (always forward biased) is of the same material (silicon or germanium) as the base-emitter junction and is maintained at the same temperature. In practice, diode D is mounted near the transistor so that the base-emitter junction and diode D are at the same temperature. Thus, the

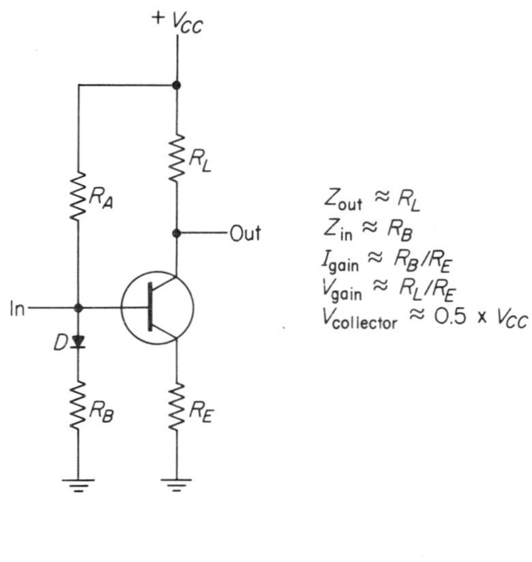

FIGURE 2-18 Bias network with diode for improved temperature stability.

voltage drops across diode D and the base-emitter junction are the same and remain the same with changes in temperature.

For example, should temperature rise, the emitter and collector currents of the transistor also tend to rise. The same temperature rise causes the forward resistance of diode D to decrease. As a result, the current flowing through R_A increases, causing an increase in the voltage drop across R_A. Since this voltage drop tends to reverse bias the emitter-base junction, the net forward bias for that junction is reduced and the emitter-collector currents are reduced to normal (or tend to reduce to normal). Should the temperature fall, the action is in the reverse direction and the emitter-collector currents are raised toward their normal values.

Thermistor temperature stabilization. In the circuit of Fig. 2-19, the basic characteristics are the same as for the Fig. 2-17 circuit except that the *negative temperature coefficient* (NTC) characteristcs of a *thermistor* are used to provide temperature compensation. The resistance of a thermistor decreases with increases in temperature and vice versa. For best results, the thermistor is mounted near the transistor so that both devices are at the same temperature.

With the Fig. 2-19 circuit, the thermistor is used to vary the emitter voltage with temperature variations to minimize the effects of these variations upon the emitter current. Resistors R_A and R_B form a voltage divider to apply

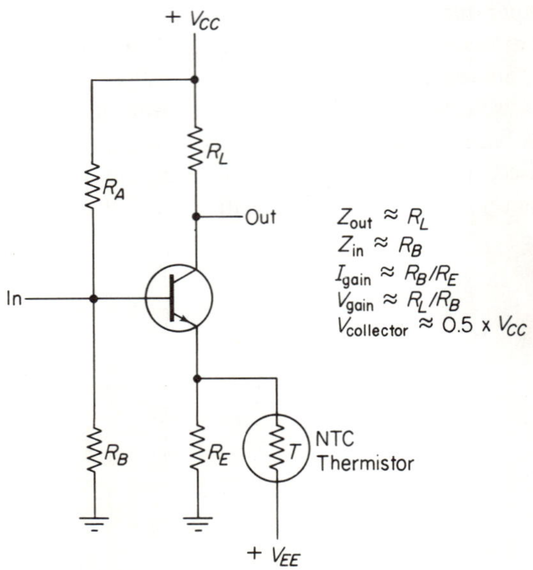

FIGURE 2-19 Bias network with thermistor for improved temperature stability.

a portion of the supply voltage V_{CC} in a direction to forward bias the emitter-base junction.

Resistor R_E and the thermistor form a second voltage divider across V_{EE}. The direction of the voltage drop across R_E is such as to reverse bias the base-emitter junction. Since the forward bias applied to the base is larger than the reverse bias applied to the emitter, however, the net result is that the emitter-base junction is forward biased.

Should the temperature rise, the emitter and collector currents tend to rise. The same rise in temperature reduces the resistance of the thermistor. This action permits more current to flow through the voltage divider. The increase in current flow increases the voltage drop across R_E, thus increasing the reverse bias being applied to the emitter-base junction. As a result, the net forward bias of the junction is reduced, thus reducing the emitter and collector currents toward their normal values. Should the temperature decrease, the action is reversed, thus preventing a decrease in emitter and collector currents.

Positive and negative base current supply. In the circuit of Fig. 2-20, the basic characteristics are the same as for the Fig. 2-17 circuit. The circuit of Fig. 2-20 is used in those special applications, however, that require a negative and positive voltage, each with respect to ground, to control base current. The values for R_L, R_E, and R_B in Fig. 2-20 are the same as for the Fig. 2-17 circuit. The value of R_A in the Fig. 2-20 circuit is different, due to the large amount of current through R_B.

Positive and negative emitter-current supply. In the circuit of Fig. 2-21, the characteristics are somewhat different than in all the previous bias networks. The Fig. 2-21 circuit is used in those special applications where it is

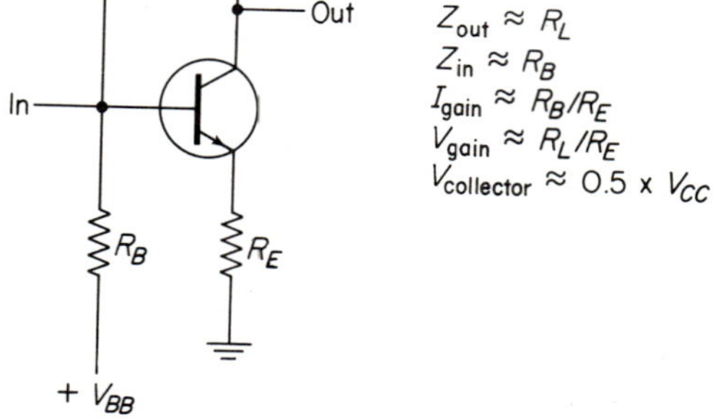

FIGURE 2-20 Bias network with positive and negative supply voltages for control of base current.

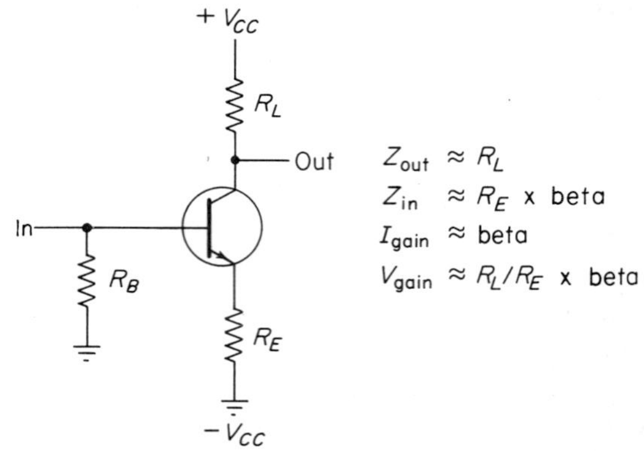

FIGURE 2-21 Bias network with positive and negative supply for emitter-collector currents.

necessary to supply collector-emitter current from both a positive and negative source. Since the transistor is *NPN*, the collector is connected to the positive source through R_L, while the emitter is connected to the negative source through R_E. If both sources are approximately equal, there is generally no voltage gain. The collector and emitter currents are approximately equal

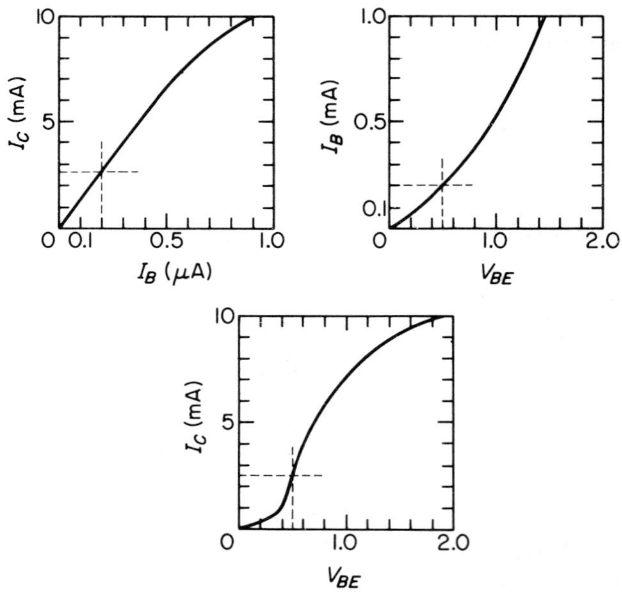

FIGURE 2-22 Typical transfer characteristic curves. (Lenk, *Handbook of Simplified Solid-State Circuit Design*, Prentice-Hall.)

(ignoring the base current). Thus, if R_L drops the positive source to half (say from 20 to 10 V), then R_E will be approximately twice the resistance of R_L. This will produce a voltage loss, all other factors being equal.

2-10. INTERPRETING TWO-JUNCTION TRANSISTOR DATASHEETS

Most of the basic design information for a particular transistor can be obtained from the datasheet. There are some exceptions to this rule. For extreme high-frequency work and in digital work where switching characteristics are of particular importance, it may be necessary to test a transistor under simulated operating conditions.

In any event, it is always necessary to interpret datasheets and each manufacturer has its own system. It would be impractical to discuss all datasheet formats here. Instead, we shall discuss the typical information found on datasheets and see how this information affects simplified design.

Figure 2-23 is the datasheet for a 2N332 transistor. This transistor is an industrial type, suitable for low-power *AF* or *RF* (up to about 10 MHz) as well as switching.

2-10.1 Maximum voltage

The first specifications are those of maximum voltage. In Fig. 2-23, the maximum collector voltage is listed as V_{CBO} of 45 V. Actually this is a test voltage rather than an operating design voltage. (V_{CBO} usually indicates collector-base breakdown voltage, with the emitter circuit open, as discussed in Sec. 2-13. Transistors do not operate in this way in circuits.)

For design purposes, the 45-V figure can be considered as the *absolute maximum* collector voltage. Keep the following in mind on maximum collector voltage: Except for *RF* circuits used in transmitters, most transistors are operated with their collector at some voltage value (V_C) less than the source voltage (V_{CC}). For example, in a typical class A circuit, the collector voltage is half the source voltage, at the normal operating point. The collector voltage will rise to or near the source voltage when the transistor is at or near cutoff. Therefore,

Never design any circuit where the collector is connected to a source higher than the maximum voltage rating, even through a resistance.

The next maximum voltage design problem to be considered is the type of source voltage. A battery power source will not deliver more than its rated voltage; however, any electronic power source is subject to some voltage variation. Thus,

Always allow for some variation in source voltage when an electronic power supply is used.

TYPICAL TRANSISTOR SPECIFICATION 2N332

ABSOLUTE MAXIMUM RATINGS (25°C)

Voltages:		
Collector to base (emitter open)	V_{CBO}	45 V
Emitter to base (collector open)	V_{EBO}	1 V
Collector current	I_C	25 mA
*Power**		
Collector dissipation (25°C)	P_C	150 mW
Collector dissipation (125°C)	P_C	50 mW
Temperature range:		
Storage	T_{STG}	−65 to 200°C
Operating	T_A	−55 to 175°C

ELECTRICAL CHARACTERISTICS (25°C)

(Unless otherwise specified, $V_{CB} = 5$ V; $I_E = -1$ mA; $f = 1$ kc)

Small-signal characteristics:		*min*	*nom*	*max*	
Current transfer ratio	h_{fe}	9	15	20	
Input impedance	h_{ib}	30	53	80	Ω
Reverse voltage transfer ratio	h_{rb}	0.25	1.0	5.0	$\times 10^{-4}$
Output admittance	h_{ob}	0.0	0.25	1.2	μΩ
Power gain ($V_{CE} = 20$ V; $I_E = -2$ mA; $f = 1$ kc; $R_G = 1$ kΩ; $R_L = 20$ kΩ)	G_e		35		dB
Noise figure	NF		28		dB
High-frequency characteristics:					
Frequency cutoff ($V_{CB} = 5$ V; $I_E = -1$ mA)	f_{ab}		15		Mc
Collector-to-base capacity ($V_{CB} = 5$ V; $I_E = -1$ mA; $f = 1$ Mc)	C_{ob}		7		μμF
Power gain (common emitter) ($V_{CB} = 20$ V; $I_E = -2$ mA; $f = 5$ Mc)	G_e		17		dB
dc characteristics:					
Collector breakdown voltage ($I_{CBO} = 50$ μA; $I_E = 0$; $T_A = 25°$C)	BV_{CBO}	45			V
Collector cutoff current ($V_{CB} = 30$ V; $I_E = 0$; $T_A = 25°$C)	I_{CBO}		0.02	2	μA
($V_{CB} = 5$ V; $I_E = 0$; $T_A = 150°$C)	I_{CBO}			50	μA
Collector saturation resistance ($I_B = 1$ mA; $I_C = 5$ mA)	R_{SC}		80	200	Ω
Switching characteristics: ($I_{B1} = 0.4$ mA; $I_{B2} = -0.4$ mA; $I_C = 2.8$ mA)					
Delay time	t_d		0.75		μs
Rise time	t_r		0.5		μs
Storage time	t_s		0.05		μs
Fall time	t_f		0.15		

*Derate 1 mW/°C increase in ambient temperature.

FIGURE 2-23 Typical transistor datasheet (2N332). (Lenk, *Handbook of Simplified Solid-State Circuit Design*, Prentice-Hall.)

Another factor that affects maximum voltage is temperature. Note that in Fig. 2-23 the maximum voltage is specified at 25°C. Usually, breakdown will occur at a lower voltage when temperature is increased. The topic of how operating temperatures affect transistor design is discussed further in Sec. 1-7 and 2-12.

In Fig. 2-23, the maximum base-emitter voltage is listed as V_{EBO} of 1 V. Again, this is a test voltage rather than an operating design voltage. (Refer to Sec. 2-13.) Usually, the base-emitter junction has some current flowing at all times. The voltage drop across the junction is about 0.2 to 0.3 for germanium and 0.5 to 0.7 for silicon transistors. The lower voltage drops produce some current flow, while the higher drops produce heavy current flow. As discussed in Sec. 2-9, bias circuits are designed to produce the desired drop. Either the higher or lower drops can be used, depending on the desired results.

In general, the lower drops produce less current drain and lower no-signal power dissipation. The higher drops may result in operation on a more linear portion of the transistor characteristics.

In practical work, it is often necessary to select a bias (base-emitter voltage) on the basis of *input signal* rather than on some arbitrary point of the transistor's characteristic curve. Keep in mind that the input signal to a transistor can come from an external source or a previous stage or can be *feedback* in the case of an oscillator or operational amplifier. Thus,

Always consider any input signal that may be applied to the base-emitter junction, in addition to the normal operating bias.

2-10.2 Collector current

In Fig. 2-23, the collector current is listed as I_C of 25 mA at 25°C. Collector current increases with temperature (and temperature increases as current increases). Thus,

Do not operate any transistor at or near its maximum current rating.

Of course, if you could be absolutely certain that the transistor will dissipate any temperature increases (this is a practical impossibility), the circuit can be designed to operate near the maximum current.

In practical design applications, it is the *power dissipated* in the collector circuit (rather than a given current) that is of major concern. For example, assume that the collector is operated at 45 V and 25 mA. This results in a power dissipation of over 1 W, far above the 150 mW specified for 25°C.

2-10.3 Power and temperature range

The power dissipation capabilities of a transistor in any circuit are closely associated with the temperature range. As shown in Fig. 2-23, power dissipation is 150 mW at 25°C and 50 mW at 125°C and must be *derated*

1 mW for each degree (°C) increase in ambient temperature. Refer to Sec. 1-7 and 2-12.

2-10.4 Small-signal characteristics

Small-signal characteristics can be defined as those where the ac signal is small compared to the dc bias. For example, h_{fe} or *forward current transfer ratio* (also known as ac beta or dynamic beta) is properly measured by noting the change in collector alternating current for a given change in base alternating current, without regard to static base and collector currents.

Small-signal characteristics do not provide a truly sound basis for practical design. As discussed in related chapters, the performance of a transistor in a working circuit can be controlled by the circuit component values (within obvious limits, of course). There are two basic reasons for this approach.

First, not all manufacturers list the same small-signal characteristics on their datasheets. To further complicate matters, manufacturers call the same characteristic by different names (or even use the same name to identify different characteristics).

Second, the small-signal characteristics listed in datasheets are based on a set of *fixed operating conditions.* If the conditions change (as they must in any practical circuit), the characteristics will change. For example, beta changes drastically with temperature, frequency, and *operating point.* Therefore,

Use small-signal characteristics as a starting point for simplified design, not as hard and fast design rules.

2-10.5 High-frequecny characteristics

High-frequency characteristics are especially important in the design of *RF* networks. As discussed in Chapter 6, networks (such as *RF* stages in a transmitter) provide the dual function of frequency selection (tank circuit) and impedance matching between the transistor and a load.

The high-frequency information provided in many datasheets (such as Fig. 2-23) is not adequate for simplified *RF* circuit design. To match impedances properly, both the resistive (real part) and reactive (imaginary part) components must be considered. The reactive component (either inductive or capacitive) changes with frequency. Thus, it is necessary to know the reactance values over a wide range of frequencies and not at some specific test frequency (unless you happen to be designing for that frequency only).

The best way to show how resistance and reactance vary in relation to frequency for a particular transistor is by means of curves. Fortunately, manufacturers who are trying to sell their transistors for high-frequency use generally provide a set of curves showing the characteristics over the anticipated frequency range.

2-10.6 Direct-current characteristics

Direct-current characteristics, while important in the design of basic bias circuits, do not have too critical an effect on the operation of the complete working circuit. The dc characteristics shown in Fig. 2-23 are primarily test values rather than design parameters. The important point to remember regarding such dc characteristics is that they serve as a starting point for bias design and that they will change with temperature.

2-10.7 Switching characteristics

Switching characteristics are important in the design of all pulse circuitry (multivibrators, gates, etc.). The switching times shown in Fig. 2-23 are defined in Fig. 2-24. These time factors (delay, storage, rise, fall) determine the operating limits for switching circuits. For example, if a transistor gate with a 20-ns rise time is used to pass a 15-ns pulse, the pulse will be hopelessly distorted. Similarly, if there is a 1.5-μs delay added to a 1-μs pulse, an absolute minimum of 2.5-μs is required before the next pulse can occur. This results in a maximum pulse repetition rate (prr) or pulse repetition frequency (prf) of 400 kHz ($1/2.5^{-6} = 400{,}000$ Hz). Actually, the prf must be lower since there is some "off" time between pulses.

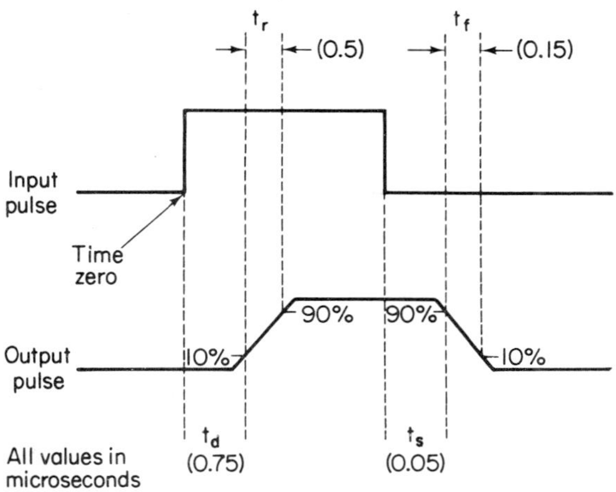

FIGURE 2-24 Definition of switching-time characteristics. (Lenk, *Handbook of Simplified Solid-State Circuit Design*, Prentice-Hall.)

2-11. DETERMINING TWO-JUNCTION TRANSISTOR PARAMETERS AT DIFFERENT FREQUENCIES

Transistor datasheets do not always specify all the parameters needed. Likewise, the datasheets specify the parameters at some given operating frequency when it is desired to know the parameter at another frequency.

The following paragraphs describe methods for converting from one parameter to another at different frequencies.

2-11.1 Basic parameter relationships

The two major parameters are h_{fb} (alpha, the common base ac short-circuit forward current gain) and h_{fe} (beta, the common emitter ac short-circuit forward current gain). The following glossary defines terms used in this paragraph and in later chapters.

GLOSSARY OF TWO-JUNCTION TRANSISTOR TERMS

- h_{fb} Common-base ac forward current gain (alpha).
- h_{fbo} Value of h_{fb} at 1 kHz.
- h_{fe} Common-emitter ac forward current gain (beta).
- h_{feo} Value of h_{fe} at 1 kHz.
- f_{ab} Common-base current-gain cutoff frequency. Frequency at which h_{fb} has decreased to a value 3 dB below h_{fbo} ($h_{fb} = 0.707 \times h_{fbo}$).
- f_{ae} Common-emitter current-gain cutoff frequency. Frequency at which h_{fe} has decreased to a value 3 dB below h_{feo} ($h_{fe} = 0.707 \times h_{feo}$).
- f_T Gain-bandwidth product. Frequency at which $h_{fe} = 1$ (0 dB).
- G_{pe} Common-emitter power gain.
- f_{max} Maximum frequency of oscillation. Frequency at which $G_{pe} = 1$ (0 dB).
- $K\theta$ Excess phase-shift factor. Factor that is a function of excess phase shift of current in the base of a transistor.

h_{fbo} (the value of h_{fb} at 1 kHz) remains constant as frequency is increased, until a top limit is reached. At that point, h_{fb} begins to drop rapidly. The frequency at which a significant drop in h_{fb} occurs provides a basis for comparison of the expected high-frequency performance of different transistors. This frequency is known as f_{ab} and is defined as that frequency at which h_{fb} is 3 dB below h_{fbo}. Expressed in magnitude, h_{fb} at f_{ab} is 70.7% of h_{fbo}.

A curve of h_{fb} versus frequency for a transistor with an f_{ab} of 1 MHz is shown in Fig. 2-25. This curve has the following significant characteristics:

1. At frequencies below f_{ab}, h_{fb} is nearly constant and approximately equal to h_{fbo}.
2. h_{fb} begins to decrease significantly in the region of f_{ab}.
3. Above f_{ab}, the rate of decrease in h_{fb} with increasing frequency approaches 6 dB per octave.

The curve of common-base current gain versus frequency for any

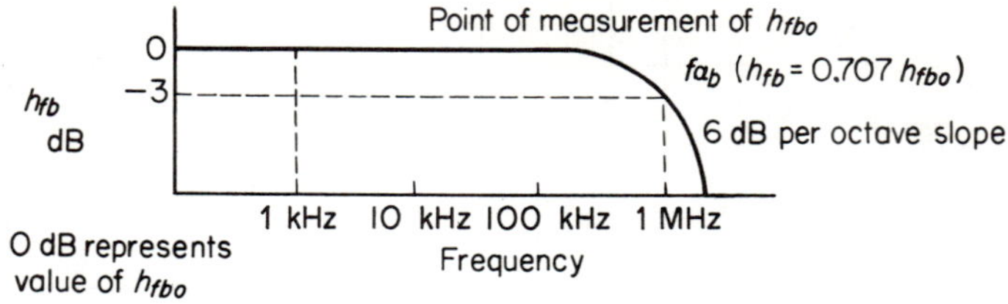

FIGURE 2-25 Common base current gain versus frequency (courtesy Motorola).

transistor has these characteristics and the same general appearance as the curve of Fig. 2-25.

2-11.2 Common-emitter parameter $\mathbf{f_{ae}}$

The common-emitter parameter that corresponds to f_{ab} is f_{ae}, the common-emitter current-gain cutoff frequency. f_{ae} is the frequency at which h_{fe} (beta) has decreased 3 dB below h_{feo}. A typical curve of h_{fe} versus frequency for a transistor with an f_{ae} of 100 kHz is shown in Fig. 2-26.

These characteristics allow such a curve to be constructed for a particular transistor by knowing only h_{feo} and f_{ae}. From the curve, h_{fe} at any frequency can be determined. If f_{ae} is not known, a curve can also be constructed if h_{feo} and h_{fe} at any frequency above f_{ae} are known. Thus, to find h_{fe} at any frequency, it is necessary to know only h_{feo} and either f_{ae} or h_{fe} at some frequency greater than f_{ae}.

Sometimes h_{feo} is needed and only h_{fbo} is given or vice versa. The quantities h_{fbo} and h_{feo} are related by the following:

$$h_{feo} = \frac{h_{fbo}}{1} - h_{fbo} \tag{2-1}$$

$$h_{fbo} = \frac{h_{feo}}{h_{feo}} + 1 \tag{2-2}$$

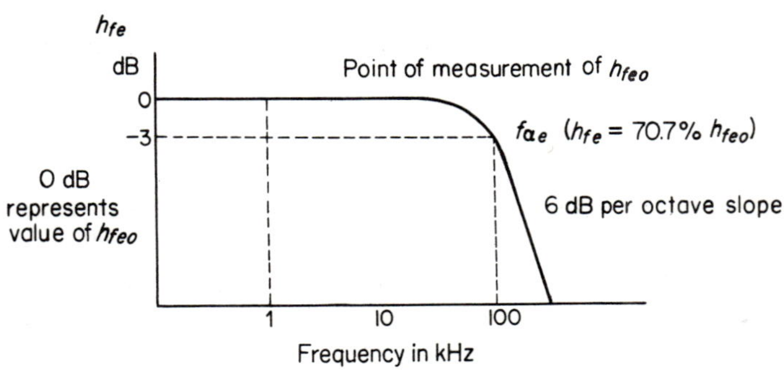

FIGURE 2-26 Common-emitter current gain versus frequency (courtesy Motorola).

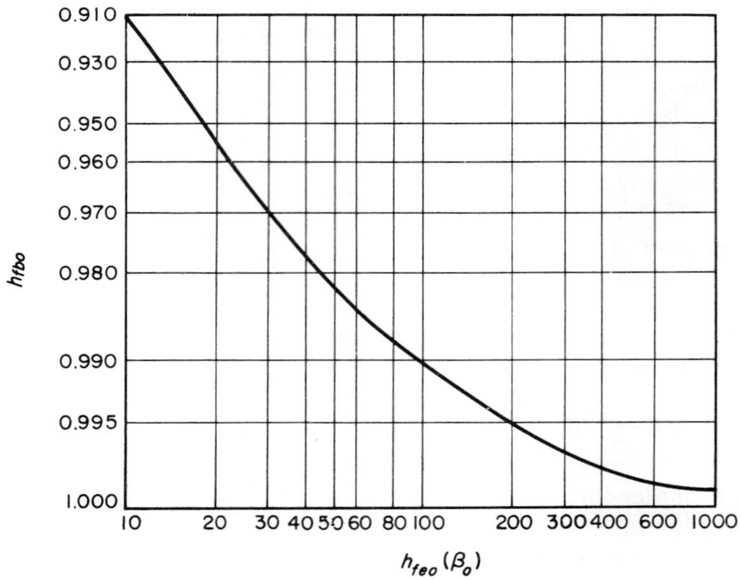

FIGURE 2-27 Relationship of h_{feo} and h_{fbo} (courtesy Motorola).

Equations 2-1 and 2-2 are plotted in Fig. 2-27. To simplify the problem, the low-frequency current scales of Fig. 2-31 through 2-34 contain both an h_{fbo} and h_{feo} scale and may be entered with either quantity.

2-11.3 Relationships between f_{ae} *and* f_{ab}

Figures 2-31 through 2-34 provide a means of converting from f_{ae} to f_{ab} and vice versa. The nomograms in these illustrations are based on a $K\theta$ (excess phase-shift factor) of 0.5 to 1.0. Most transistors have a $K\theta$ in the 0.8 to 1.0 range. Certain germanium transistors have a $K\theta$ of about 0.6. Alloy transistors have a $K\theta$ of 0.82.

The quantity f_{ae} is usually a much lower frequency than f_{ab} for the same transistor. In general, a common-base circuit can be operated at higher frequencies than a common-emitter circuit, using the same transistor.

2-11.4 Determining h_{fe} *at different frequencies*

As previously stated, h_{fe} is probably the most important parameter and is specified on most transistor datasheets; however, h_{fe} is usually specified for one frequency only. This limits the usefulness of the datasheet for practical purposes. Figures 2-27 through 2-34 can be used to determine h_{fe} at various frequencies. The following rules summarize how to find h_{fe} at a particular frequency.

1. When the frequency is less than f_{ae}, then h_{fe} is approximately equal to h_{feo}.
2. When the frequency is equal to (or near) f_{ae}, then h_{fe} is approximately equal to $0.7 \times h_{feo}$.
3. When the frequency is greater than f_{ae}, consider h_{fe} to be decreasing at 6 dB per octave at the frequency, and use Fig. 2-28 to find h_{fe}. The nomogram of Fig. 2-28 is based on entering the f (frequency) scale with the frequency at which it is desired to find h_{fe} and the f_T scale

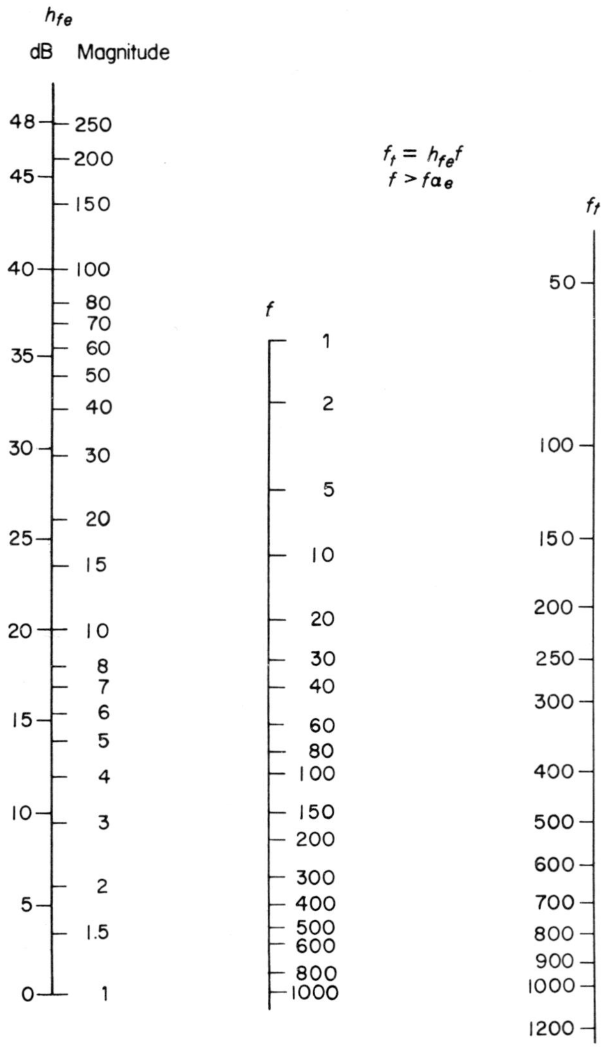

FIGURE 2-28 Finding h_{fe} when frequency is greater than f_{ae} (courtesy Motorola).

with the gain bandwidth product. f_T is defined as that frequency at which h_{fe} is 1 (or 0) dB.

The value of f_T is sometimes specified indirectly on high-frequency transistor datasheets. This is done by specifying h_{fe} at some frequency above f_{ae}. f_T is then obtained by multiplying the magnitude of h_{fe} by the frequency of measurement. This relationship arises from the 6 dB per octave characteristic of the h_{fe} versus frequency curve (Fig. 2-25 and 2-26) above f_{ae}. Since 6 dB represents a current-gain magnitude of 2, h_{fe} is halved each time frequency is doubled and vice versa. Thus, the product of h_{fe} and the frequency of the sloping portion of the curve yields f_T.

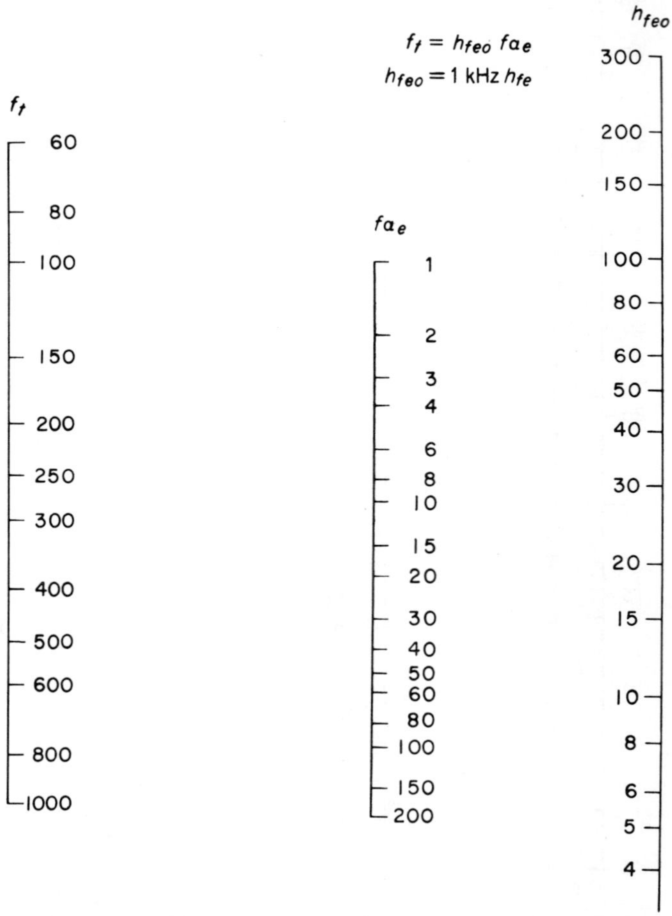

FIGURE 2-29 Finding f_T when f_{ae} and h_{feo} are known (courtesy Motorola).

f_T is also equal to the product of h_{feo} and f_{ae} expressed by

$$f_T = h_{feo} \times f_{ae} \tag{2-3}$$

With h_{feo} known, Eq. 2-3 provides a simple means of finding f_{ae} when f_T is known or vice versa. Equation 2-3 is represented by the nomogram of Fig. 2-29.

As a guide, remember that f_T is approximately equal to or slightly less than f_{ab} (the common-base cutoff frequency).

4. If the datasheet specifies h_{fbo} instead of h_{feo}, use Fig. 2-27 to find h_{feo}.

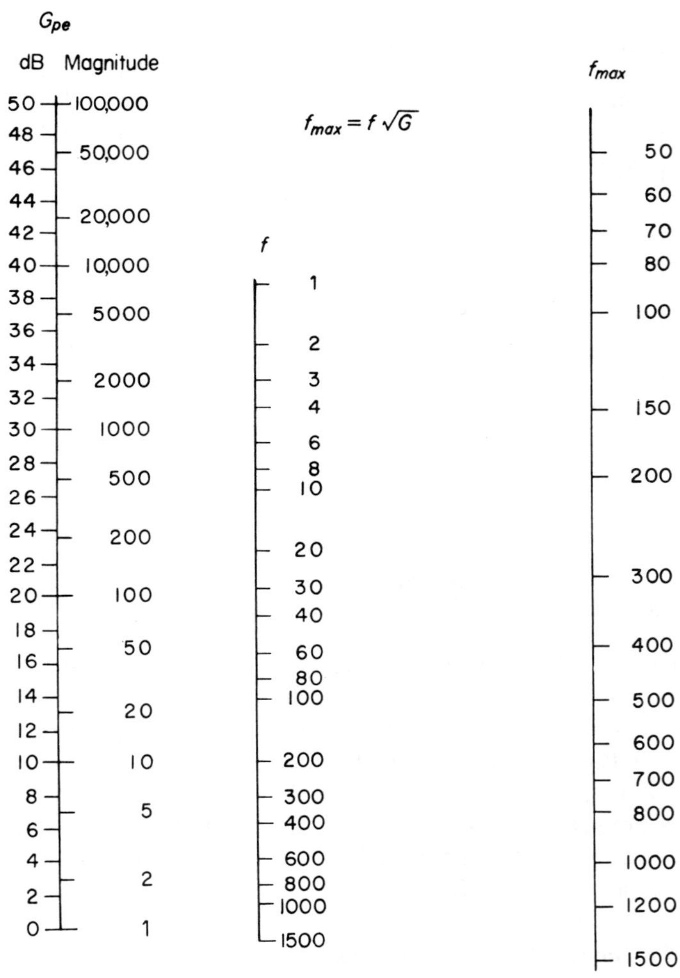

FIGURE 2-30 Finding f_{max} when f and power gain are known (courtesy Motorola).

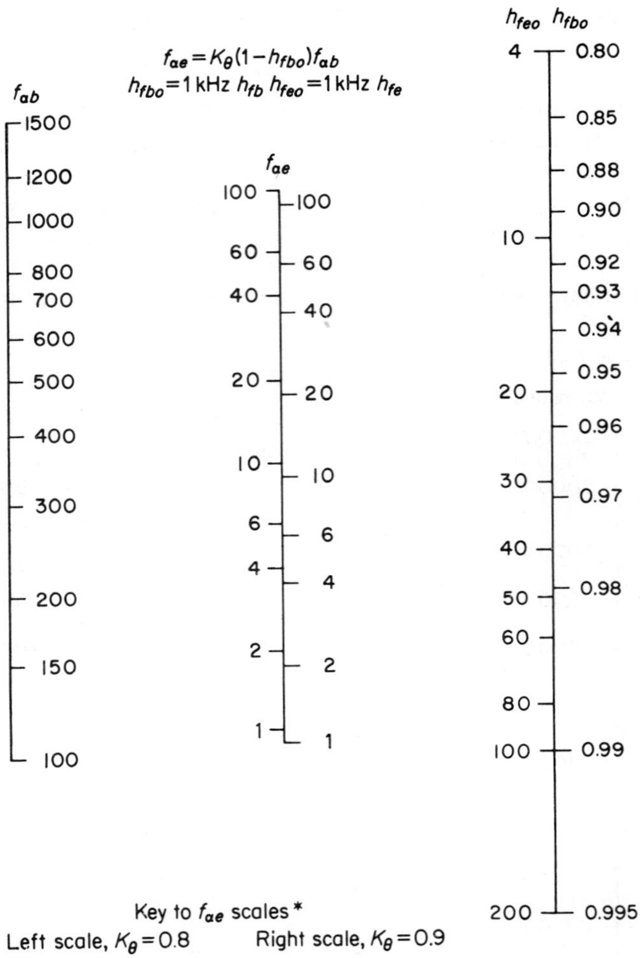

FIGURE 2-31 Finding f_{ae} when f_{ab} is known (courtesy Motorola).

If the datasheet does not specify f_T, use Fig. 2-29 to find f_T (with h_{feo} and f_{ae} known) or Fig. 2-32 to find f_T (with f_{ab} known). Use Fig. 2-34 instead of Fig. 2-32 to find f_T if the transistor is a MADT germanium transistor.

If f_{ae} is not specified, use Fig. 2-29 to find f_{ae} (with f_T and h_{feo} known) or Fig. 2-31 to find f_{ae} (with f_{ab} known). Use Fig. 2-33 instead of Fig. 2-31 to find f_{ae} if the transistor is a MADT type.

2-11.5 Using the curves and nomograms

The following general instructions apply to the curves and nomograms of Fig. 2-25 through 2-34.

All power-gain and current-gain scales (except h_{fbo} and h_{feo}) are cali-

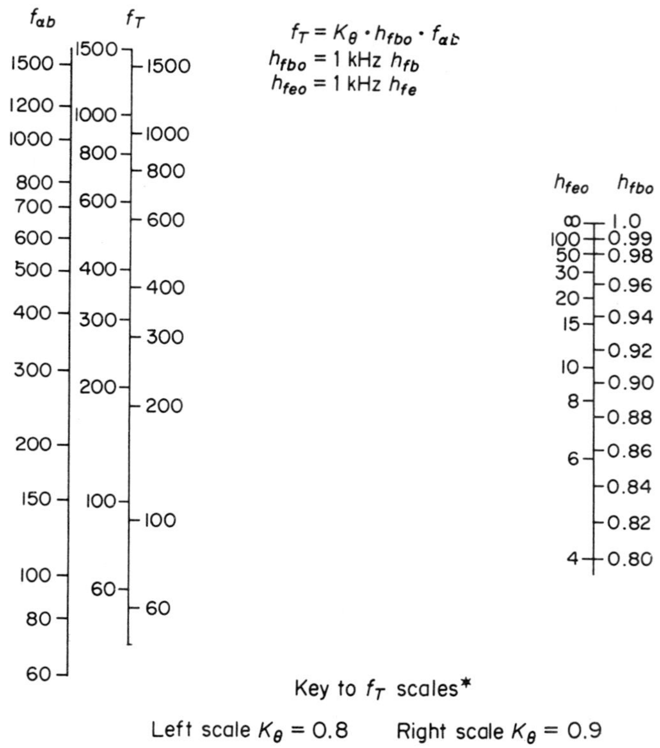

FIGURE 2-32 Relationship of f_T, f_{ab} and either h_{fbo} or h_{feo} (courtesy Motorola).

brated in both actual magnitudes (ratios) and decibel values for convenience.

The nomograms assume no shift in operating point. Known parameters used to find an unknown must be measured at the *same collector voltage and collector current* as the desired unknown.

Frequency scales on the nomograms are calibrated in numbers only, without units. Also, all nomograms contain two frequency scales. Decimal points may be shifted on the frequency scales of any nomogram as long as they are shifted the same amount on both scales. That is, both frequency scales of a nomogram must be multiplied by 10 to the same power. This permits the same nomogram to be used for both high- and low-frequency transistors.

The nomograms assume that both power gain and current gain decrease with increasing frequency at a rate of 6 dB per octave at high frequencies.

2-11.6 Example of using the nomograms

The following is an example of how the nomograms can be used to find various parameters of a transistor. The datasheet of a 2N337 transistor shows

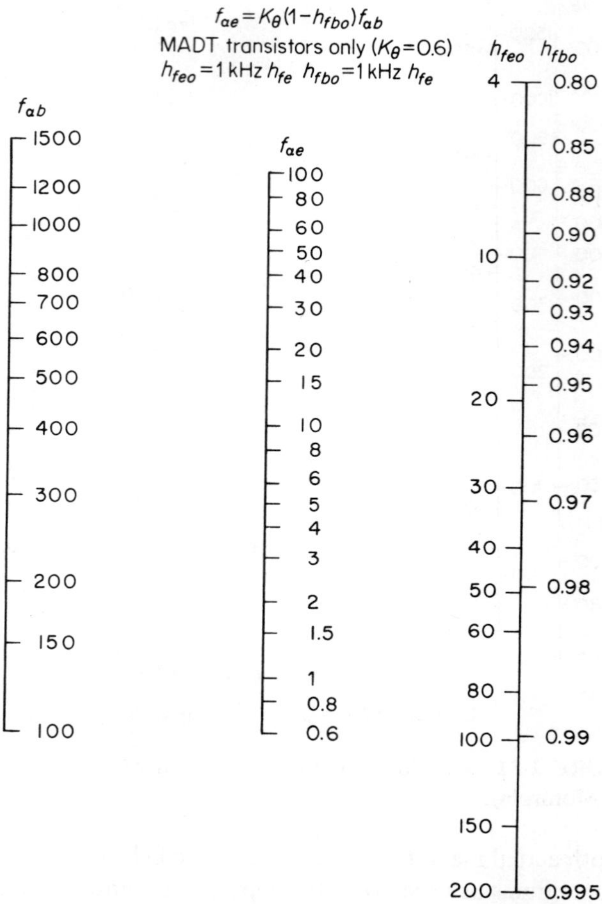

FIGURE 2-33 Finding f_{ae} when f_{ab} is known (for MADT type transistors) (courtesy Motorola).

a "typical" current transfer ratio h_{fe} of 55 (magnitude) and a "typical" alpha cutoff frequency of 30 MHz. (The datasheet lists alpha cutoff frequency as f_{hfb}.) Assume that it is desired to find the h_{fe} at 2.5 MHz. Also, assume that the transistor has a phase shift factor $K\theta$ of approximately 0.9.

Using Fig. 2-32 with a known f_{ab} of 30 MHz (use the 300 mark on the f_{ab} scale, and move decimal one place left) and an h_{feo} (h_{fe} at 1 kHz) of 55, find an f_T of 28 MHz (shown as 280 MHz on right-hand f_T scale).

Using Fig. 2-28 with an f_T of 28 MHz (280 MHz) and the desired frequency of 2.5 MHz (25 on the f scale), find an h_{fe} of approximately 12 (magnitude) or 23 dB. This is close to the "typical" 24-dB figure of h_{fe} at 2.5 MHz, also shown on the datasheet.

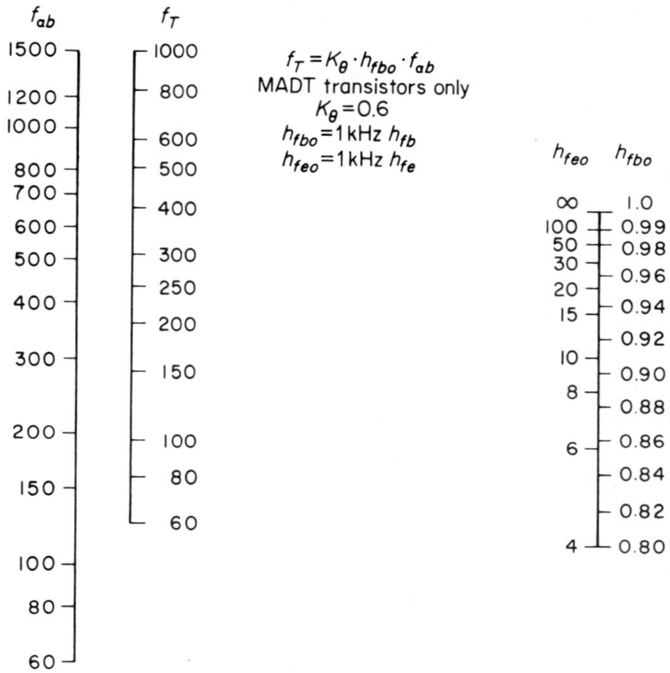

FIGURE 2-34 Relationship of f_T, f_{ab} and either h_{fbo} or h_{feo} (for MADT type transistors) (courtesy Motorola).

Using Fig. 2-29 with f_T of 28 MHz (280 MHz) and h_{feo} of 55, find f_{ae} to be 0.5 MHz (5 MHz).

As an alternate method, use Fig. 2-31 to find f_{ae} using the known f_{ab} and h_{feo}. Once f_{ae} is found, use Fig. 2-29 to find f_T.

2-11.7 Determining maximum operating frequency of a transistor

Although common-emitter *current gain* is equal to 1 at f_T, there may still be considerable *power gain* at f_T due to different input and output impedance levels. Thus, f_T is not necessarily the highest useful frequency of operation for a transistor. An additional parameter, the maximum frequency of oscillation ($f_{\max}$), can be found.

The term $f_{\max}$ is the frequency at which common-emitter *power gain* is equal to 1 and is related to f_T by

$$f_{\max} \approx \frac{f_T}{25.12 \times r_b' \times C_c}$$

where r_b' is base resistance

C_c is collector capacitance.

A plot of common-emitter power gain versus frequency has the same characteristics as shown in Fig. 2-25 and 2-26. This leads to another gain-bandwidth product:

$$f_{max} = f \times \sqrt{\text{power gain}}$$

where f is the frequency of measurement and *power gain is expressed in magnitude* and not decibels.

f_{max} can be found by measuring power gain at some frequency on the 6 dB per octave portion of the power gain versus frequency curve and multiplying the square root of the power gain (expressed in magnitude) with the frequency of measurement.

The nomogram of Fig. 2-30 permits the f_{max} to be found when the common-emitter power gain (G_{pe}) is known (either in magnitude or decibels).

The parameters are voltage and current dependent, and operating point must be considered in all cases. For example, the high-frequence h_{fe} measured at one collector voltage and current must not be used to calculate f_T directly at another voltage and/or current, without considering the added effects of the different operating point.

The parameter f_{ae} for present high-frequency transistors usually lies in the region between 10 and 50 MHz. The term h_{fe}, measured at any frequency above this region, is assumed on the 6 dB per octave portion of h_{fe} versus frequency curve and is used to calculate f_T directly.

Power gain measured at any frequency above 50 MHz is assumed on the 6 dB per octave portion of the power gain versus frequency curve and is used to calculate f_{max} directly.

2-12. DETERMINING TWO-JUNCTION TRANSISTOR PARAMETERS AT DIFFERENT TEMPERATURES

The following paragraphs describe methods for determining transistor parameters at temperatures other than those specified on the datasheet.

2-12.1 Collector-base leakage current I_{cbo} *versus temperature*

Figure 2-35 shows the typical change of I_{cbo} with temperature for a number of different power transistors. Five typical curves (three germanium and two silicon) are shown. The dotted lines indicate the increases in I_{cbo} using the following rules of thumb:

I_{cbo} *doubles with every 10°C increase in temperature for germanium transistors.*

I_{cbo} *doubles with every 15°C increase in temperature for silicon transistors.*

Figure 2-36 shows a much closer prediction of the change in I_{cbo} with temperature. This is a normalized I_{cbo} curve with a V_{cb} of 2 V and is based on

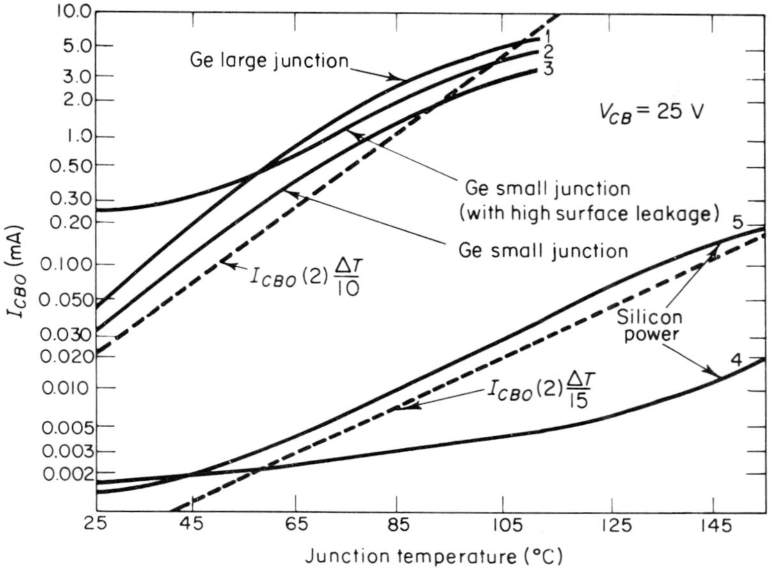

FIGURE 2-35 Change of I_{cbo} with junction temperature (courtesy Motorola).

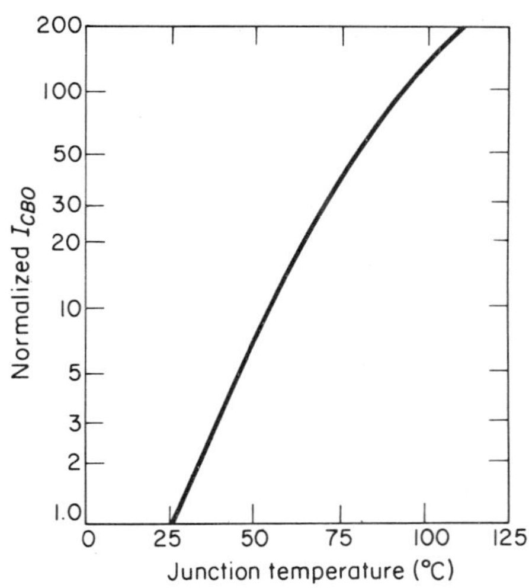

FIGURE 2-36 Normalized I_{cbo} versus junction temperature (V_{cb} = 2) (courtesy Motorola).

the similarity of the slopes of the I_{cbo} curves for germanium transistors as indicated by the low surface-leakage devices of Fig. 2-35.

Collector leakage also increases with the voltage applied to the collector. For example, the 2N332 datasheet lists an I_{cbo} of 2 μA at 25°C and of 50 μA at 150°C. The 25°C figure is with a voltage of 30 V (V_{cb}), however, while the 150°C figure is with 5 V. If the temperature is raised from 25 to 150°C with 30 V at the collector, the I_{cbo} is about 500 μA.

Always consider the possible effects of a different collector voltage when approximating I_{cbo} at temperatures other than those on the datasheet.

2-12.2 Current gain versus temperature

Figure 2-37 shows the variation of collector current versus base current for a germanium power transistor over a temperature range of −40 to +100°C. The dotted line indicates the actual base current readings obtained at 100°C. Because I_{cbo} is flowing through the base in the opposite direction, however, we must combine the I_{cbo} current with the meter reading (subtract I_{cbo} from I_b) to obtain the actual gain component of base current, as shown by the solid line. This solid line curve is more useful in determining the increase in transistor gain characteristics with temperature because the solid line eliminates the highly variable I_{cbo} component. The curve of Fig. 2-37 shows that

Current gain doubles when the temperature is raised from 25 to 100°C, for germanium transistors.

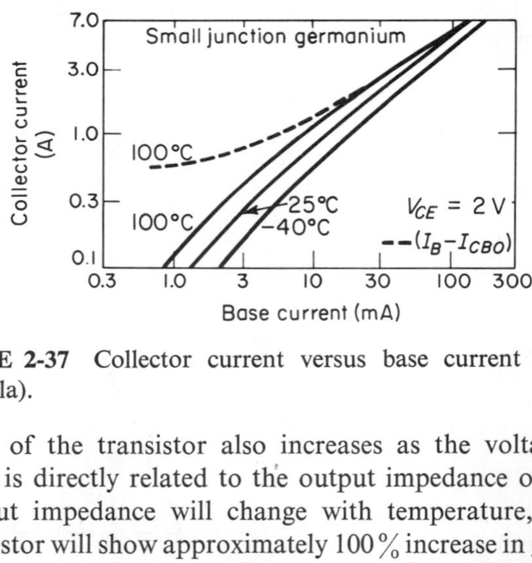

FIGURE 2-37 Collector current versus base current (courtesy Motorola).

The gain of the transistor also increases as the voltage across it is increased and is directly related to the output impedance of the transistor. Because output impedance will change with temperature, the same germanium transistor will show approximately 100% increase in gain at a voltage level equal to one-half of BV_{ceo} (collector-emitter breakdown) over the same temperature increase of 25 to 100°C. If such information is not stated on the datasheet, the curve of Fig. 2-36 is approximately true and can be used for most germanium power transistors operated as class B.

Figure 2-38 shows the variation of collector current versus base current for a typical diffused silicon power transistor over the temperature span of −40 to +175°C. The curve of Fig. 2-38 shows that

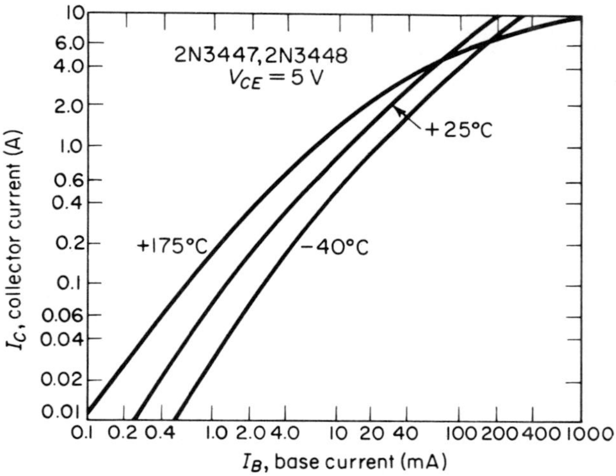

FIGURE 2-38 Typical I_C versus I_B for silicon power transistors (courtesy Motorola).

Current gain doubles when the temperature is raised from 25 to 175°C, for silicon transistors.

At higher voltages, the change will be somewhat greater; however, the effect of output impedance changes on silicon transistors is less than on germanium transistors.

It is obvious that silicon transistors are less temperature-sensitive than germanium transistors. Datasheets usually specify a maximum operating temperature or ambient temperature. If this temperature is not given or is unknown, the rule of thumb is

Do not exceed 100°C for germanium transistors or 200°C for silicon transistors.

2-12.3 Base-to-emitter voltage V_{be} *versus temperature*

Figure 2-39 shows the transconductance (base voltage versus collector current) curve versus temperature of a typical germanium power transistor operating as class B. The solid curves show the change in V_{be} (at a V_{ce} of 2 V) from 25 to 100°C. In the range of 30- to 300-mA collector current, V_{be} changes at the rate of approximately 2 mV/°C.

Although there is a change in V_{be} with an increase in collector voltage (as shown by the dotted lines of Fig. 2-39), the slope of the V_{be} curve is generally not greatly affected at 25°C over the collector-current ranges that apply to class B operation. If the slope of the V_{be} curve is changed at higher temperatures because of a temperature effect on the output impedance at high voltage levels, however, this effect must be considered.

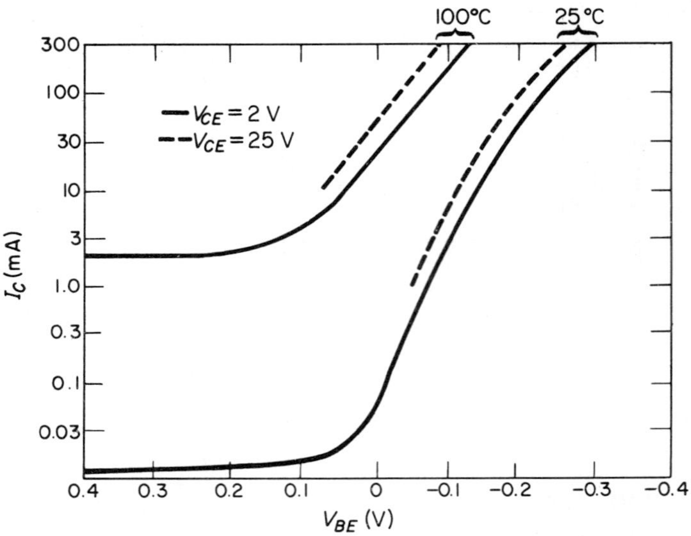

FIGURE 2-39 Transconductance versus temperature (with V_{BE} equal to −0.2 to +0.8) (courtesy Motorola).

The dotted lines of Fig. 2-39 show the change in V_{be} with a change in temperature for a high output impedance transistor when operating at a collector voltage equal to half the BV_{ceo} rating. This indicates that the rate of change for this transistor is increased at 2.2 mV/°C. Such would be a useful temperature coefficient for V_{be} operating as class B.

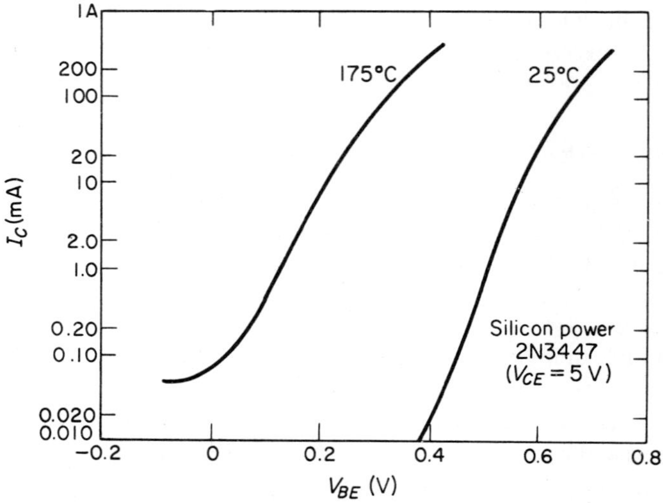

FIGURE 2-40 Transconductance versus temperature (courtesy Motorola).

Figure 2-40 shows the effects of temperature on the transconductance of a typical diffused silicon power transistor. The curves of Fig. 2-40 show that V_{be} changes 2.2 mV/°C over a temperature range of 25 to 175°C in class B operation. This change in V_{be} with temperature is shown at a collector voltage of 5 V. Because of the very high output impedance of silicon power transistors operating as class B, 2.2 mV/°C would be valid except at very high voltage levels. This may not be true with class A operation (due to higher current), depending on transistor characteristics.

2-13. PRACTICAL TWO-JUNCTION TRANSISTOR TESTING

This section describes transistor characteristics and test procedures from the practical, technician-level standpoint. That is, the data in this section permit the technician to tell all the *important* two-junction transistor characteristics and to *understand the basis for such tests.* It is assumed that the reader is familiar with operation and use of conventional electronic test equipment. Such information is found in *Handbook of Electronic Test Equipment* by John D. Lenk, Prentice-Hall, Inc., Englewood Cliffs, N.J., 1971.

2-13.1 Basic two-junction transistor tests

Transistors are subjected to a variety of tests during manufacture. It is neither practical nor necessary to duplicate all of these tests in the field. There are only four basic tests required in practical applications: *gain, leakage, breakdown, and switching time.* Unless a transistor is used for pulse or digital work, the switching characteristics are not of great importance.

In the final analysis, the only true test of a transistor is in the circuit with which the transistor is to be used. Except in special circumstances, however, a transistor will operate properly in-circuit provided

1. The transistor shows the proper gain.
2. The transistor does not break down under the maximum operating voltages.
3. Leakage, if any, is within tolerance.
4. In the case of pulse circuits, the switching characteristics (such as delay time and storage time) are within tolerance.

There are two exceptions to this rule. Transistor characteristics change with variations in operating frequency and temperature, as discussed in other sections of this chapter.

For example, a transistor may be tested at 1 MHz and show more than enough gain to meet circuit requirements. At 10 MHz, however, the gain of the same transistor may be zero. This can be due to a number of factors. Any transistor will have some capacitance at the input and output. As frequency increases, the capacitive reactance changes and, at some frequency, the transistor becomes unsuitable for the circuit (insufficient gain, unstable oscillation etc.).

In the case of temperature, the current flow in any junction of the transistor will increase with increases in temperature. A transistor may be tested for leakage at a normal ambient temperature and show a leakage well within tolerance. When the same transistor is used "in-circuit," the temperature increases, increasing the leakage to an unsuitable level.

It is usually not practical to test transistors over the entire range of operating frequencies and temperature with which the transistor will be used. Instead, the transistor is generally tested under the conditions specified in the datasheet. Then, using the equations and graphs described in other sections of this chapter, the transistor characteristics can be *predicted* at other frequencies and temperatures. In the case of some high-frequency *RF* applications, the transistor must be tested at the intended operating frequency in a special test circuit that is similar (if not identical) to the operating circuit.

2-13.2 Two-junction transistor leakage tests

Since *NPN* and *PNP* transistors can be considered as two diodes connected back-to-back, the procedures for transistor leakage tests are similar to those of diodes. In theory, there should be no current flow across a diode junction when the junction is reverse biased. Any current flow under these conditions is the result of current leakage. In the case of a transistor, the collector-base junction is reverse biased and should show no current flow. In practical applications, however, there will be some collector-base current flow, particularly as the collector voltage is operated near its limits and as the operating temperature of the transistor is increased.

Collector leakage. Collector leakage current is designated as I_{cbo} or I_{bo} on most datasheets. Collector leakage can also be termed "collector cutoff current" on other datasheets, where a nominal and/or maximum current is specified for a given collector-base voltage and ambient temperature. Collector-base leakage is normally measured with the emitter open but can also be measured with the emitter shorted to the base or connected to the base through a resistance.

Figure 2-41 shows the basic circuits for the collector-base leakage test. As shown, the circuits of Fig. 2-41(a), (b), and (c) are for *PNP* transistors, while the circuits of Fig. 2-41(d), (e), and (f) are for *NPN* transistors. Although any of the circuits can be used, those of Fig. 2-41(a) and (d) (emitter open) are the most popular.

The procedure is the same for all the circuits shown in Fig. 2-41. The voltage source is adjusted to a given value (thus providing a given reverse bias), and the current (if any) is read on the meter. This current must be below a given maximum for a given reverse bias.

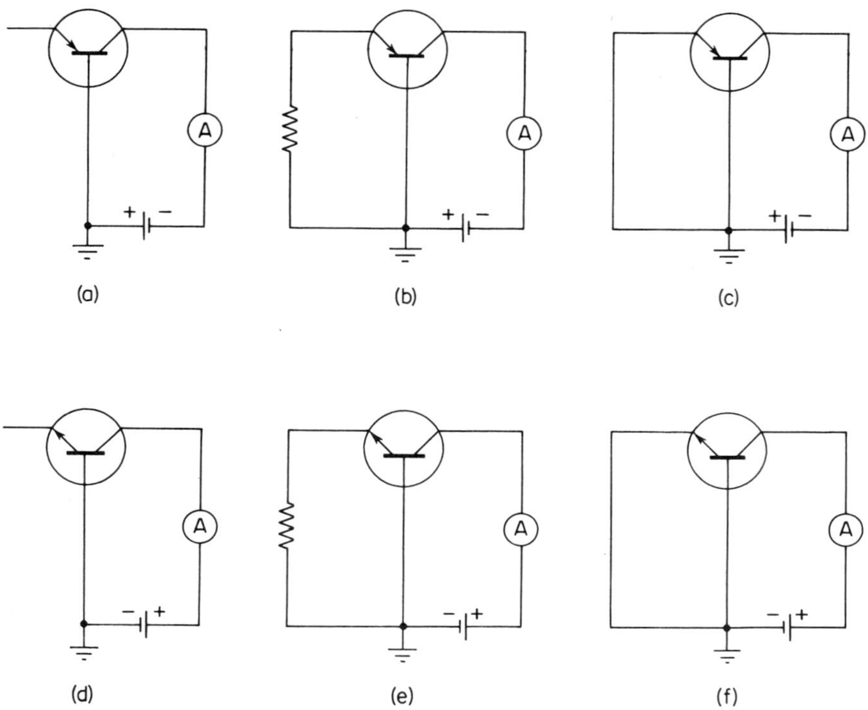

FIGURE 2-41 Collector-base leakage test circuits. (Lenk, *Practical Semiconductor Databook for Electronic Engineers and Technicians*, Prentice-Hall.)

Temperature is often a critical factor in leakage measurements. Using a 2N332 transistor as an example, the maximum collector leakage current at 25°C is 2 μA (with 30 V applied between collector and base). When the collector-base voltage is lowered to 5 and the temperature is raised to 150°C, however, the maximum collector leakage is 50 μA.

Emitter leakage. Some datasheets also specify emitter-base current leakage. This is not usually the case, however, since the emitter-base junction is usually forward biased in most circuits. Should it become necessary to test the emitter-base current leakage (I_{eo} or I_{ebo}), the circuits of Fig. 2-41 can be used, except that the collector and emitter connections are interchanged; that is, the emitter-base junction is reverse biased and the collector is left open. The procedures for testing emitter-base leakage are identical to those for collector-base voltage.

Testing two-junction transistor leakage with an ohmmeter. It is possible to make a quick check of transistor leakage with an ohmmeter. For the purpose of this test, a transistor is considered as two diodes connected back-to-back. Each diode should show low forward resistance and high reverse

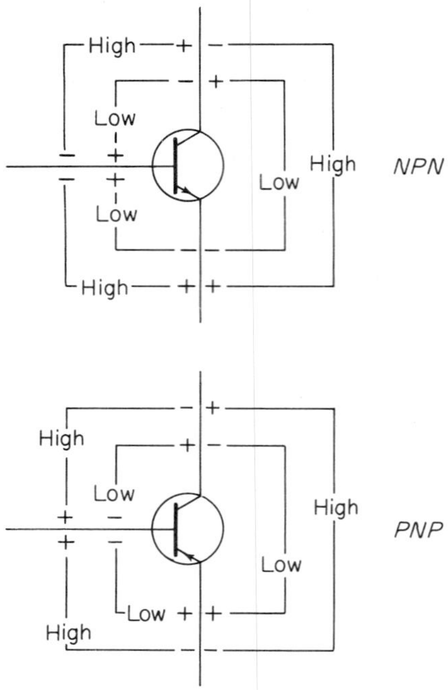

FIGURE 2-42 Transistor leakage tests with an ohmmeter.

resistance. These resistances can be measured with an ohmmeter as shown in Fig. 2-42. The same ohmmeter range is used for each pair of measurements (base-to-emitter, base-to-collector, collector-to-emitter). On low-power transistors, there may be a few ohms indicated from collector to emitter. Avoid using the $R \times 1$ range of an ohmmeter with a high internal-battery voltage. Either of these conditions can damage a low-power transistor.

If both forward and reverse readings are very high, the transistor is open. Similarly, if any of the readings show a short or very low resistance, the transistor is shorted or leaking badly. Also, if the forward and reverse readings are the same (or nearly equal), the transistor is defective.

A typical forward resistance is 300 to 700 Ω. Typical reverse resistances are 10 to 60 kΩ. Actual resistance values depend on ohmmeter range and battery voltage rather than the transistor characteristics. Thus, the *ratio of forward-to-reverse resistance* is the best indicator. Almost any transistor will show a ratio of 30 : 1. Many transistors show ratios of 100 : 1 or greater.

CAUTION: Transistors should not be tested for any characteristic unless all the characteristics are known. The transistor can be damaged if this rule is not followed. Even if no damage occurs, the test results can be inaccurate. Never test a transistor with voltages, or currents, higher than the rated values. The *maximum* current rating is often overlooked. For example, if a transistor is designed to operate with a maximum of 45 V at the collector, it could be assumed that a 9-V battery is safe for all measurements involving the collector. Assuming that the internal (emitter-to-collector) resistance of the transistor is 90 Ω and the maximum rated emitter-collector current is 25 mA, however, the maximum current will be exceeded. With 9 V connected directly between the emitter and collector, the emitter-collector current is 100 mA, four times the maximum rated 25 mA. This will cause the junctions to overheat and damage the transistor.

2-13.3 Two-junction transistor breakdown tests

The circuits and procedures for transistor breakdown tests are similar to those for leakage tests. The most important breakdown test is to determine the collector-base breakdown voltage. In this test, the collector and base are reverse biased, with the emitter open, and the voltage source is adjusted to a *given value of leakage current*. The voltage is then compared with the minimum collector breakdown voltage specified for the transistor.

For example, the minimum collector breakdown voltage specified for a 2N332 transistor is 45 V (with 50 μA flowing and an ambient temperature of 25°C). If 50 μA flows with less than 45 V, the transistor collector-base junction is breaking down.

Another breakdown test specified on some transistor datasheets is the collector-emitter breakdown voltage. In this test, the collector and emitter are reverse biased, with the base open. The voltage source is then adjusted for a given value of leakage current through both the emitter-base and collector-base junctions. The collector-emitter breakdown voltage test determines the condition of both junctions simultaneously.

Breakdown voltage is designated as BV_{cbo} (collector-base, emitter open), BV_{ces} (emitter shorted to base), or BV_{ceo} (collector-to-emitter, base open) on most datasheets. Breakdown is normally measured with the emitter (or base) open but can also be measured with the emitter shorted to the base or connected to the base through a resistor or with the emitter and base reverse biased.

Figure 2-43 shows the basic circuits for breakdown tests. The circuits shown are for *PNP* transistors. The same circuits can be used for *NPN* transistors when the voltage source polarity is reversed.

In all cases, the voltage source is adjusted for a given leakage current flow. Then the voltage is compared with a minimum specified voltage.

2-13.4 Two-junction transistor gain tests

The dynamic gain of a transistor is determined by the amount of change in output for a given change in input. Two-junction transistors are tested for *current gain*. The change in output current for a given change in input current is measured, without changing the output voltage.

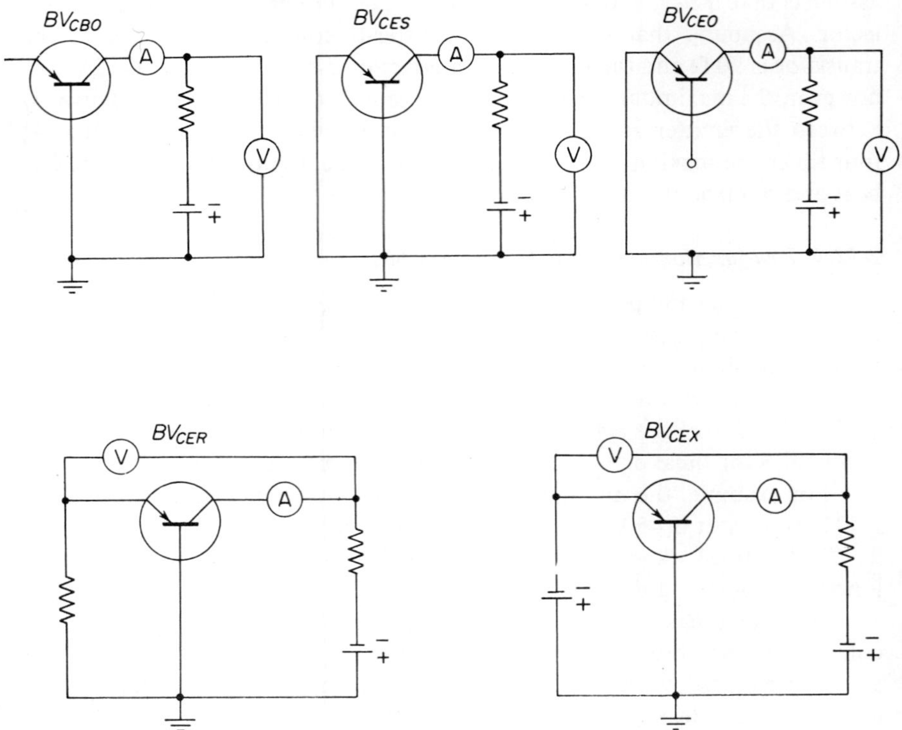

FIGURE 2-43 Breakdown voltage test circuits. (Lenk, *Practical Semiconductor Databook for Electronic Engineers and Technicians*, Prentice-Hall.)

When a transistor is connected in a common-base circuit, the collector forms the output circuit and the emitter forms the input circuit. Common-base current gain is known as alpha, indicated by the lowercase Greek letter (α).

Most datasheets specify gain with the transistor connected in a common-emitter circuit rather than a common-base circuit. In the common-emitter circuit, the base is the input and the collector is the output. Current gain for a common-emitter circuit is known as beta, indicated by the lowercase Greek letter (β).

In addition to alpha and beta, datasheets use several other terms to specify gain. The term "forward current transfer ratio" and the letters h_{fe} are the most popular means of indicating current gain for two-junction transistors, even though some manufacturers use "collector-to-base current gain."

Hybrid system. As discussed in Chapter 1, the h in the letters h_{fe} refers to the hybrid (mixture) of tı ınsistor equivalent model circuits. Transistor test circuits are structured along the same lines. In the hybrid system, the transistor and the test or operating circuits are considered as a "black box" with an input and an output rather than individual components.

When lowercase letters h_{fe} (or sometimes improperly H_{fe}) are used in transistor specifications, this indicates that the current gain is measured by noting the change in collector alternating current, for a given change in base alternating current. This is also known as "ac beta" or "dynamic beta."

When capital letters H_{FE} are used in transistor specifications, the current gain is measured by noting the collector direct current for a given base direct current. This is also known as "dc beta."

Direct versus alternating gain measurements. Direct-current gain measurements apply under a wider range of conditions and are easier to make. Alternating-current gain measurements require more elaborate test circuits, and the test results will vary with the frequency of the alternating current used for the test. Alternating-current measurements are more realistic, however, since transistors are normally used with ac signals.

There are a number of circuits for both ac and dc gain tests and a number of test procedures. Likewise, there are many commerical transistor testers as well as adapters that permit transistors to be tested with oscilloscopes. Some testers permit transistors to be tested while still connected in the circuit. It is impractical to cover the use and operation of all such testers and oscilloscope adapters in this book. Also, detailed operating instructions are provided with the testers. These instructions must be followed in all cases. Instead of attempting to duplicate the operating instructions, the following paragraphs describe the *operating principles* of the tests.

2-13.4.1 *Basic Transistor Gain Tests*

Alpha tests. Figure 2-44 shows the basic circuits for alpha measurement of *PNP* and *NPN* transistors. Both emitter current I_e and collector current I_c are measured under static conditions. Then the emitter current I_e is changed a given amount by varying the resistance of R_1 or by changing the emitter-base source voltage. The collector voltage must remain the same.

The difference in collector current I_c is noted, and the value of alpha is calculated using the equation shown. For example, assume that the emitter current I_e is changed 4 mA and that this results in a change of 3 mA in collector current I_c. This means a current gain of 0.75.

Beta tests. Figure 2-45 shows the basic circuits for beta measurement of *PNP* and *NPN* transistors. Both base current I_b and the collector current I_c are measured under static conditions. Then, without changing the collector voltage, the base current I_b is changed by a given amount and the difference in collector current I_c is noted.

For example, assume that when the circuit is first connected, the base current I_b is 7 mA and the collector current I_c is 43 mA. When the base

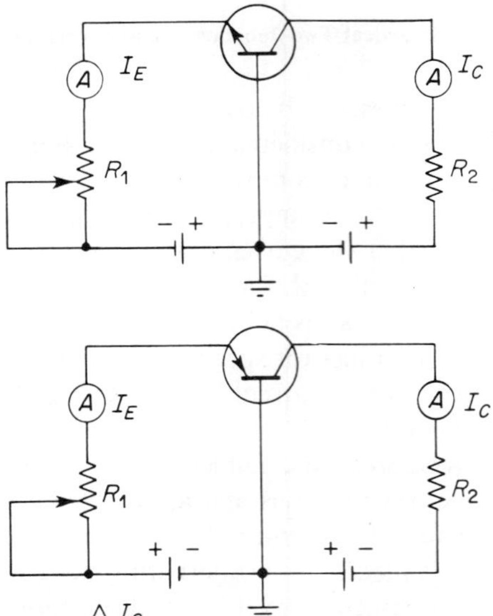

$$\text{Alpha} = \frac{\Delta I_C}{\Delta I_E}$$

FIGURE 2-44 Basic dc alpha test circuits. (Lenk, *Practical Semiconductor Databook for Electronic Engineers and Technicians*, Prentice-Hall.)

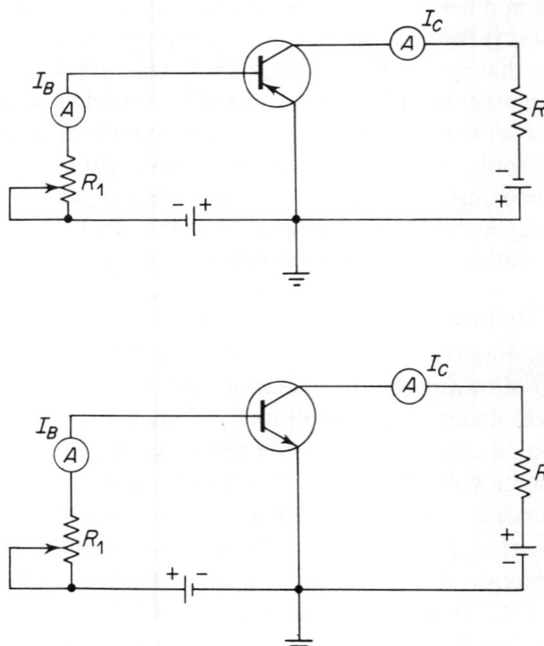

$$\text{Beta} = \frac{\Delta I_C}{\Delta I_B}$$

FIGURE 2-45 Basic dc beta test circuits. (Lenk, *Practical Semiconductor Databook for Electronic Engineers and Technicians*, Prentice-Hall.)

current I_b is increased to 10 mA (a 3-mA increase), the collector current I_c is increased to 70 (a 27-mA increase). This represents a 27-mA increase in collector current I_c for a 3-mA increase in base current I_b, or a current gain of 9.

Precautions in transistor gain tests. Certain precautions should be observed if transistors are to be tested using *noncommercial* test circuits. The most important of these are as follows:

1. The collector and emitter (or base) load resistances (represented by R_1 and R_2 in Fig. 2-44 and 2-45) should be of such value that the maximum current limitations of the transistor are not exceeded. In the case of power transistors, the wattage rating of the load resistance should be large enough to dissipate the heat.
2. Where there is a large collector leakage current, this must be accounted for in test conditions. Measure leakage using the same voltages and currents as for gain tests. Then subtract any significant leakage currents from the gain test currents.
3. The effect of meters used in the test circuits must also be taken into account.

2-13.4.2 *Basic Two-Junction Transistor ac Gain Tests*

There are several types of circuits used for the ac or dynamic test of transistors. Some of the commercial transistor testers use the same basic circuits shown in Fig. 2-44 and 2-45, except that an alternating-current signal is introduced into the input and the gain is measured at the output. Usually, these testers provide a 60- or 1000-Hz signal for injection into the input. Where it is desirable to test transistors at higher frequencies, some tester circuits permit an external high-frequency signal to be injected.

Feedback gain tester. One of the most common methods used in shop-type (nonlaboratory) transistor testers for ac gain measurement is the feedback circuit. A typical feedback gain test circuit is shown in Fig. 2-46. In this circuit, the transistor under test is inserted into an audio-oscillator configuration. The amount of feedback is adjusted by means of a calibrated control until the circuit begins to oscillate. Oscillation is indicated by a tone on the loudspeaker. The transistor current gain at the oscillation starting point can be read directly from the dial calibration.

The setting of the feedback control and the transistor gain determine the point at which oscillation starts. For example, if a large feedback is required to produce oscillation, the transistor gain is low. If the circuit will oscillate with very little feedback, the transistor gain is high. Thus, the dial connected to variable feedback resistor R_1 can be calibrated directly in terms of gain.

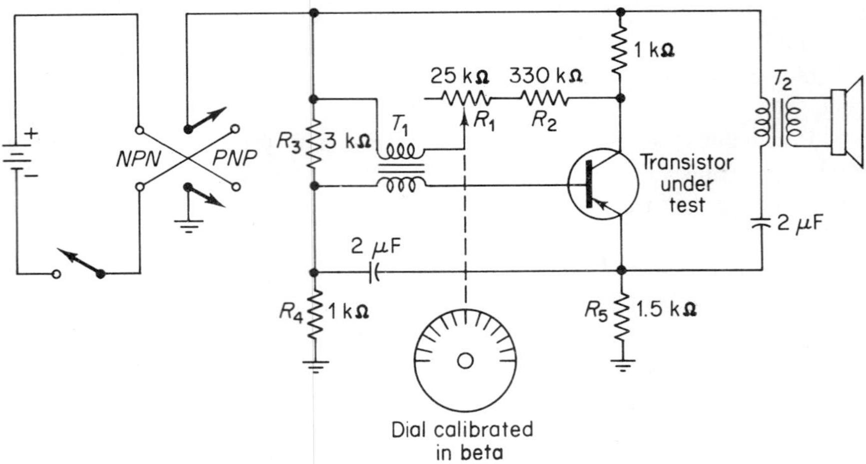

FIGURE 2-46 Feedback type gain (beta) test circuit. (Lenk, *Practical Semiconductor Databook for Electronic Engineers and Technicians*, Prentice-Hall.)

2-13.4.3 *Two-Junction Transistor Curve Tracers*

The most practical means of measuring transistor characteristics in the laboratory is to display the transistor characteristics as *curve traces* on an oscilloscope. Since the oscilloscope screen can be calibrated in voltage and current, the transistor characteristics can be read from the screen directly. If a number of curves are made with an oscilloscope, they can be compared with the curves drawn on transistor datasheets. (Some datasheet curves are direct reproductions of those obtained by tracing curves on an oscilloscope.)

There are a number of oscilloscopes (or oscilloscope adapters) manufactured specifically to display transistor characteristic curves. Some curve tracers are for one type of transistor, while other instruments will display the characteristics for many types (FETs, UJTs, etc.).

Figure 2-47 shows the functional block diagram of a typical transistor curve tracer. Such a tracer introduces changes in base current in the form of equal-value steps, that is, steps of selectable known amounts. These steps occur at the same rate as the collector supply voltage is swept between 0 V and some peak value and back to zero, producing a separate curve corresponding to each different value of base current.

When curves show collector current versus collector voltage (for different values of base current), the change in collector current induced by one step of base current is proprotional to the vertical distance between adjacent curves. This change can be read directly from the scale.

Which vertical line is chosen for the scale will depend on what collector

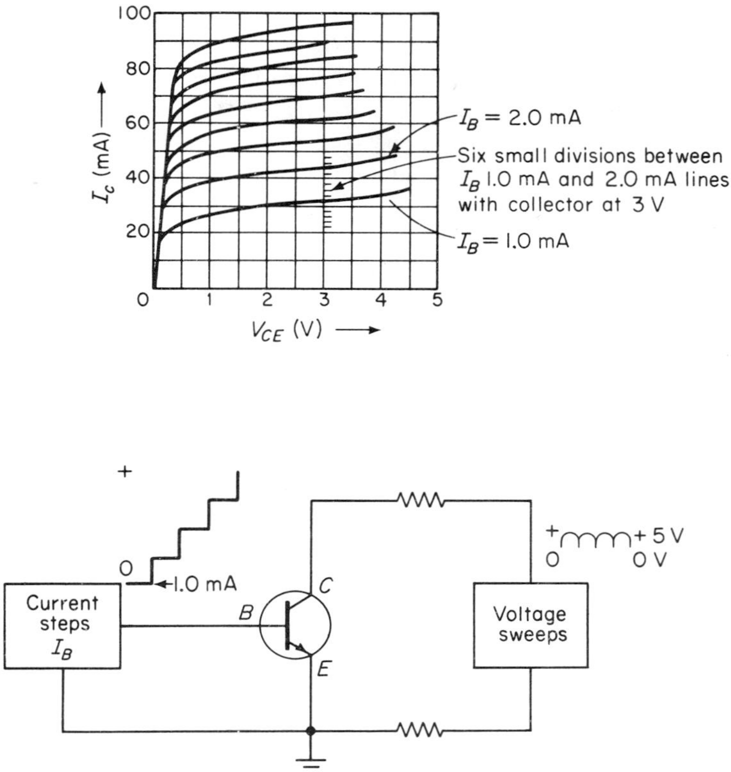

FIGURE 2-47 Basic test connections and typical oscilloscope displays for transistor curve tracers (courtesy Tektronix).

voltage is specified because each vertical line corresponds to a particular collector voltage.

Basic gain test using curves. The basic technique for measuring beta using a curve tracer is as follows (see Fig. 2-47):

1. The transistor is connected into a grounded-emitter circuit. Precise 1-mA current steps are applied to the base. Voltage sweeps from 0 to approximately 5 V are applied to the collector. The oscilloscope vertical deflection is obtained from the collector voltage sweeps, while horizontal deflection is obtained from the resultant collector current. All this is accomplished by setting of switches on the curve tracer.
2. On some curve tracers, each sweep must be initiated individually, while other instruments produce a series of 10 (or more) curves in sequence automatically.

3. Once the curves are made, they must be interpreted, as follows.
4. Choose the vertical line corresponding to the specified collector voltage. For example, the 3-V vertical line is chosen in Fig. 2-47.
5. Note (on that line) the distance between the two curves that appears above and below the specified base current (or specified collector current). For example, assume that the transistor of Fig. 2-47 is to be operated with a collector current of 30 to 40 mA, with 3 V at the collector. The two most logical curves will then be the 1- and 2-mA base current curves. This distance between these two curves (on the 3-V line) are approximately six small divisions.
6. Translate that distance into the *difference* in collector current according to the current per division of the scale. One small division on the vertical (collector current) scale equals 2 mA so the difference in collector current is 12 mA (6 × 2 mA = 12 mA).
7. Divide that collector current difference by the base current difference that caused it, depending on the current per step. Since the base current per step is 1 mA and the difference in collector current is 12 mA, the gain (beta) is 12. Had the base current per step been 0.1 mA and all other conditions the same, the beta would be 120 (12/0.1 = 120).

2-13.4.4 *Testing Two-Junction Transistor Gain with an Ohmmeter*

It is possible to make a quick check of transistor gain with an ohmmeter. The basic circuit is shown in Fig. 2-48. Normally, there will be little or no current flow between emitter and collector until the base-emitter junction is forward biased. This fact can be used to provide a basic gain test of a two-junction transistor.

In this test, the R_1 range should be used. Any internal-battery voltage can be used provided it does not exceed the maximum collector-emitter breakdown voltage.

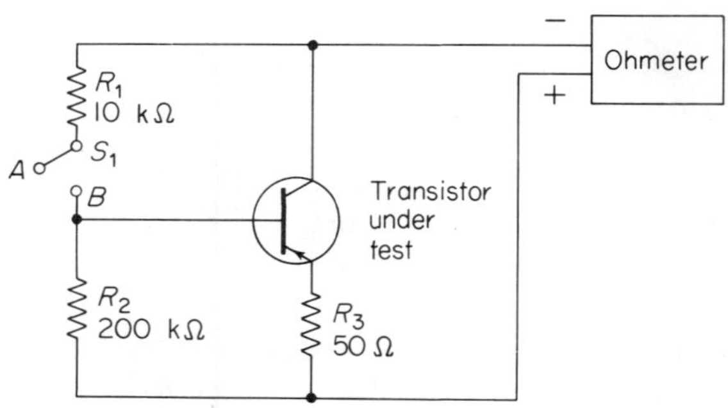

FIGURE 2-48 Simple transistor gain test using an ohmmeter.

In position A of switch S_1, there is no voltage applied to the base and the base-emitter junction is not forward biased. Thus, the ohmmeter should read a high resistance. When switch S_1 is set to B, the base-emitter circuit is forward biased (by the voltage across R_1 and R_2) and current flows in the emitter-collector circuit. This is indicated by a lower resistance reading on the ohmmeter. A 10 : 1 resistance ratio is typical for an audio-frequency transistor.

2-13.4.5 *Testing Two-Junction Transistor* RF *Gain*

The tests described thus far can be used to establish the gain of a two-junction transistor operating at low frequencies; however, the test will not establish the gain in a high-frequency (*RF*) circuit. The only true test of circuit gain (either voltage gain or power gain) is to operate the device in a working circuit and to measure actual gain. The most practical method is to operate the device in the circuit with which the device is to be used; however, it may be convenient to have a "standard" or "universal" circuit for gain test.

Very often, the datasheets for transistors to be used in *RF* work show the circuits used to test gain. With the transistor operating in such a circuit, simply introduce an *RF* signal (at the frequency of interest) to the input, and measure both input and output voltage with an *RF* voltmeter (or meter with an *RF* probe).

If power-gain measurement is desired, use input and output load resistances (noninductive or composition resistors). Measure the *RF* voltages across the input and output resistances and then calculate the power gain using $P = E^2/R$. The ratio of output power divided by input power is the power gain.

2-13.5 Two-junction transistor switching tests

Transistors to be used in pulse or digital applications must be tested for switching characteristics. For example, when a pulse is applied to the input of a transistor, there is a measurable time delay before the pulse starts to appear at the output. Likewise, after the pulse is removed, there is additional time delay before the transistor output returns to its normal level. These "switching times" or "turn-on" and "turn-off" times are usually in the order of a few microseconds (or possibly nanoseconds) for high-speed pulse transistors.

The switching characteristics of transistors designed for computer or industrial work are listed on the datasheets. Each manufacturer lists his own set of specifications; however, there are four terms (rise time, fall time, delay time, and storage time) common to most datasheets for transistors used in pulse work. These switching characteristics are of particular importance where the pulse durations are short. For example, assume that the turn-on time of a transistor is 10 μs and that a 5-μs pulse is applied to the transistor input. There will be no pulse output, or the pulse will be drastically distorted.

Pulse and square wave definitions. The following terms are commonly used in describing transistor switching characteristics. The terms are illustrated in Fig. 2-49. The input pulse represents an ideal input waveform for comparison purposes. The other waveforms in Fig. 2-49 represent typical output waveforms in order to show the relationships. The terms are defined as follows:

Rise time t_r: the time interval during which the amplitude of the output voltage changes from 10 to 90% of the rising portion of the pulse.

Fall time t_f: the time interval during which the amplitude of the output voltages changes from 90 to 10% of the falling portion of the pulse.

Time delay t_d (or delay time): the time interval between the beginning of the input pulse (time zero) and the time when the rising portion of the output pulse attains an arbitrary amplitude of 10% above the base line.

Storage time t_s: the time interval between the end of the input pulse (trailing edge) and the time when the falling portion of the output pulse drops to an arbitrary amplitude of 90% from the base line.

Pulse width (or pulse duration) t_w: the time duration of the output pulse measured between two 50% amplitude levels of the rising and falling portions of the waveform.

Tilt (also known as sag or droop): a measure of the full amplitude (flat top) portion of a pulse. The tilt measurement is usually expressed as a percentage of the amplitude of the rising portion of the pulse.

Overshoot: a measure of the overshoot generally occurring above the 100% amplitude level. This measurement is also expressed as a percentage of the pulse rise.

Preshoot: a measure of the preshoot generally occurring below the base line (or zero line). This measurement is usually expressed as a percentage of the 100% amplitude level.

These definitions are for guide purposes only. When pulses are very irregular (such as excessive tilt, droop, sag, preshoot, overshoot, ringing, and oscillation), the definitions may become ambiguous.

Testing two-junction transistors for switching time. An oscilloscope having wide-frequency response, good transient characteristics, and a dual trace can be used to check the high-speed switching characteristics of transistors used in pulse or computer work. The oscilloscope vertical channel must be voltage calibrated in the normal manner, while the horizontal channel should be time calibrated (rather than sweep-frequency calibrated).

As shown in Fig. 2-49, the transistor is tested by applying a pulse to the base of the transistor under test. The same pulse is applied to one of the

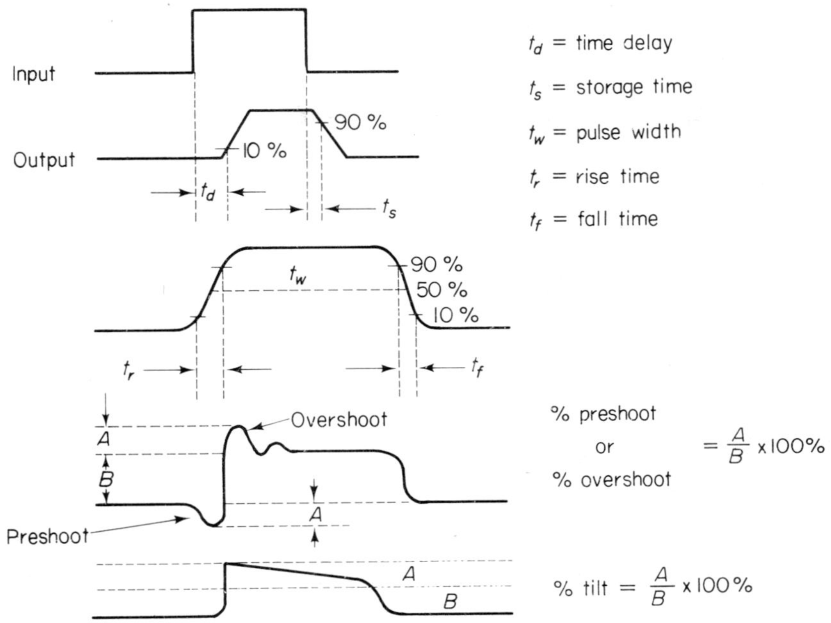

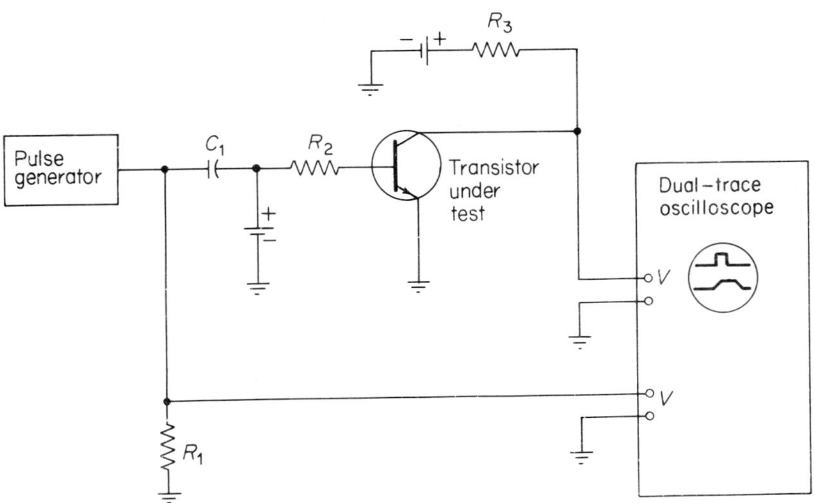

FIGURE 2-49 Pulse definitions and test connections used in two-junction transistor switching tests.

oscilloscope vertical inputs. The transistor collector output is applied to the other oscilloscope vertical input (inverted 180 deg by the common-emitter circuit). The two pulses are then compared as to rise time, fall time, delay time, storage time, etc. The transistor output pulse characteristics can then be compared with transistor specifications.

Rule of thumb for switching tests. Since rise-time and fall-time measurements are of special importance in switching tests, the relationship between oscilloscope rise time and the rise time of the transistor must be taken into account. Obviously, the accuracy of rise-time measurements can be no greater than the rise time of the oscilloscope. Also, if the transistor is tested by mean of an external pulse from a pulse generator, the rise time of the pulse generator must also be taken into account.

For example, if an oscilloscope with a 20-ns rise time is used to measure the rise time of a 15-ns pulse, the measurement will be hopelessly inaccurate. Similarly, if a 20-ns pulse generator and a 15-ns oscilloscope are used to measure the rise time, the fastest rise time for accurate measurement will be something greater than 20 ns.

There are two basic rules of thumb that can be applied to rise-time measurements.

The first method is known as the *root of the sum of the squares.* It involves finding the square of all the rise times associated with the test, adding these squares together, and finding the square root of this sum. For example, using the 20-ns pulse generator and 15-ns oscilloscope, the calculation is

$$20 \times 20 = 400, \qquad 15 \times 15 = 225, \qquad 400 + 225 = 625,$$

$$\sqrt{625} = 25 \text{ ns}$$

One major drawback to this rule is that the coaxial cables required to interconnect the test equipment are subject to "skin effect." As frequency increases, the signals tend to travel on the outside or skin of the conductor. This decreases conductor area and increases resistance. In turn, this increases cable loss. The losses of cables do not add properly when applied to the root-sum-squares method, except as an approximation.

The second rule or method states that *if the equipment or pulse being measured has a rise time 10 times slower than the test equipment, the error is 1%.* This is small and can be considered as negligible. *If the equipment being measured has a rise time 3 times slower than the test equipment, the error is slightly less than 6%.*

3

Field Effect Transistors

The field effect transistor (or FET) is sometimes known as the solid-state device that acts like a vacuum tube. This is because the characteristics of an FET are similar but definitely not identical to those of a pentode vacuum tube. There are two basic types of FETs, known as the *junction field effect transistor* (or JFET) and the *metal–oxide semiconductor field effect transistor* (or MOSFET). The MOSFET is also known as the *insulated gate field effect transistor* (or IGFET) due to its physical structure. Note that the terms MOSFET and IGFET are used interchangeably throughout this handbook.

The principles on which the JFET and MOSFET operate (current controlled by an electric field) are very similar. The primary difference between the two is in the method by which the control element is made. This difference, however, results in a considerable difference in device characteristics.

3-1. JUNCTION FIELD EFFECT TRANSISTORS (JFET)

The basic JFET is essentially a bar of doped silicon that acts like a resistor, as shown in Fig. 3-1(a). The terminal into which current is injected is called the *source*. The source terminal is similar to the cathode of a vacuum tube. The opposite terminal is called the *drain* terminal and can be likened to a vacuum-tube plate. In an FET, however, the polarity of the voltage applied to the drain and source can be interchanged.

Current flow between source and drain is related to the drain-source voltage by the resistance of the material between the two terminals. In Fig.

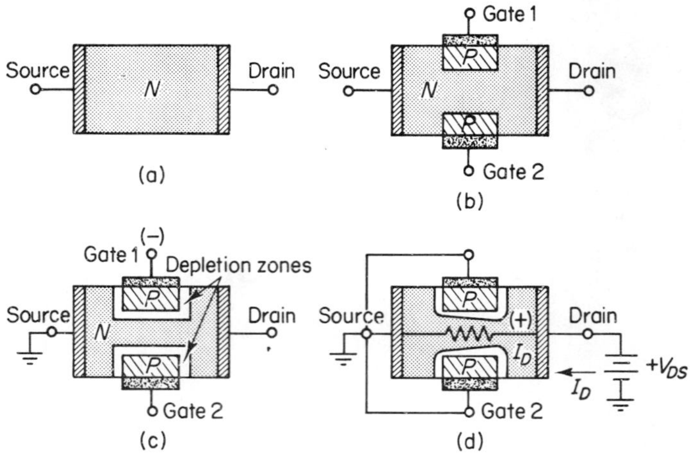

FIGURE 3-1 Development of JFETs (courtesy Motorola).

3-1(b), *P*-type regions have been diffused into the *N*-type substrate. This leaves an *N*-type channel between the source and drain. (It is also possible to make a complementary JFET with a *P*-type channel by reversing all the material types.) In the configuration of Fig. 3-1, the *P*-type regions are used to control the current flow between the source and the drain and are thus called *gate* regions.

A depletion region surrounds the *PN* junction when the junctions are reverse biased, as shown in Fig. 3-1(c). When the reverse voltage is increased, the depletion regions spread into the channel until they meet. This creates an almost infinite resistance between the source and drain.

Assume that a JFET is connected as shown in Fig. 3-1(d). The gate voltage (corresponding to a vacuum-tube grid-bias voltage) is zero (grounded). Current flows from source to drain because of an external voltage V_{DS} (or drain-source voltage). This voltage corresponds to the *B*+ voltage of a vacuum tube.

A reverse bias is set up along the surface of the gate, parallel to the channel, because of the drain-current flow in the channel. As the V_{DS} increases, the depletion regions again spread into the channel because of the voltage drop in the channel that reverse biases the junctions. The depletions will continue to grow with each increase in V_{DS} until the depletion regions meet. When this occurs, any further increase in V_{DS} is offset by an increase in the depletion region. In effect, there is an increase in channel resistance that prevents any further increase in drain current. The drain-source voltage that causes this current-limiting condition is called the *pinch-off voltage* (or V_P). Futher increases in drain-source voltage produce only a slight increase in drain current.

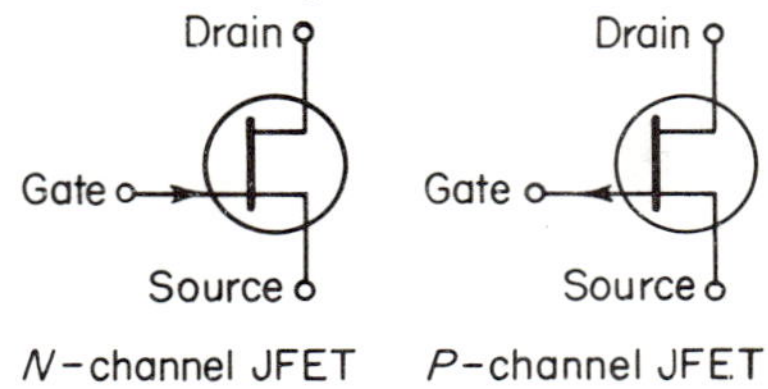

FIGURE 3-2 *P*-channel and *N*-channel JFET symbols (courtesy Motorola).

The symbols for a *P*-channel JFET and an *N*-channel JFET are shown in Fig. 3-2. Note that the arrow points away from the substrate (or bulk) in a *P*-channel device. This appears to be inconsistent with two-junction (bipolar) transistors where arrows point toward the substrate in *PNP* devices. It will be seen that the symbols are consistent, however, when it is realized that *P*-channel FETs are formed of *N*-type substrate material (and *N*-channel FETs use *P*-type material).

The drain-current (I_D) versus drain-source voltage (V_{DS}), with zero gate-source voltage (V_{GS}), is shown in Fig. 3-3(a). In the low-current region, the drain current is linear in relation to V_{DS}. As the I_D increases, the "channel" begins to deplete and the slope of the I_D curve decreases. When the V_{DS} equals the pinch-off voltage V_P, the I_D reaches saturation and stays relatively constant until drain-to-gain *avalanche* voltage (V_A) is reached.

If a reverse voltage (corresponding to a vacuum-tube negative grid bias) is applied to the gates, channel pinch-off occurs at a lower I_D as shown in Fig. 3-3(b). This is because the depletion region spread caused by the reverse-biased gates added to that produced by V_{DS}, thus reducing the maximum current for any value of V_{DS}.

The geometry of a typical JFET is single-ended, as shown in Fig. 3-4. This is because of the difficulty in diffusing impurities into both sides of a semiconductor wafer. In Fig. 3-4, the substrate is of *P*-type material onto which an *N*-type channel is grown. A *P*-type gate is then diffused into the *N*-type channel.

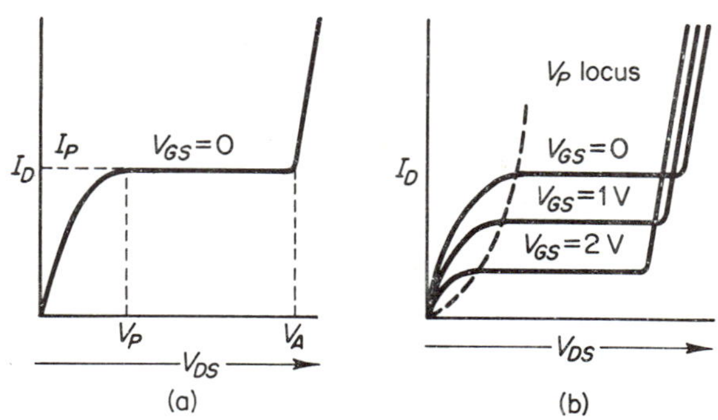

FIGURE 3-3 Drain-current characteristics (courtesy Motorola).

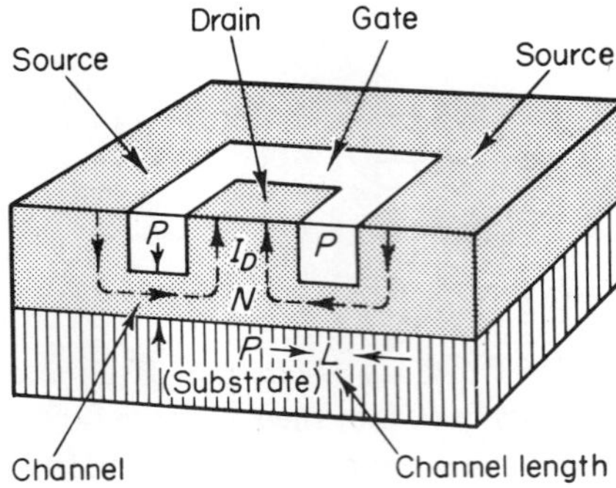

FIGURE 3-4 Typical JFET construction (courtesy Motorola).

The substrate, which functions as gate 2 of Fig. 3-1, is of relatively low resistance material to obtain maximum gain. For the same purpose, gate 1 is of very low resistance material, allowing the depletion region to spread mostly into the *N*-type channel.

3-2. METAL–OXIDE SEMICONDUCTOR TRANSISTORS (MOSFET)

The MOSFET (or IGFET) operates with a slightly different control mechanism than the JFET. Figure 3-5 shows the development of an *N*-channel MOSFET. The substrate is a high resistance *P*-type material. Two separate low resistance *N*-type regions (source and drain) are diffused into the substrate as shown in Fig. 3-5(b). The surface of the structure is covered with an insulating oxide layer as shown in Fig. 3-5(c).

Holes are cut into the oxide layer, allowing metallic contact to the source and drain. The gate metal area is overlayed on the oxide, covering the entire channel region. Simultaneously, metal contacts to the drain and source

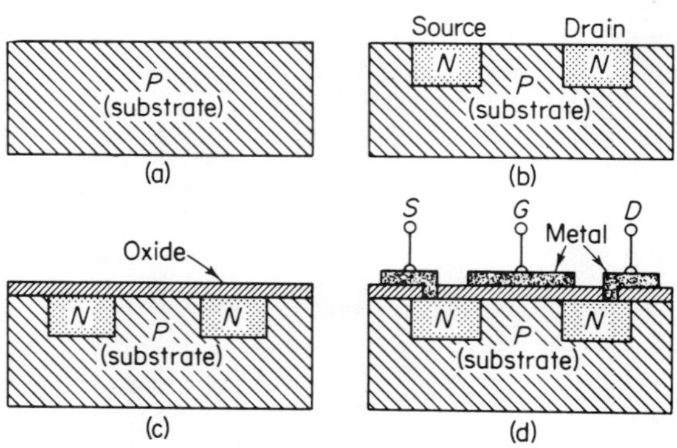

FIGURE 3-5 Typical *N*-channel MOSFET construction (courtesy Motorola).

are made as shown in Fig. 3-5(d). The contact to the metal area covering the channel is the gate terminal. There is no physical penetration of the metal through the oxide into the substrate. Since the drain and source are isolated by the substrate, any drain-to-source current in the absence of gate voltage is very low.

The metal area of the gate, in conjunction with the insulating oxide layer and the semiconductor channel, forms a *capacitor*. The metal area is the top plate; the substrate material is the bottom plate.

Operation of the gate is as follows (see Fig. 3-6): Positive charges at the metal side of the metal–oxide capacitor induce a corresponding negative

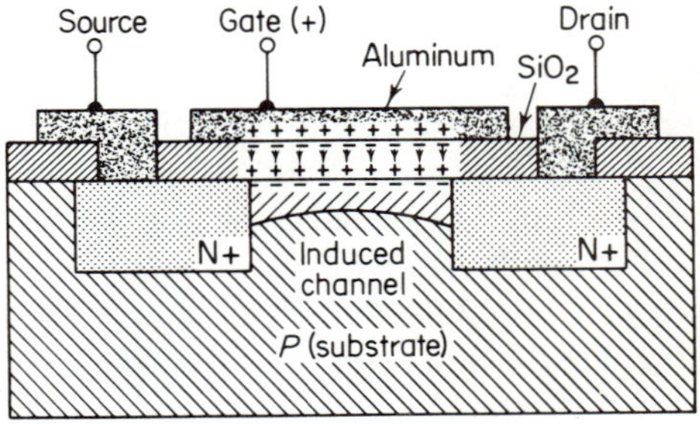

FIGURE 3-6 Operation of typical MOSFET (channel enhancement) (courtesy Motorola).

charge at the semiconductor side. As the positive charge at the gate is increased, the negative charge "induced" in the semiconductor increases until the region beneath the oxide becomes an *N*-type semiconductor region and current can flow between source and drain through the induced channel. Drain-current flow is *enhanced* by the gate voltage and can therefore be modulated by the gate voltage. Channel resistance is directly related to the gate voltage. (It is also possible to make a complementary MOSFET with a *P*-type channel by reversing all the material types.)

The FET just described is called an *enhancement*-type MOSFET. A *depletion* MOSFET is formed when an *N*-type channel is diffused between the source and drain so that drain current can flow when the gate voltage is zero. Such an arrangement is shown in Fig. 3-7 and 3-8. The structure shown in Fig. 3-7 is both an enhancement and a depletion MOSFET. When positive

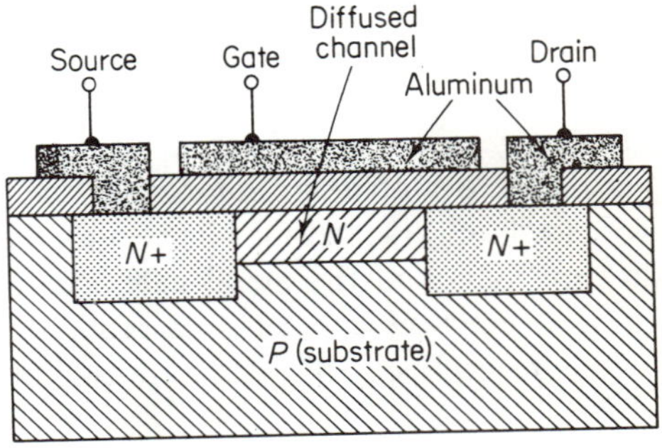

FIGURE 3-7 Typical depletion mode MOSFET construction (courtesy Motorola).

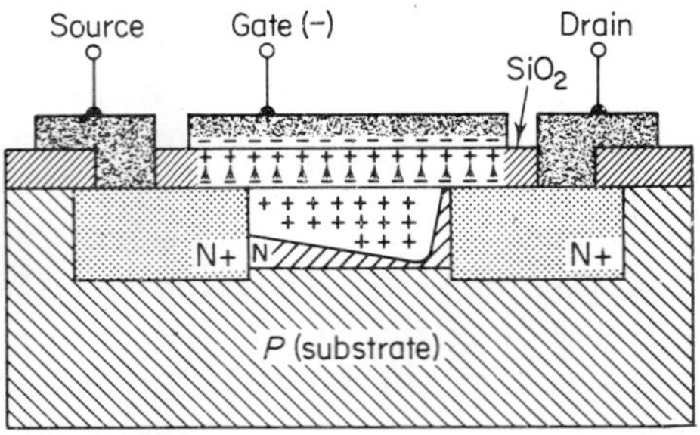

FIGURE 3-8 Operation of typical MOSFET (channel depletion) (courtesy Motorola).

gate voltages are applied, the structure enhances in the same manner as that shown in Fig. 3-5 and 3-6. When negative gate voltages are applied, the channel begins to deplete of carriers as shown in Fig. 3-8.

It should be noted that, with both the JFET and the depletion MOSFET, drain-current flow depletes the *channel area nearest the drain terminal first.*

The symbols for a *P*-channel MOSFET and an *N*-channel MOSFET are shown in Fig. 3-9. Note that the arrow points toward the substrate for *N* channel and away from the substrate for *P* channel. In practical devices, the substrate terminal may or may not be connected to the source terminal.

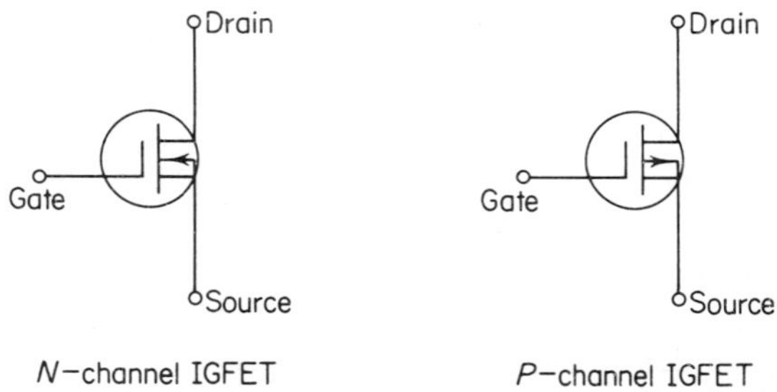

FIGURE 3-9 *P*-channel and *N*-channel IGFET symbols (courtesy Motorola).

3-3. EQUIVALENT MOSFET CIRCUIT

An equivalent circuit for a MOSFET is shown in Fig. 3-10. Here, $C_{g(ch)}$ is the distributed gate-to-channel capacitance, representing the oxide capacitance. C_{gs} is the gate-source capacitance of the metal gate area overlapping the source, while C_{gd} (sometimes known as the Miller capacitance) is the gate-drain capacitance of the metal gate area overlapping the drain. $C_{d(sub)}$ and $C_{s(sub)}$ are junction capacitances from drain to substrate and source to substrate, respectively. Y_{fs} is the transadmittance between drain-current and

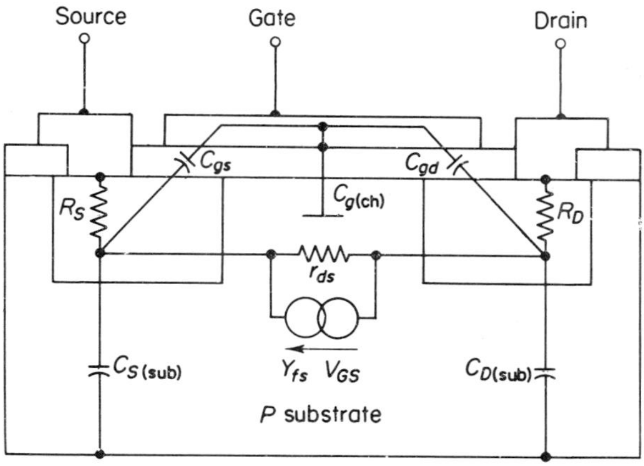

FIGURE 3-10 Equivalent circuit of enhancement mode MOSFET (courtesy Motorola).

gate-source voltage. The modulated channel resistance is r_{ds}. R_D and R_S are the bulk resistance of the drain and source.

The input resistance of the MOSFET is exceptionally high because the gate behaves as a capacitor with a very low leakage. A typical input resistance is $10^{14}\ \Omega$. The output impedance is a function of r_{ds} (which is related to gate voltage) and the drain and source bulk resistances (R_D and R_S).

To turn the MOSFET on, the gate-channel capacitance $C_{g(ch)}$, the Miller capacitance C_{gd}, and the drain-substrate capacitance $C_{d(sub)}$ must be charged. The resistance of the substrate determines the peak charging current of $C_{d(sub)}$.

3-4. FET OPERATING MODES

Both JFETs and MOSFETs operate on the principle of a "channel" current controlled by an electric field. The control mechanisms for the two are different, resulting in considerably different characteristics. The main difference between the two is in the gate characteristics. The input of the JFET acts like a reverse-biased diode; whereas the input of a MOSFET is similar to a small capacitor.

FETs can be operated in three modes: *depletion only*, *enhancement only*, and a combination of *enhancement and depletion.*

The basic differences between these three modes can most easily be understood by examining the transfer characteristics of Fig. 3-11.

The *depletion-only* FET is classified as Type A and has considerable drain-current flow for zero gate voltage. No forward gate voltage is used.

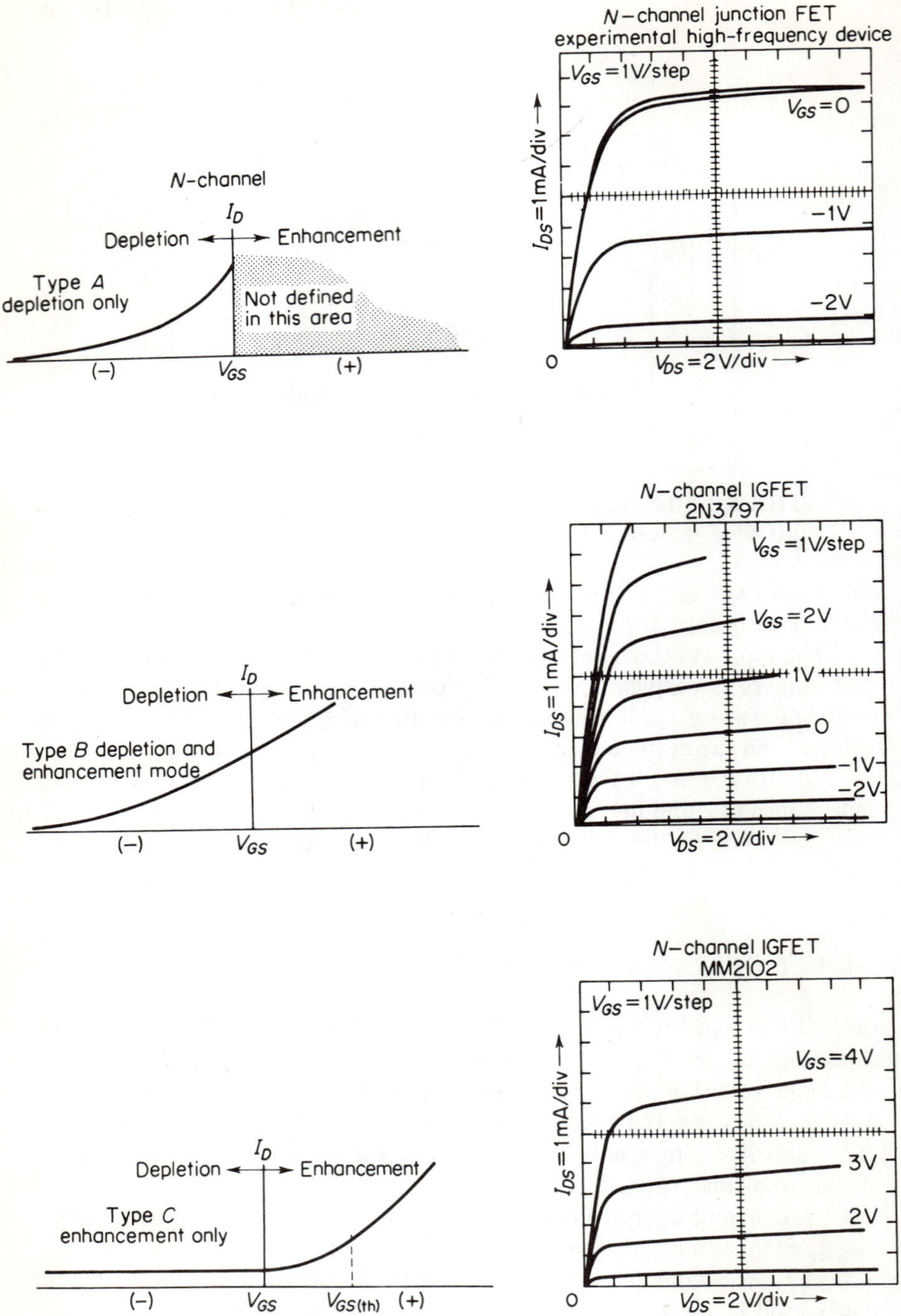

FIGURE 3-11 Transfer characteristics and associated oscilloscope traces for Types A, B and C FETs (courtesy Motorola).

Maximum drain current (I_D) flows when the gate-source voltage (V_{GS}) is zero. Drain current is reduced by applying a reverse voltage to the gate terminal. That is, I_D decreases when V_{GS} is increased.

The *depletion/enhancement* FET is classified as Type B and also has considerable drain-current flow for zero V_{GS} (but not as much as Type A). Drain current is increased by application of a forward gate voltage and reduced by application of a reverse gate voltage. For a JFET used as Type B, drain current can be increased by gate voltage only until the gate-source *PN* junction becomes forward biased. At this point, a further increase in forward gate voltage will not produce an increase in drain current.

The *enhancement-only* FET is classified as Type C and has little or no current flow for zero gate voltage. Drain current does not occur until a forward gate voltage is applied. This voltage is known as the *threshold* voltage and is indicated in Fig. 3-11 as $V_{GS(\text{th})}$. Once the threshold voltage is reached, the transfer characteristics for a Type C FET are similar to those of a Type B.

3-5. PRACTICAL MOSFETS

Typically, MOSFETs are available in *single-gate* and *dual-gate* configurations.

Figure 3-12 shows the symbol, fabrication schematic, and chip layout of a typical single-gate MOSFET. The device shown is representative of single-gate MOSFETs manufactured by RCA. Such devices are contained in a standard TO-72 metal package, the outlines of which are shown in Fig. 3-13. Note that there are four terminal leads. These leads accommodate the drain, source, gate, and bulk or substrate. Not all MOSFETs are provided with a lead to the substrate. In some cases, the substrate is connected to the source by means of an internal lead. In the MOSFET of Fig. 3-12 and 3-13, the fourth lead provides for electrical connection to the substrate and metal case.

Figure 3-14 shows the fabrication schematic of a device that is representative of the MOSFETs manufactured by Motorola. The fabrication is similar to that for typical MOS devices, except that a silicon nitride passivation process is used; that is, the surface of the substrate is covered with the conventional insulating oxide layer, plus a nitride layer.

The oxide layer serves as a protective coating and to insulate the channel from the gate; however, the oxide is subject to contamination by sodium ions that are found in varying quantities in all environments. Such contamination results in long-term instability and changes in device characteristics. Silicon nitride is impervious to sodium ions, and this is used to shield the oxide layer from contamination.

Holes are cut into both the oxide and nitride allowing metallic contact to the source and drain in the normal manner. Of course, there is no physical

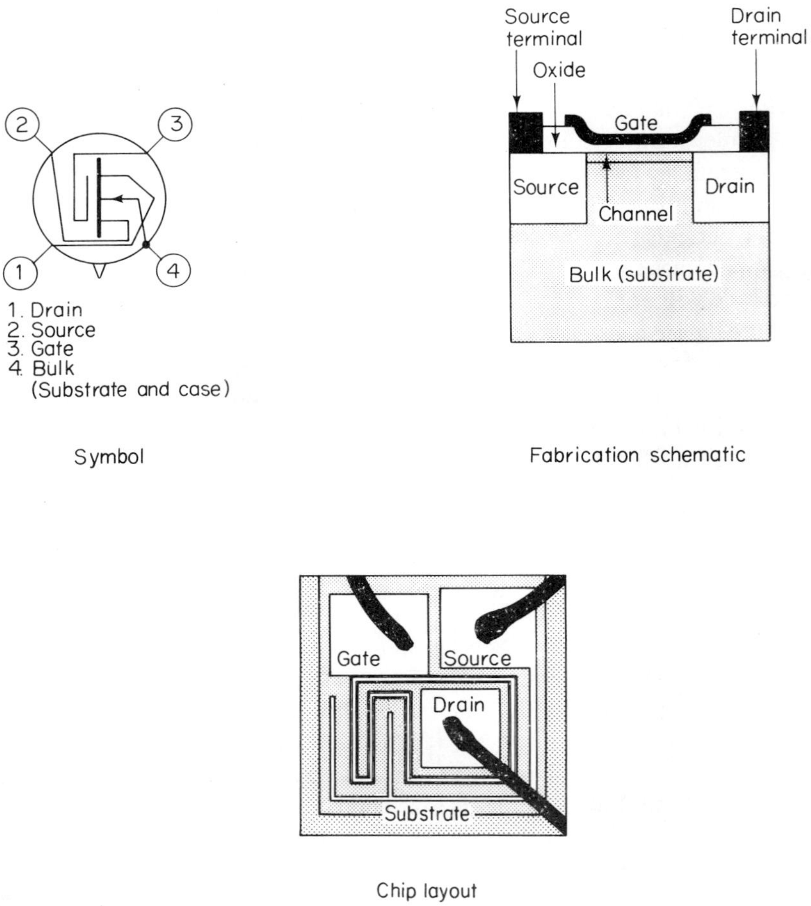

FIGURE 3-12 Symbol, fabrication schematic, and chip layout of a typical single-gate MOSFET (courtesy RCA).

penetration of the gate metal through the oxide and nitride layers to the substrate. Note that the silicon nitride passivation process also reduces susceptibility to damage from static-charge buildup during handling of the MOSFET. The subject of static-charge buildup is discussed further in later paragraphs of this chapter.

Figure 3-15 shows the symbol, fabrication schematic, and chip layout of a typical dual-gate MOSFET. The device shown is representative of dual-gate MOSFETs manufactured by RCA and features a series arrangement of two separate channels. Each channel has independent control. This arrangement results in substantially lower feedback capacitance, greater gain, remote AGC capability in *RF* amplifier applications, substantially better cross-modulation

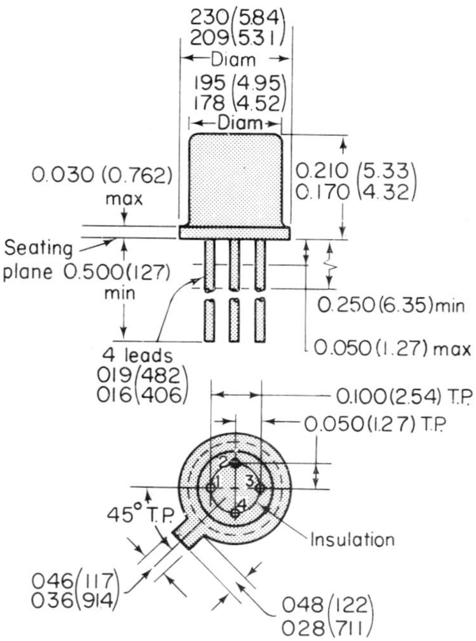

FIGURE 3-13 Dimensional outline for standard JEDEC TO-72 package (courtesy RCA).

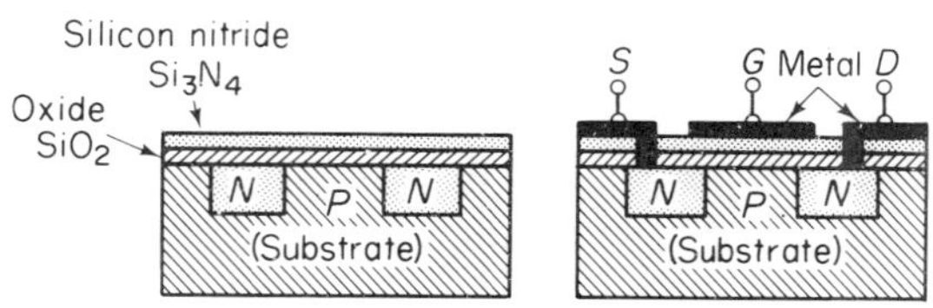

FIGURE 3-14 Fabrication schematic of MOSFET with silicon-nitride passivation (courtesy Motorola).

characteristics, and lower spurious response than are provided by single-gate types.

The availability of two independent control gates also offers unique advantages for chopper, clipper, and gated-amplifier service and for applications involving the combination of two or more signals, such as mixers, product detectors, color demodulators, and balanced modulators.

Note that the dual-gate device is also contained in a standard TO-72 metal package with four terminal leads. Three of the leads accommodate the drain, gate 1, and gate 2, respectively. The fourth lead provides electrical connection to the source, substrate, and metal case.

With the series dual-gate MOSFET, the drain current is controlled by voltages on both channels. There is a common region between channel 1 and channel 2 that acts as a drain for channel 1 and a source for channel 2. This

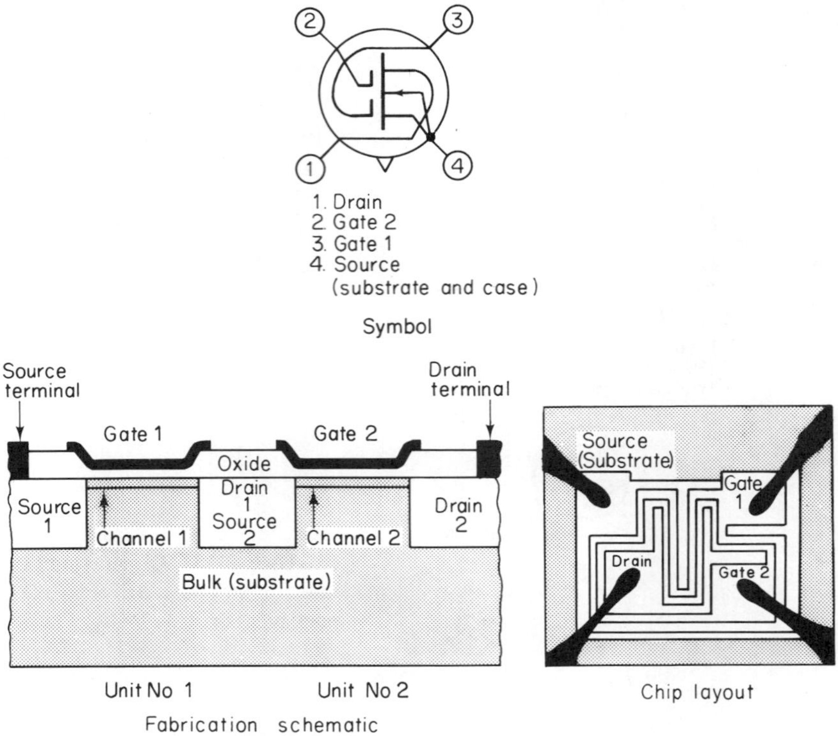

FIGURE 3-15 Symbol, fabrication schematic, and chip layout of typical dual-gate MOSFET (courtesy RCA).

region does not have a terminal lead and is not accessible from outside th package (case).

3-5.1 Transfer characteristics for basic MOSFETs

Figure 3-16 and 3-17 show the transfer characteristics for typical single gate and dual-gate MOSFETs, respectively. The transfer characteristic showi in Fig. 3-16 is for an *N*-channel, depletion-type, single-gate MOSFET. Not that there is a heavy drain-current flow (about 14 mA) at zero gate voltage When the gate voltage is at about −2.5 V, the drain current drops to zero.

The transfer characteristic shown in Fig. 3-17 is for an *N*-channel depletion-type, dual-gate MOSFET. Note that the drain current is controlled by voltages on *both* gates. For example, with zero volts on gate 2, the drain current cannot exceed about 7 mA, even with gate 1 at +1 V. If gate 2 is reduced to −1 V, then the drain current cannot exceed about 2 mA. If gate 2 is raised to +2 V, then the drain current can be increased to about 14 mA, with gate 1 at (or near) zero.

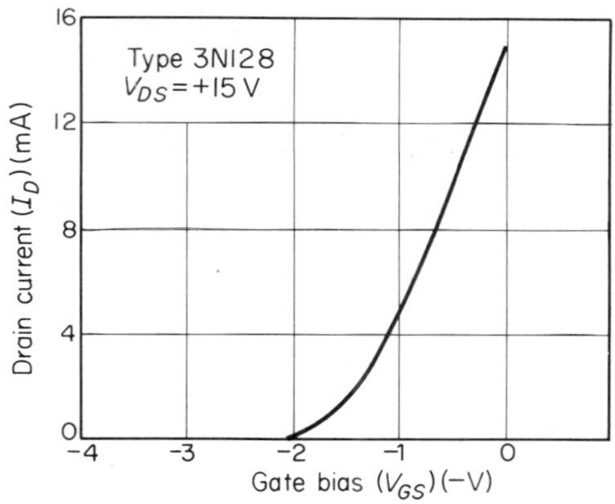

FIGURE 3-16 Transfer characteristic for typical single-gate MOSFET (courtesy RCA).

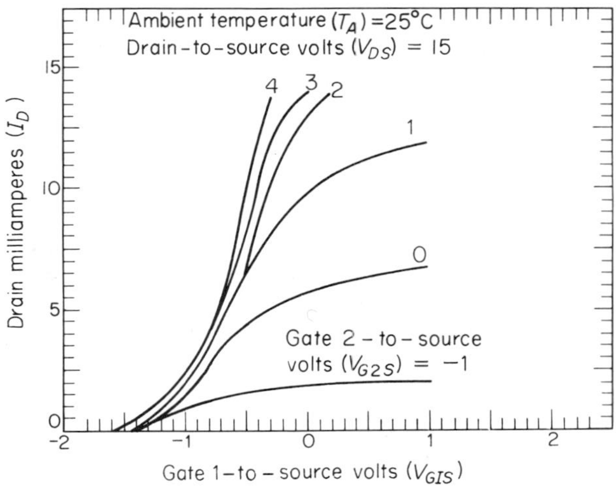

FIGURE 3-17 Transfer characteristic for typical dual-gate MOSFET (courtesy RCA).

3-6. ADVANTAGES AND DISADVANTAGES OF FETS

The FET has several advantages over a conventional transistor. The FET is relatively free of noise and is more resistant to the degrading effects of nuclear radiation because carrier lifetime effects are comparatively unimportant to FET operation. The FET is inherently more resistant to burnout than is a conventional two-junction (or bipolar) transistor.

There are additional advantages for certain design considerations. For example, the high input impedance (typically several megohms) is very useful in impedance transformations, chopper, and switching applications. Since the

FET is a voltage-controlled device, the FET can readily be "self-biased." This frequently permits a more simple circuit than is possible with a bipolar transistor. The FET also has a nonlinear region of operation that can be used for automatic gain control applications.

The JFET has a very high output resistance, making it useful as a constant-current source. Figure 3-18 illustrates a comparison of some JFET key parameters and their relative magnitudes to those of vacuum tubes, bipolar transistors, and MOSFETs. (The special capabilities and limitations of MOSFETs are discussed in later sections of this chapter.)

Characteristics	Vacuum Tube	JFET	MOSFET	Bipolar
Input impedance	High	High	Very high	Low
Noise	Low	Low	Unpredictable	Low
Warm-up time	Long	Short	Short	Short
Size	Large	Small	Small	Small
Power consumption	Large	Small	Small	Small
Ageing	Noticeable	Not noticeable	Noticeable	Not noticeable
Bias voltage temp coefficient	Low, not predictable	Low predictable	High not predictable	Low predictable
Typical gate grid current	1 nA	0.1 nA	10 pA	—
Gate/grid current change with temp	High unpredictable	Medium predictable	Low unpredictable	—
Reliability	Low	High	High	High
Sensitivity to overload	Very good	Good	Poor	Good

FIGURE 3-18 Comparison of FET characteristics (courtesy Motorola).

When compared with bipolar transistors, the chief shortcoming of any FET is its relatively small gain-bandwidth product. Although the JFET is free from carrier-transit-time limitations, parasitic capacitances limit the FET at higher frequencies.

3-7. FET OPERATING REGIONS

Both MOSFETs and JFETs have three distinct characteristic regions, only two of which are operational. Figure 3-19(a), which is the output transfer characteristic of a depletion-type FET, illustrates the three different regions. Below the pinch-off voltage V_P, the FET operates in the *ohmic* or *resistance*

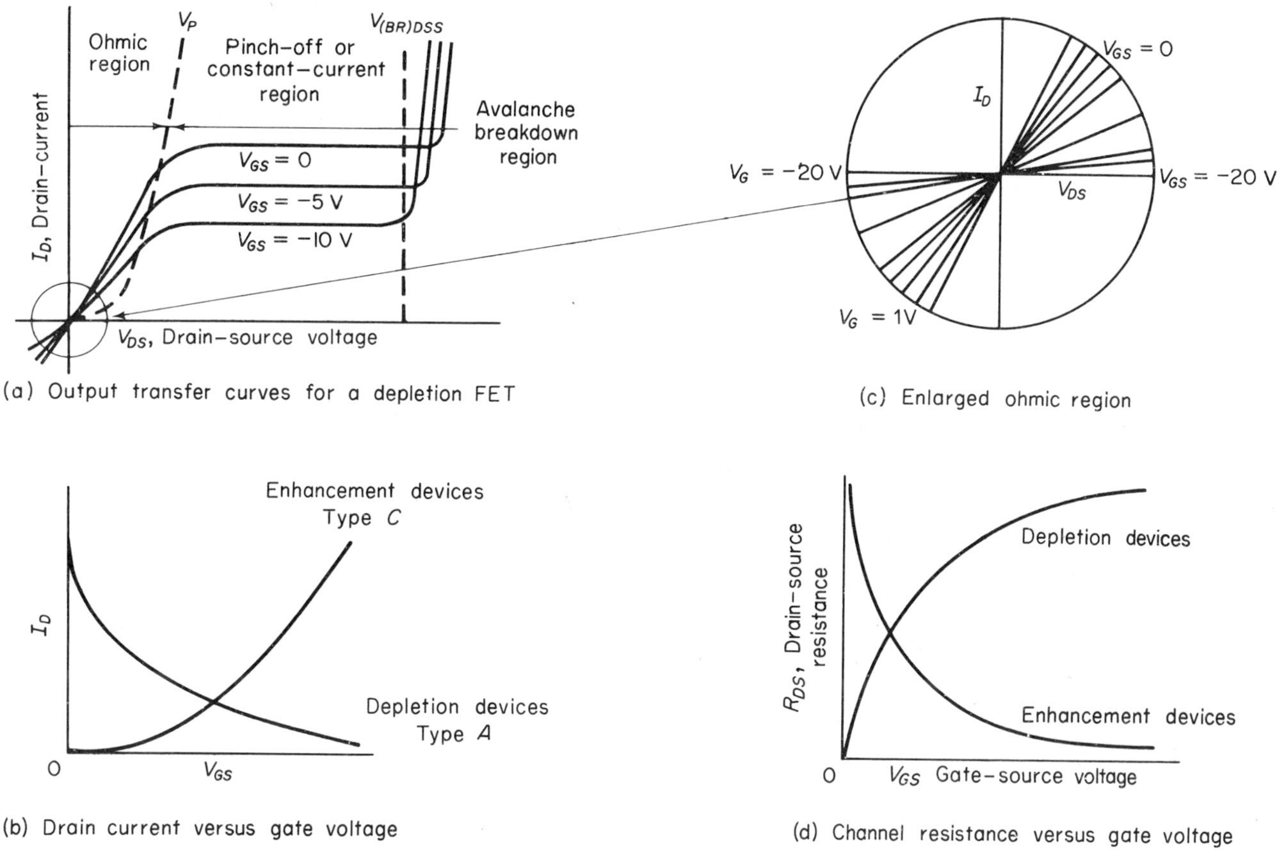

FIGURE 3-19 Characteristics of FET operating regions (courtesy Motorola).

region. Above the pinch-off voltage V_P, up to the drain-source breakdown voltage $V_{(BR)DSS}$, the FET operates in the *constant-current region*. The third region, above the breakdown voltage, is the *avalanche region* where the FET is not operated. For this reason, the avalanche region is referred to by some manufacturers as the *forbidden region*.

The drain-source resistance r_{ds} at any point on these curves is given by the slope of the curve at that point. Above the pinch-off voltage, changes in the drain-source voltage V_{DS} result in small changes in drain current I_D. This produces a very high drain-source resistance and is characteristic of a constant-current source. Also, the actual operating drain current is variable and is dependent on the gate-source voltage. This results in a voltage-controlled current source.

The I_D–V_{GS} curve, shown in Fig. 3-19(b), illustrates how the drain current varies with changes in gate-source voltage. For depletion FETs, the drain current decreases as the gate-source voltage is increased. For enhancement devices, the drain current is enhanced or increased as the gate-source voltage is increased.

If the FET is operated with a drain-source voltage below the pinch-off voltage, or preferably a voltage below several hundred millivolts, the slopes of the curves vary considerably as the gate-source voltage is varied. This is shown in Fig. 3-19(c). Since the slope varies, the drain-source resistance varies. This is operation in the *ohmic region*.

In the ohmic region, the drain-source channel is actually a voltage-variable or voltage-controlled resistor. As shown in Fig. 3-19(d), the drain-source resistance decreases with increasing gate-source voltage, for enhancement FETs. For depletion FETs, the converse is true. MOSFETs are generally not operated in the ohmic region; however, MOSFETs are capable of operation below the pinch-off voltage.

Note that the curves in the ohmic region [Fig. 3-19(c)] are relatively symmetrical. This means that ac as well as dc signals can be handled. In other words, the drain-source channel is bilateral and not unilateral.

Note also that FETs are generally operated above the pinch-off region (in the constant-current region) for linear applications; whereas the ohmic region is used only for voltage-variable resistor applications.

Figure 3-20 illustrates I_D as a function of V_{DS} for a typical *N*-channel depletion-type MOSFET. Three I_D curves are shown: $V_{GS} = 0$, $V_{GS} > 0$, and $V_{GS} < 0$. Changes in the conductivity pattern are shown in the simplified conductivity profile for each region of operation.

In the ohmic region, the I_D–V_{DS} curve has the characteristic of a resistance. The shape of this curve is a function of V_{DS}; that is, the slope of the curve is governed by V_{GS}. The V_{DS}/I_D characteristic, or resistance, is controlled by the gate voltage.

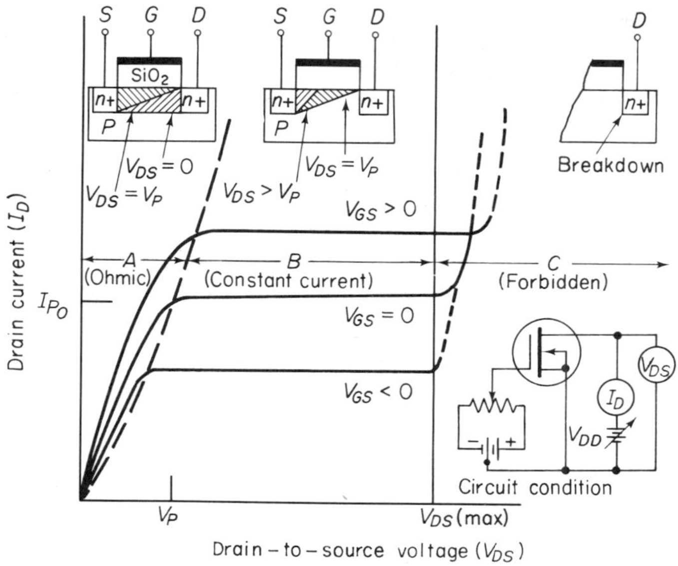

FIGURE 3-20 Regions of operation for N-channel depletion MOSFET (courtesy RCA).

As V_{DS} is increased, it produces an electrostatic stress in the channel. In turn, the stress modifies channel conductivity as shown. The channel is completely pinched off beyond V_P. Increasing V_{DS} serves only to maintain I_D at a constant level.

In the constant-current (or "amplifier") region, the I_D is held at a constant level for a given, fixed gate voltage V_{GS}. A change in V_{GS} produces a change in I_D. Thus, in the constant-current region, the device exhibits the transconductance characteristic that is essential to amplifier operation (that is, G_m or transconductance equals dI_D/dV_{GS}).

In the forbidden region, an increase of V_{DS} beyond the rated maximum could produce avalanching in the drain-to-substrate diffusion (diode). For this reason, FETs of any type should not be operated in the forbidden region (above the maximum rated V_{DS}).

3-8. ZERO-TEMPERATURE-COEFFICIENT POINT

An interesting characteristic of all FETs is their ability to operate at a zero-temperature-coefficient (0TC) point. This means that if the gate source is biased at a specific voltage and is held constant, the drain current will not vary with changes in temperature.

Typically, JFETs show the 0TC characteristic over a wide range of temperature, approximately 150°C. MOSFETs are limited to a much narrower range, approximately 50°C. For that reason, 0TC is not generally considered as an important MOSFET characteristic.

Also, it is not always practical to operate an FET at the 0TC point. For example, assume that the required V_{GS} to produce 0TC is 0.3 V and the FET is to operate as an amplifier with 0.5-V input signals. A part of the input signal will be clipped. Or, assume that the circuit is to be self-biased with a source resistor. An increase in bias resistance to produce 0TC could reduce gain. Thus, in practical circuits, FETs are often not biased for 0TC; however, the 0TC characteristic should be understood, and it is worth discussing.

The I_D–V_{GS} curves of Fig. 3-21 show that the various curves of different temperatures intersect at a common point. If the FET is operated at this value of I_D and V_{GS} (shown as I_{DZ} and V_{GS}), 0TC operation will result.

The 0TC point varies from one FET to another and is dependent on I_{DSS}, the zero-gate-voltage drain current, and V_P. The equations shown in Fig. 3-21 provide good approximations of the 0TC point. For example, if the pinch-off voltage V_P is 1 V, the 0TC mode will be obtained if the gate-source voltage V_{GS} is 0.37 V ($1 - 0.63 = 0.37$).

It is sometimes assumed that the forward transadmittance of the FET does not vary with temperature, particularly if the FET is biased at the 0TC

$$I_{DZ} \approx I_{DSS}\left(\frac{0.63}{V_P}\right)^2 \approx \frac{0.4\ I_{DSS}}{V_P{}^2}$$

$$V_{GSZ} \approx V_P - 0.63$$

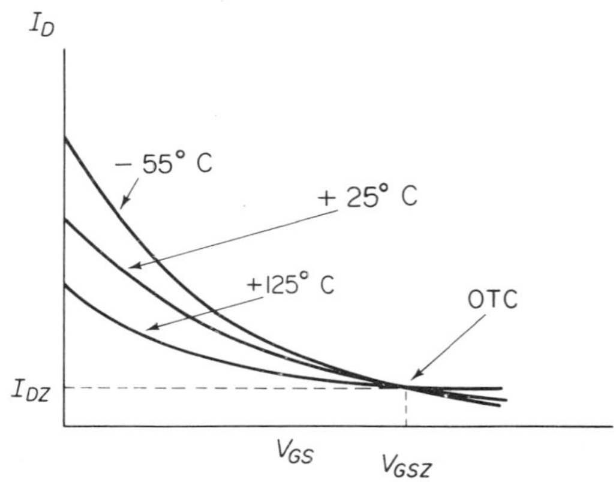

FIGURE 3-21 Zero temperature coefficient for FETs (courtesy Motorola).

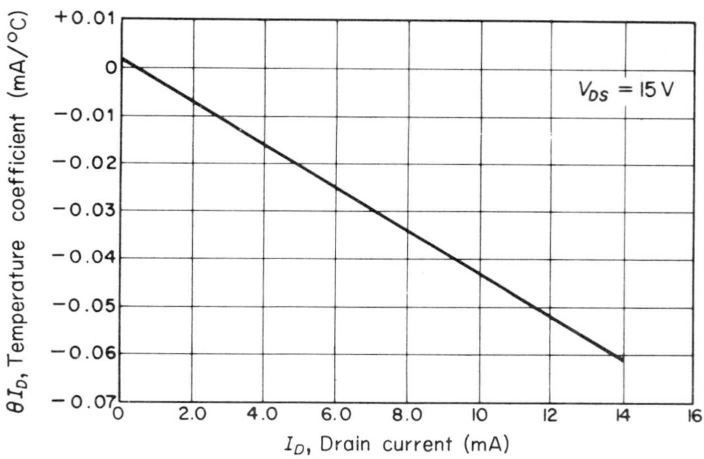

FIGURE 3-22 Drain current temperature coefficient versus drain current for typical JFET (courtesy Motorola).

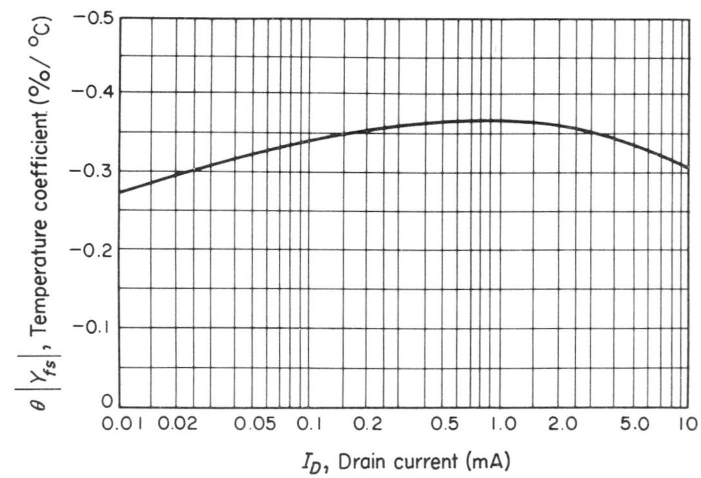

FIGURE 3-23 Forward transadmittance temperature coefficient versus drain current (courtesy Motorola).

point; however, this is not correct. The transadmittance of the FET is the slope of the I_D–V_{GS} curve. The curve of Fig. 3-21 shows that the slope varies with temperature at every point on the curve. Figures 3-22 and 3-23 illustrate the temperature coefficients for a typical JFET.

3-8.1 Practical methods for finding 0TC

The values of I_D and V_{GS} that produce 0TC can be found by using datasheet curves or by equations, as shown in Fig. 3-21; however, the values are typical approximations.

A more practical method for determining I_{DZ} requires a soldering tool, a coolant (a can of Freon), and a curve tracer (such as the Tektronix Type 575). By placing a 1000-Ω resistor across the base and emitter terminals of the curve tracer test socket, the constant-current base drive is converted to a relatively constant voltage for driving the FET gate. The curve tracer is adjusted to display the I_D–V_{GS} output family [such as shown in Fig. 3-19(a)]. By alternately bringing the soldering tool near the FET and spraying the FET with Freon, the voltage step of V_{GS} that remains motionless on the curve tracer can be observed. The I_D at this voltage step is I_{DZ}.

Typically, FETs with an I_{DSS} of about 10 to 20 mA will have an I_{DZ} of less than 1 mA. Usually, the I_{DZ} increases as I_{DSS} increases (but not always, and not in proportion). For example, the I_{DZ} of a 50-mA FET often shows an I_{DZ} below 1 mA.

3-9. HANDLING MOSFETS

Electrostatic discharges can occur when a MOSFET is picked up by its case and the handler's body capacitance is discharged to ground through the series arrangement of the bulk-to-channel and channel-to-gate capacitances of the device. For this reason, MOSFETs require proper handling, particularly when the MOSFET is out of the circuit. In-circuit, a MOSFET is just as rugged as any other solid-state component of similar construction and size.

MOSFETs are generally shipped with the leads all shorted together to prevent damage in shipping and handling (there will be no static discharge between leads). Figure 3-24 shows a typical MOSFET with a shorting spring installed. This shorting spring, or similar device, *should not be removed until* after the device is soldered into the circuit. An alternate method for shipping or storing MOSFETs is to apply a conductive foam between the leads.

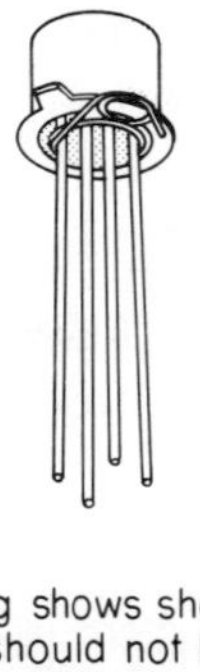

Drawing shows shorting spring, which should not be removed until after device is soldered into circuit.

FIGURE 3-24 Typical MOSFET with shorting spring installed on leads to prevent damage due to static discharge (courtesy RCA).

Such a foam is manufactured by Emerson & Cuming, Inc. and is listed as ECCOSORB LS26.

Note that polystyrene insulating "snow" is not recommended for shipment or storage of MOSFETs. Such "snow" can acquire high static charges that could discharge through the device.

If it becomes necessary to install a MOSFET in a circuit or to remove the device from the circuit for test or some other operation, consider the following points.

1. When removing or installing a MOSFET, first turn the power off. If the MOSFET is to be removed, your body should be at the *same potential* as the unit from which the device is removed. This can be done by placing one hand on the chassis *before* removing the device. If the MOSFET is to be connected to an external circuit (for test, etc.), put the hand holding the MOSFET against the tester front panel and connect the leads from the tester to the device leads. This procedure prevents possible damage from static charges on the tester (or other external unit).

   WARNING: Make certain that the chassis or external unit is at ground potential before touching it. In some obsolete or defective equipment, the chassis or panel is above ground (typically by one-half of the line voltage).
2. When handling a MOSFET, the leads must be shorted together. Generally, this is done in shipment by a shorting ring or spring as shown in Fig. 3-24. When testing MOSFETs, connect the tester lead to the MOSFET (preferably the source lead first); then remove the shorting ring.
3. When soldering or unsoldering a MOSFET, the soldering tool tip must be at ground potential (no static charge). Connect a clip lead from the barrel of the soldering tool to the chassis or tester case. The use of a soldering gun is *not recommended* for MOSFETs.
4. Remove power to the circuit before inserting or removing a MOSFET (or a plug-in module containing a MOSFET). The voltage transients developed when terminals are separated may damage the MOSFET. This same caution applies to circuits with conventional transistors; however, the chances of damage are greater with MOSFETs.

3-10. PROTECTING MOSFETS

Because of the static-discharge problem described in Sec. 3-9, manufacturers provide some form of protection for a number of their MOSFETs. Generally, this protection takes the form of a diode incorporated as part of the substrate material. This section describes such protective diodes, as found in certain RCA MOSFETs.

Before going into specific protection methods, let us analyze the breakdown mechanism associated with gate destruction. Figure 3-25 shows a typical single-gate MOSFET, including the structure [Fig. 3-25(a)], the symbol [Fig. 3-25(b)], and an oversimplified equivalent circuit [Fig. 3-25(c)]. A study of Fig. 3-25 will show the possible discharge paths within the device. Note that a substrate "diode" is formed by the *PN* junction integrated over the entire junction area (that is, the source and drain diffusions connected by the inversion layer or *N*-type channel). Figure 3-25(c) lumps the substrate-to-channel "diode" into one equivalent diode *D* that terminates at the center of the channel.

C_{in} in Fig. 3-25(c) represents the gate-to-channel capacitance; R_1 and R_2 represent the channel resistance. R_3 is the leakage resistance associated with the substrate-to-channel equivalent diode D_1. Leakage resistance across C_{in} is intentionally deleted because such resistance is many thousands of times greater than R_3.

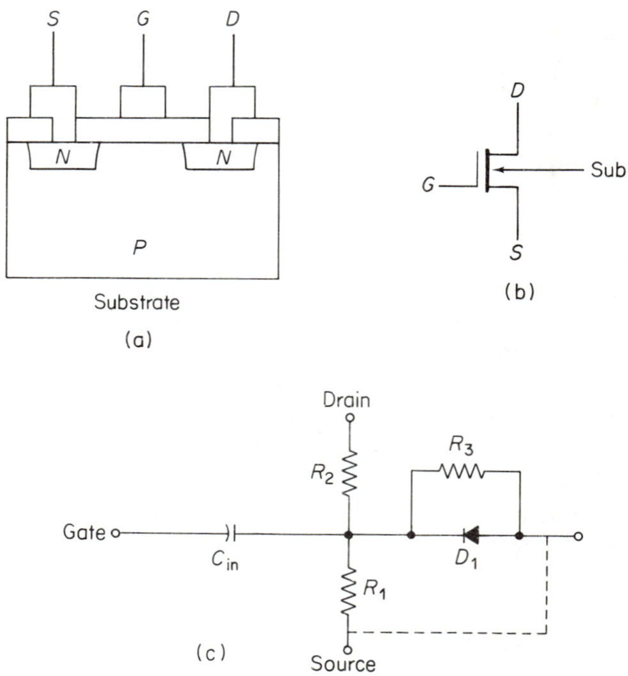

FIGURE 3-25 Structure, symbol and simplified equivalent circuit of typical single-gate MOSFET.

In a typical RCA MOSFET (the 3N128), C_{in} is less than 5 pF. The channel resistance $R_1 + R_2$, which is a function of the applied bias, can range from 10^2 to 10^{10}. R_3 is also subject to variations determined by operating conditions but can be assumed to be in the order of $10^9\ \Omega$. Thus, a dc voltage applied between the gate and any other element results in practically all this voltage being applied across C_{in}.

Typically, the gate voltage-handling capability of a MOSFET (without protection) is not over 30 V, although some MOSFETs can withstand gate-to-substrate voltages of about 100 V and not result in breakdown. With any MOSFET, however, once the oxide insulation breaks down, the device is destroyed.

3-10.1 Basic gate diode protection

The ideal situation in gate protection is to provide a signal-limiting configuration that allows for a typical sine wave to be handled without clipping or distortion. The signal-limiting devices must limit all transient voltages that exceed the gate breakdown voltage. For example, if the gate breakdown is 30 V, transient voltage must be limited to something less (say 25 V).

One possible means of securing proper limiting is to place a diode in parallel with C_{in} as shown in Fig. 3-26. This method is popular with some single-gate MOSFETs. The method has certain limitations. In terms of signal handling, the single diode of Fig. 3-26 clips the positive peaks of a sine wave when the device is operated near zero bias.

The single-diode method is not suitable for dual-gate MOSFETs where one gate is operated with a fixed dc bias. For example, a dual-gate MOSFET is frequently operated with an *RF* signal superimposed on a slightly positive "bias" on one gate and an AGC voltage on the other gate. The AGC voltage is capable of both positive and negative swings.

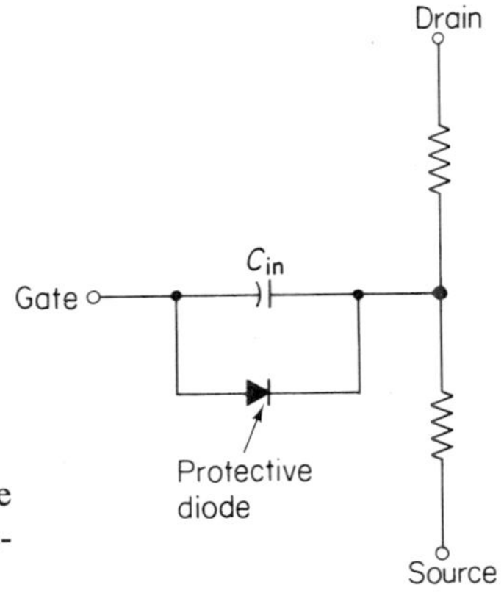

FIGURE 3-26 Static-discharge protection provided by single-gate diode.

3-10.2 Back-to-back diode protection

Whatever limiting device is used, it is important that the limiting device be an effective "open circuit" to any incoming signals, throughout the amplitude range of such signals. One of the best available methods for accomplishing this effective open circuit is the back-to-back diode arrangement (pioneered by RCA) shown in Fig. 3-27. The back-to-back method is generally used with dual-gate MOSFETs, and each gate is provided with a separate pair of back-to-back diodes.

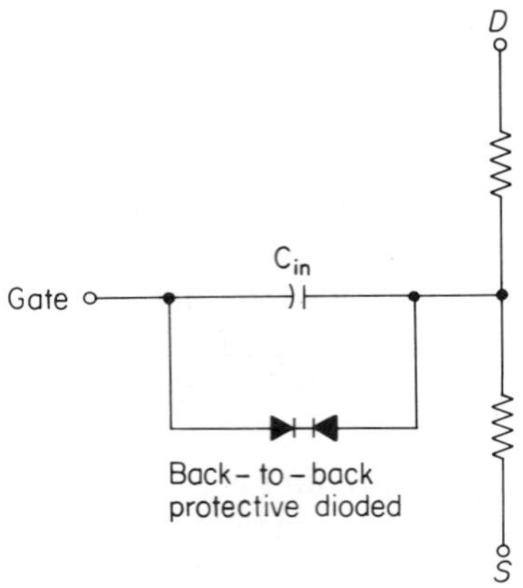

FIGURE 3-27 Static-discharge protection provided by back-to-back gate diodes.

3-10.3 Practical monolithic gate-protected MOSFETs

To keep cost and size at a minimum, it is best if the gate protection is accomplished by monolith means. One approach to integrating protective diodes into a MOSFET is shown in Fig. 3-28. With such a scheme, the diodes provide adequate protection against the transients experienced in handling, in addition to those found when the device is installed in some piece of equipment.

In this approach, the silicon substrate required for an *N*-channel, depletion-type MOSFET is the starting material. The *N*-type wells are diffused into the silicon to provide pockets for the protective devices. The surface concentrations and depth of these wells are carefully controlled because both of the factors are important in determining diode characteristics.

The *P*+ regions are diffused into the *N*-type wells to form the diodes. *P*+ regions are also diffused around the periphery to isolate the diode structure from the surface of the MOSFET and to provide a region into which the channels may be terminated. The size of the diodes is determined by the desired current-handling capability and the amount of capacitance that can be tolerated across the gate. The spacing of the diodes is determined by the area available and the desired amount of transistor control action from

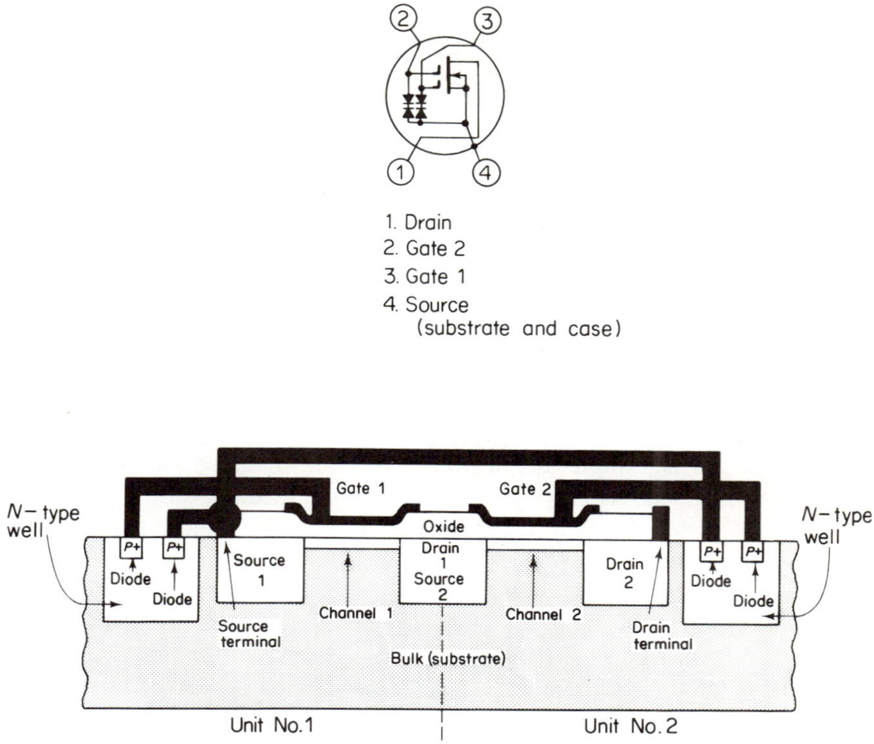

FIGURE 3-28 Practical monolithic gate-protected MOS device (courtesy RCA).

diode to diode. After the diode structures are formed, they are covered by a protective oxide. The MOSFET is then fabricated by conventional means.

The MOSFET shown in Fig. 3-28 is a dual-gate device. Such a MOSFET can be converted to a single-gate device, but with diode gate protection, by a simple modification as shown in Fig. 3-29. This modification is accomplished by connecting the two gates as shown in Fig. 3-29(a). This produces the electrical equivalent shown in Fig. 3-29(b).

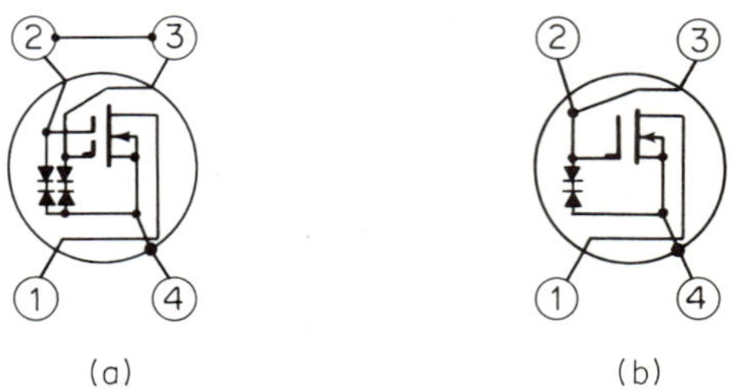

FIGURE 3-29 Dual-gate MOSFET modified to provide single-gate operation, with diode protection (courtesy RCA).

No matter what fabrication technique is used, there will be some effect on the MOSFET when protective diodes are added; however, the change in characteristics need not be significant. For example, with a typical RCA MOSFET operating as an *RF* amplifier at 200 MHz, the following changes were noted after the addition of protective diodes:

Power gain reduction, less than 0.5 dB.
Noise figure increase, less than 0.5 dB.
Input capacitance increase, from 5 to 8 pF.
Input resistance decrease, from 1 kΩ to 800 Ω.

3-11. TRIODE-CONNECTED DUAL-GATE MOSFET

As discussed in Sec. 3-10.3, it is possible to connect a dual-gate MOSFET so that it functions as a single-gate device. This is known as the *triode-connected configuration*, which has curve tracer (drain family) characteristics that look like a "real" triode MOSFET. The curves in Fig. 3-30 show that characteristics for the triode MOSFET (3N128) and the triode-connected dual-gate MOSFET (3N187) are essentially similar.

Some useful triode-connected device characteristics are provided in Fig. 3-31 in the form of comparison with dual-gate and single-gate devices. It should be noted that the difference in I_{DS} level between the 3N187 and 3N200 carries over to their diode-connected versions.

It should not be inferred from these comparisons that all single-gate applications can be handled by protected (back-to-back diodes) dual-gate

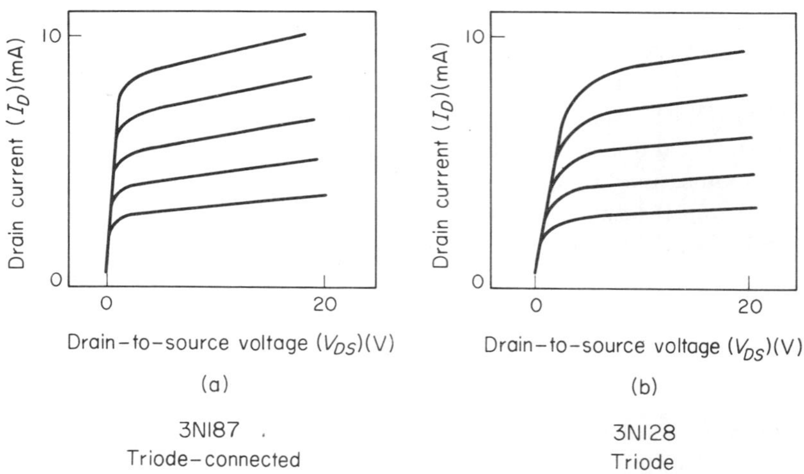

FIGURE 3-30 Comparison of $V_{DS} - I_D$ curves for triode MOSFET and triode-connected MOSFET (courtesy RCA).

Characteristic	Conditions	Triode-connected 3N187	Triode-connected 3N200	Dual-gate Circuit* 3N187	Dual-gate Circuit* 3N200	Single Gate 3N128	Units
I_{DS}	$V_{DS} = 15$ V	6.0	2.0	15	5	15	mA
g_{fs}	$V_{DS} = 15$ V, $I_D = 10$ mA, $f = 1$ kHz	7.0	8.5	12	15	9	mmho
$V_{G1S(\text{off})}$	$V_{DS} = 15$ V, $I_D = 50\ \mu$A	−2.0	−1.0	−2.0	−1.0	−1.5	V
I_{G1SS}	$V_{GS} = \pm 6$ V	2.0	2.0	1.0	1.0	10^{-4}	nA
C_{iss}	$V_{DS} = 15$ V, $I_D = 10$ mA, $f = 1$ kHz	10.0	10.0	6.0	6.0	5.5	pF
C_{rss}	$V_{DS} = 15$ V, $I_D = 10$ mA, $f = 1$ kHz	0.5	0.5	0.02	0.02	0.2	pF
C_{oss}	$V_{DS} = 15$ V, $I_D = 10$ mA, $f = 1$ kHz	2.0	2.0	2.0	2.0	1.4	pF
$R_{DS(\text{on})}$	$V_{DS} = 1$ V, $V_{GS} = 0$	160	250	100	150	300	Ω

*V_{G2S}=4V except for I_{GSS} measurement, where $V_{G2S} = 0$.

FIGURE 3-31 Comparison of typical electrical characteristics for triode-connected dual-gate, standard dual-gate, and triode (single-gate) MOSFETs (courtesy RCA).

devices. Gate leakage is increased when back-to-back diodes (or any form of protective diode) is added to a MOSFET. For those applications where the gate leakage I_{GSS} must be in the picoampere range, it is necessary to use a classic single-gate type (such as the 3N128) and take precautions against gate-insulation puncture.

3-12. FET CHARACTERISTICS

The various characteristics of FETs are discussed in detail throughout this chapter and in other chapters of this book. How these characteristics affect certain applications of specific FETs is covered in detail. For example, drain-substrate junction capacitance is of great importance to an FET use in switching circuits but is of no particular concern to an FET operating as an *RF* amplifier.

In this section, we outline all the characteristics that generally apply to FETs of all types. This provides a foundation for understanding how the characteristics affect design with FETs. Both static and dynamic characteristics are discussed.

3-12.1 Static FET characteristics

Static (or direct-current) characteristics of an FET are those that indicate the effect of a control (or gate) voltage (or signal) on the output (drain-source) current. The following paragraphs describe static characteristics found on typical FET datasheets. Figure 3-32 shows the relationship between the characteristics and typical V_{GS}–I_D (gate-source voltage versus drain current) curves.

Note that some of the characteristics must be related to the *operating mode* of the FET. For example, I_{DSS} is the drain current found with the FET connected as a common-source device but with *zero gate voltage*. As shown in Fig. 3-32, I_{DSS} for an FET in the depletion-only mode is essentially the maximum drain current (or the drain current with the device full-on). In the depletion/enhancement mode, I_{DSS} is the drain current about halfway between full-off and full-on. In the enhancement-only mode, I_{DSS} is essentially leakage current.

Control Voltage

The gate-source voltage, or V_{GS}, is considered as the control voltage (or signal). That is, the amount of V_{GS} controls the amount of source-drain current.

$V_{GS(\text{off})}$ is the gate voltage necessary to reduce the drain-source current (generally specified as I_D) to zero or to some specified value near zero. Often, $V_{GS(\text{off})}$ is defined as the gate voltage required to reduce the drain current to 0.01 or preferably 0.001 of the minimum I_{DSS} value. For example, assume that

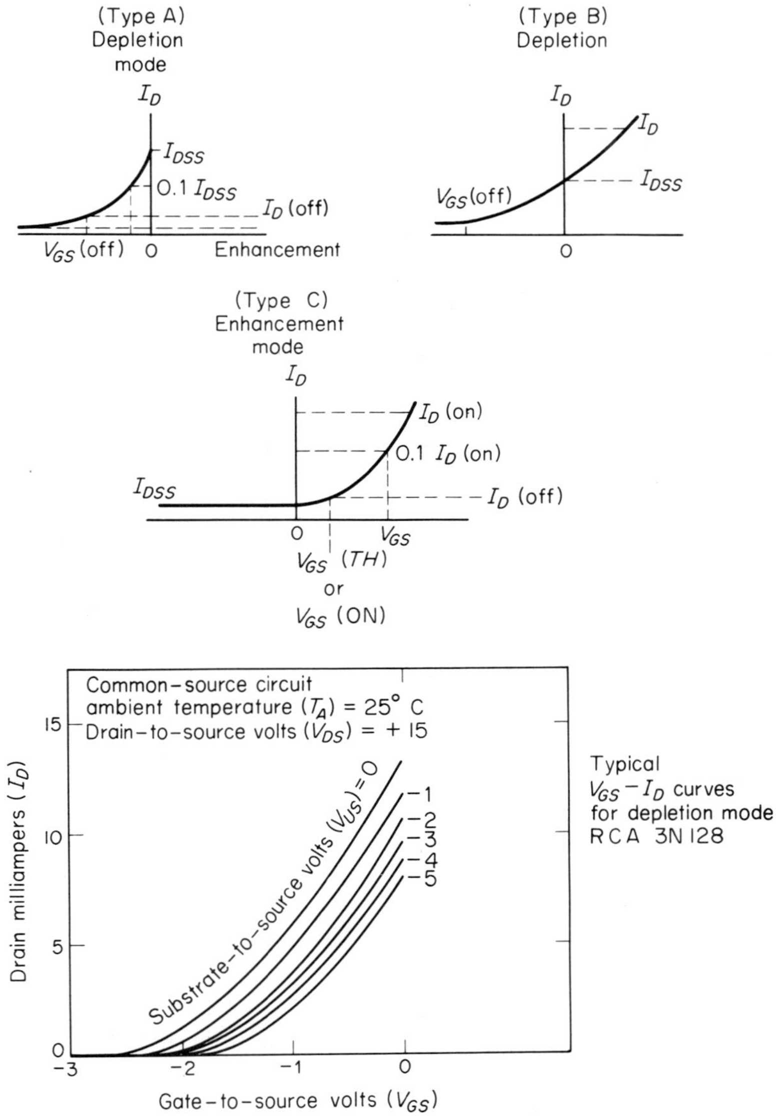

FIGURE 3-32 Static MOS characteristics.

1 mA flows when V_{GS} is at zero. This means that I_{DSS} is 1 mA. Under such conditions, $V_{GS(\text{off})}$ is the gate voltage required to produce a drain current of 0.01 mA (or 0.001 mA).

Although the term $V_{GS(\text{off})}$ can be applied to all operating modes, it is best applied to depletion and depletion/enhancement modes. In the enhancement-only mode, the term $V_{GS(\text{th})}$ or *threshold gate voltage* is more descriptive. $V_{GS(\text{th})}$ is the gate voltage of an enhancement-only mode FET where I_D *just starts to flow*. Some datasheets specify $V_{GS(\text{off})}$ as the gate voltage that produces some given value of I_D, such as 1 μA, 1 pA, or 1 nA. The term $V_{GS(\text{on})}$ is sometimes used in place of $V_{GS(\text{th})}$ for enhancement-only FETs.

No matter what term is used, the value of $V_{GS(\text{off})}$ determines the *minimum operational* drain-source voltage V_{DS}. As a rule of thumb, V_{DS} *must be a minimum of 1.5* times $V_{GS(\text{off})}$ (or $V_{GS(\text{th})}$ or $V_{GS(\text{on})}$).

Some datasheets specify $V_{GS(\text{off})}$ as minimum-typical-maximum values; whereas other datasheets specify only a maximum $V_{GS(\text{off})}$. There is a simple method of finding a workable value for $V_{GS(\text{off})}$ if the V_{GS}–I_D curves are available.

Simply lay a straightedge along the straight portion of the curve as shown in Fig. 3-32. Note where the straightedge intersects the V_{GS} axis. Double this value to find a workable V_{GS}. For example, in the curve of Fig. 3-32, the straightedge intersects the V_{GS} axis at about -2 V. Thus, a workable value for V_{GS} is -4 V. Note that the datasheet for an FET with the Fig. 3-32 curve specifies $V_{GS(\text{off})}$ as -0.5 V, -4 V typical, and -8 V maximum.

Operating Voltage

The *drain-source* voltage, or V_{DS}, is considered as the operating voltage. V_{DS} is the equivalent of collector voltage in a two-junction transistor or plate voltage in a vacuum tube. Generally, the only concern is that the maximum datasheet value for V_{DS} must not be exceeded. In most applications, however, V_{DS} must be a minimum of 1.5 times $V_{GS(\text{off})}$.

When FETs are used as switches and choppers, the terms $V_{DS(\text{on})}$ and $V_{DS(\text{off})}$ sometimes appear on the datasheets. $V_{DS(\text{on})}$ is similar to saturation voltage in conventional transistors. $V_{DS(\text{off})}$ is the drain-source voltage at which the I_D increases very little for an increase in drain-source voltage, with V_{GS} held at zero. Sometimes, the term V_P or pinch-off voltage is used instead. V_P generally applies to JFETs, however, rather than to MOSFETs.

The various values of V_{DS} are dependent on source resistance R_S, drain resistance R_D, or drain-to-source resistance R_{DS}. All of these are static, direct-current values and rarely appear on datasheets. The dynamic r_{ds} values discussed in Sec. 3-12.2 are far more important.

Although FETs do not usually have an operating voltage between the drain and gate, the term V_{DG} usually appears on most FET datasheets. Generally, this is a maximum voltage and is usually the same as V_{DS} maximum.

Operating Current

The *drain-source* current, or I_D, is considered as the operating current. I_D is the equivalent of collector current in a two-junction transistor or plate current in a vacuum tube. Generally, the only concern is that the maximum datasheet value for I_D must not be exceeded.

In addition to I_D, the terms $I_{D(\mathrm{on})}$ and $I_{D(\mathrm{off})}$ are used on some datasheets. $I_{D(\mathrm{on})}$ is an arbitrary current value (usually near the maximum rated current) that locates a point in the enhancement operating mode, as shown in Fig. 3-32. $I_{D(\mathrm{off})}$ is the current value that flows when $V_{GS(\mathrm{off})}$ is applied. On some FETs operating in the enhancement-only mode, the datasheet value of $I_{D(\mathrm{off})}$ is the current when $V_{GS(\mathrm{th})}$ is applied.

Voltage Breakdown

Voltage breakdown is a particularly important characteristic of FETs. There are a number of specifications to indicate the maximum voltage that can be applied to the various elements. These include

$V_{(BR)GSS}$—gate-to-source breakdown voltage.
$V_{(BR)DGO}$—drain-to-gate breakdown voltage.
$V_{(BR)DSX}$—drain-to-source breakdown voltage.
$V_{(BR)DSS}$—drain-to-source breakdown voltage (alternate).

$V_{(BR)GSS}$ is the breakdown voltage from gate to source, with the drain and source shorted. Under these test conditions, the gate-channel junction (*PN* diode for a JFET; oxide layer for a MOSFET) also meets the breakdown specification since the drain and source are the connections to the channel. For the designer, this means that the drain and source may be interchanged, for symmetrical devices, without fear of the individual junction breakdown. For MOSFETs, the gate is insulated from the drain-source and channel by an oxide layer. Thus, the gate breakdown voltage is dependent on the thickness and purity of the layer. $V_{(BR)GSS}$ for a MOSFET is the voltage that will physically puncture the oxide layer.

$V_{(BR)DGO}$ is essentially the same specification as $V_{(BR)GSS}$, except that $V_{(BR)DGO}$ represents breakdown from gate to drain. $V_{(BR)DGO}$ is more properly applied to JFETs but may appear on a MOSFET datasheet.

Drain-to-source breakdown voltage for an FET is determined by the operating mode. For enhancement-only mode operation (Type C), with the gate connected to the source (in the cutoff condition) and the substrate floating, there is no effective channel between drain and source, and the applied drain-source voltage appears (in effect) across two back-to-back connected, reverse-biased diodes. These diodes are represented by the source-to-substrate and substrate-to-drain junctions.

During drain-to-source breakdown tests, drain current remains at a very low (typically picoampere) level as drain voltage is increased until the drain voltage reaches a value that causes reverse (*reach-through* or *punch-through*) breakdown of the diodes. This condition, generally represented on datasheets as $V_{(BR)DSS}$, is indicated by an increase in I_D.

For Types A and B FETs, the $V_{(BR)DSS}$ designation is sometimes replaced by $V_{(BR)DSX}$. The main difference between the two designations is the replacement of the last subscript S with the subscript X. The S normally indicates that the gate is shorted to the source. The X indicates that the gate is biased to cutoff or beyond. To obtain cutoff in Types A and B FETs, a depletion bias voltage must be applied to the gate.

In some datasheets, there may be ratings and specifications indicating the maximum voltages that may be applied between the individual gates and the drain or source, between drain and source, and so on. Not all of these specifications are found on every datasheet. Some of the specifications provide the same information in slightly different form. By understanding the various breakdown mechanisms, however, it should be possible to interpret the intent of each specification and rating.

For example, in MOSFET specifications, the *breakdown* voltage is generally interchangeable with the term *avalanche* voltage (V_A). Such avalanche voltage is the V_{DS} that will cause the I_D to go "full-on," sometimes to the point of destroying the device. Note that avalanche occurs at a lower value of V_{DS} when the gate is reverse biased than when the gate is at zero bias. This condition is caused by the fact that the reverse-bias gate voltage adds to the drain voltage, thus increasing the effective voltage across the junction. As a rule of thumb, *the maximum V_{DS} that may be applied is the drain-gate breakdown voltage, less the V_{GS}.*

Gate Leakage

Gate leakage is a very important characteristic of an FET since such leakage is directly related to input resistance. When gate leakage is high, input resistance is low and vice versa. Gate leakage is usually specified as I_{GSS} (reverse-bias, gate-to-source current with drain shorted to source) and is a measure of the static short-circuit input impedance.

Normally, input resistance of an FET is very high, particularly for a MOSFET, since the leakage current across a capacitor is very small. For example, at a temperature of 25°C, typical MOSFET input resistance R_{GS} is in the order of $10^{14}\ \Omega$. Of course, JFET input resistance values are lower since their input is essentially a diode junction rather than a capacitor.

Input resistance may decrease when temperature is raised to the typical maximum of +175°C. FETs are not drastically affected by temperature (particularly MOSFETs), however, and their input resistance remains high even at maximum operating temperature.

Some datasheets specify gate leakage current as I_{GDO} (leakage between gate and drain with the source open). Other datasheets use the term I_{GSO} (leakage between gate and source with the drain open). When these characteristics are used, the gate leakage current is lower than when I_{GSS} is specified. Consequently, I_{GDO} and I_{GSO} does not represent the "worst case" condition, making I_{GSS} the preferred specification.

Device Dissipation

Datasheets list device dissipation (P_D) or transistor dissipation (P_T) or some similar term at a given ambient temperature, together with a derating factor. Typically, P_D or P_T is the maximum power that can be dissipated with the device at 25°C without exceeding the maximum allowable internal temperature. Most FETs cannot stand temperatures above about 200°C without damage. In many cases, the maximum temperature is 150 to 175°C.

As an example, a typical FET has a maximum temperature of +175°C, a maximum dissipation of 330 mW at 25°C, and a derating factor of 2.2 mW/°C. Thus, at 125°C, the power dissipation in the FET must not exceed 110 mW, which can be calculated as follows:

$$125°\text{C} - 25°\text{C} = 100°\text{C}, \qquad 100 \times 2.2\ \text{mW} = 220\ \text{mW},$$

$$330\ \text{mW} - 220\ \text{mW} = 110\ \text{mW}$$

Temperature Ranges

Critical temperatures for FETs are usually the *ambient temperature* during both operation and storage as well as *lead temperature* during soldering. As a guideline, a typical ambient temperature range for FETs during both operation and storage is −65 to +175°C. Lead temperatures should not exceed +265°C during soldering and this temperature should not be closer than $\frac{1}{32}$ in. from the seating surface for more than 10 s.

Protective Diode Voltage

On those FETs with protective diodes (MOSFETs), the datasheets list the protective diode clamp voltage. This voltage is often referred to as the *knee* voltage, or V_{knee}. The input signal voltages cannot exceed this value.

Forward Gate Current

Some FET datasheets list a forward gate current characteristic, such as I_{GS} or $I_{G(f)}$. This is not to be confused with I_{GSS}, which is leakage current. I_{GS} or $I_{G(f)}$ is the maximum recommended forward current through the gate terminal. Although I_{GS} or $I_{G(f)}$ is primarily a JFET specification, the forward gate current characteristic can be a limiting factor in some MOSFET applications.

I_{GS} or $I_{G(f)}$ is caused by a large forward bias current on the gate. When such a bias must be used, the gate current must be limited to the I_{GS} or $I_{G(f)}$ value (or degeneration of the device could occur). A resistor in series with the gate will limit current but the resistor value will determine the variation of gate bias, as affected by the gate leakage current, and will affect input resistance (or impedance) of the device.

Dual-Gate Characteristics

In addition to the static characteristics described thus far, dual-gate FETs have certain other characteristics. For example, with dual-gate MOSFETs, both gates control I_D. As a result, there are a number of datasheet specifications that describe how the voltage on one gate will affect I_D, with the other gate held at some specific voltage, or zero volts. Likewise, there are specifications that describe how gate leakage or gate current flow is affected by gate voltages. For practical design situations, the most important dual-gate characteristics include *cutoff voltage*, *breakdown voltage*, and *gate current*.

With a dual-gate MOSFET, the gate 1-to-source cutoff voltage (with gate 2 connected to the source) is identified as $V_{GIS(\text{off})}$; the gate 2-to-source cutoff voltage (with gate 1 connected to the source) is identified as $V_{G2S(\text{off})}$. Note that dual-gate MOSFETs are not always symmetrical; that is, the gate voltages in different combinations produce different drain currents. For example, with the RCA 40822 dual-gate MOSFET, $V_{G1S(\text{off})}$ is tested with a V_{DS} of +15 V and a V_{G2S} of +4 V. Under these conditions, a typical cutoff voltage of −2 V at gate 1 will produce an I_D of 50 μA [which is considered to be $I_{D(\text{off})}$ for this depletion-type device]. With the same MOSFET, $V_{G2S(\text{off})}$ is tested with a V_{DS} of +15 V, and a V_{G1S} of 0 V. Under these conditions, a −2 V at gate 2 will produce the I_D of 50 μA.

The gate 1-to-source breakdown voltage (with gate 2 connected to the source) is identified as $V_{(BR)G1SS}$; gate 2-to-source breakdown voltage is $V_{(BR)G2SS}$. The gate 1 leakage current is identified as I_{G1SS}; gate 2 leakage current is I_{G2SS}. With the RCA 40822, $V_{(BR)G1SS}$ is tested using a V_{DS} and a V_{G2S} of 0 V. Under these conditions, a V_{G1S} of 9 V will produce an I_{G1SS} of 100 μA.

With a dual-gate MOSFET, gate leakage should be symmetrical, even though the gate-voltage versus drain-current characteristics may not be symmetrical. Thus, for the RCA 40822 with a V_{DS} and V_{G1S} of 0 V, a V_{G2S} of 9 V will also produce an I_{G2SS} of 100 μA. Further, gate leakage or gate current should be the same when gate voltages are reversed in polarity. Thus, dual-gate MOSFET datasheets often specify forward and reverse breakdown voltages as well as gate leakage currents.

As is the case with single gate, dual-gate FETs have a maximum recommended gate current. This sets the limit of gate voltage. For example, the RCA 40822 has a maximum recommended gate current of 50 nA. A maximum of 50-nA gate current will flow when V_{DS} and one gate are at 0 V and the opposite gate is at 6 V (either +6 V or −6 V). Thus, 6 V is the maximum recommended gate voltage. Gate currents for dual-gate FETs are usually identified as I_{G1SSF} and I_{G2SSF} (for forward) or I_{G1SSR} and I_{G2SSR} (for reverse).

Dual-gate FETs have a zero-bias drain-current specification I_{DS}, which corresponds to an I_{DSS} for single-gate units. The measurement for I_{DS} is usually based on 0 V at one gate, however, with a fixed voltage at the opposite gate. For example, with the RCA 40822, I_{DS} is measured with V_{DS} at +15 V, V_{G1S} at 0 V, and V_{G2S} at +4 V. Typically, the I_{DS} is between 5 and 30 mA for this depletion mode device.

3-12.2 Dynamic FET characteristics

Unlike the static characteristics, the dynamic characteristics (ac or signal) of FETs apply equally to Types A, B, and C; however, conditions and presentations of the dynamic characteristics depend mostly on the intended *application*. Figure 3-33 indicates the dynamic characteristics generally used to describe an FET for various applications.

Audio	*RF–IF*	Switching	Chopper
Y_{fs} (1 kHz)	Y_{fs} (1 kHz)	C_{iss}	C_{iss}
C_{iss}	C_{iss}	C_{rss}	C_{rss}
C_{rss}	C_{rss}	$C_{D(\text{sub})}$	$C_{D(\text{sub})}$
Y_{os} (1 kHz)	G_{PS}	$R_{ds(\text{on})}$	
NF	*MAG*		$R_{ds(\text{on})}$
	MUG		
	$Re(Y_{fs})(HF)$	t_d, t_s	
	$Re(Y_{is})(HF)$	t_r, t_f	
	$Re(Y_{os})(HF)$		
	NF		

FIGURE 3-33 Important dynamic characteristics generally used to describe a MOS device for various applications.

FORWARD TRANSADMITTANCE (TRANSCONDUCTANCE)

Forward transadmittance Y_{fs} is defined as the magnitude of the common-source forward transfer admittance. Y_{fs} is the most important dynamic characteristic for FETs, no matter what application. Y_{fs} serves as a basic design parameter in audio and *RF* and is a widely accepted device figure of merit.

Because FETs have many characteristics similar to those of vacuum tubes, the symbol g_m is sometimes used instead of Y_{fs}. This is further confused since the *g* notation school also uses a number of subscripts. In addition to g_m, some datasheets show g_{fs}, while others go even further out with g_{21}.

No matter what symbol is used, Y_{fs} defines the relationship between an input signal voltage and an output signal current, with the drain-source voltage held constant or

$$Y_{fs} = \frac{\Delta I_D}{\Delta V_{GS}}$$

with V_{DS} held constant.

Y_{fs} is expressed in *mhos* (current divided by voltage). In most datasheets, Y_{fs} is specified at 1 kHz with a V_{DS} the same as that for which $I_{D(\text{on})}$ or I_{DSS}

is obtained. At 1 kHz, Y_{fs} is almost entirely real. Thus, Y_{fs} at 1 kHz $= Y_{fs}$. At higher frequencies, Y_{fs} includes the effects of gate-to-drain capacitance and may be misleadingly high. For high-frequency operation, the real part of transadmittance $R_e(Y_{fs})$, as discussed in later paragraphs, should be used.

Since the I_D–V_{GS} curves of an FET are nonlinear, Y_{fs} will vary considerably with changes in I_D. This variation for a typical N-channel MOSFET (the 3N152) is illustrated in Fig. 3-34. Note that the curve ends sharply where Y_{fs} is at its maximum. This is the point where $I_D = I_{DSS}$.

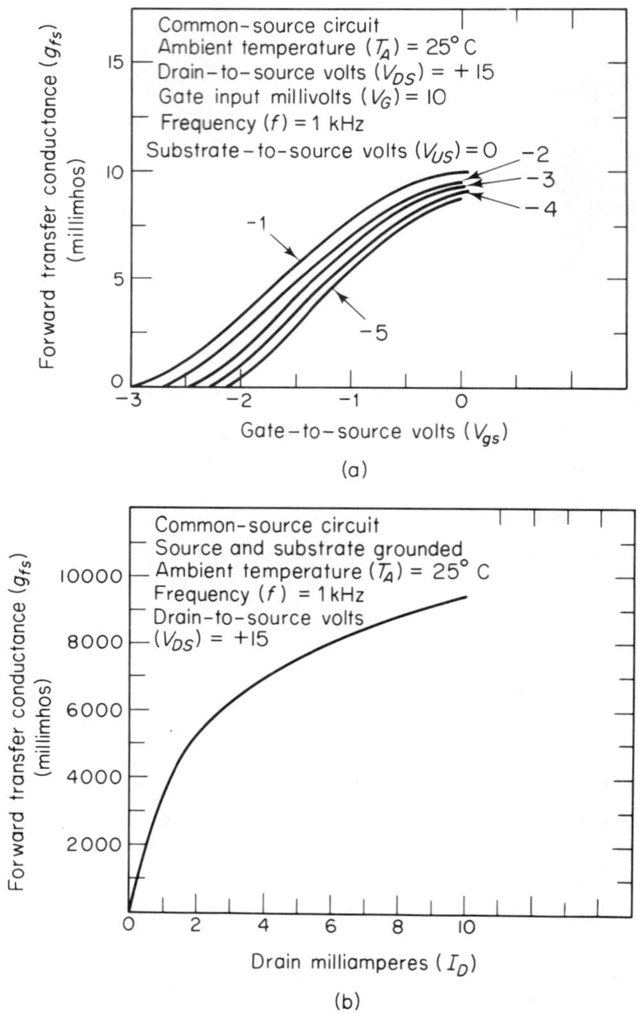

FIGURE 3-34 (a) Forward transconductance versus gate bias voltage (b) drain current for 3N152 (courtesy RCA).

Three Y_{fs} measurements are often specified for FETs. One of these measurements, with two gates tied together, provides a Y_{fs} value for the condition where a signal is applied to both gates simultaneously. The other two measurements provide the Y_{fs} for the two gates individually. Generally, with the two gates tied together, Y_{fs} is higher and more gain may be realized in a given circuit. Because of the increased capacitance, however, the gain-bandwidth product is much lower.

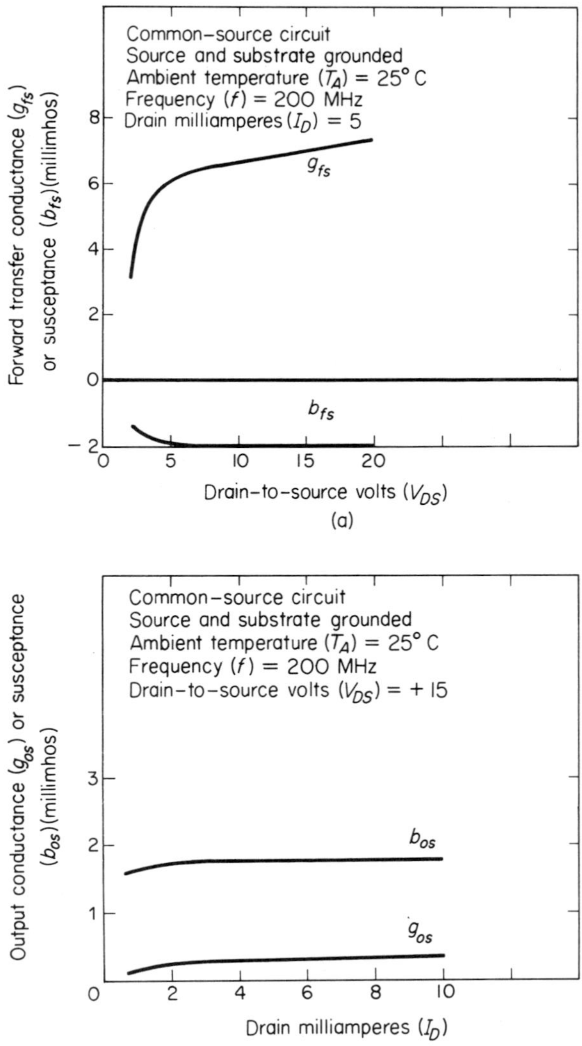

FIGURE 3-35 (a) Forward transadmittance versus drain-to-source voltage (b) drain current for 3N154 (courtesy RCA).

For FETs used at radio frequencies, an additional value of Y_{fs} is often specified at or near the highest frequency of operation. This value is measured at the same voltage conditions as those used for $I_{D(on)}$ or I_{DSS}.

Because of the importance of the imaginary component at radio frequencies, the high-frequency Y_{fs} specification is generally a complex representation; that is, both forward transconductance (g_{fs}, g_{21}, etc.) and forward susceptance (b_{fs}, b_{21}, etc.) are shown, as in Fig. 3-35, which illustrates the curves for a 3N154 at a test frequency of 200 MHz.

Some datasheets list the real part of Y_{fs} (or forward transconductance) as $Re(Y_{fs})$ or $Re(Y_{fs})(HF)$. Such $Re(Y_{fs})$ is defined as the common-source forward transfer conductance (drain current versus gate voltage). For high-frequency applications, $Re(Y_{fs})$ is considered a significant figure of merit. The dc operating conditions are the same as for Y_{fs}, but the test frequency is typically 100 to 200 MHz.

Rather than listing a value for the real part of Y_{fs} at a specific frequency, some datasheets show both the forward transconductance (g_{fs}) and forward susceptance (b_{fs}) over the useful frequency range of the device. Such a representation is shown in Fig. 3-36 by means of curves. No matter what datasheet system is used, *an increase in the real part of Y_{fs} produces an increase in the voltage gain* produced by an FET amplifier stage.

In comparing $Re(Y_{fs})$ with Y_{fs} (on those datasheets that list both values), the minimum values of the two are quite close, considering the difference in frequency at which the measurements are made. At high frequencies, about 30 MHz and above, Y_{fs} will increase due to the effect of gain-drain capacitance C_{gd} so that Y_{fs} will be misleadingly high.

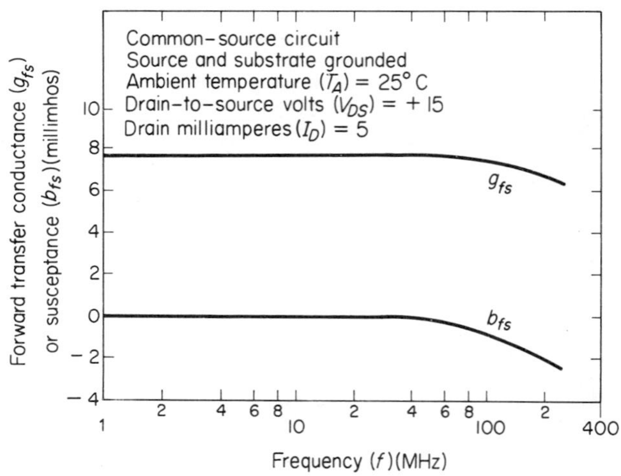

FIGURE 3-36 Forward transadmittance versus frequency for 3N152 (courtesy RCA).

REVERSE TRANSADMITTANCE

Reverse transadmittance Y_{rs} (or Y_{12}) is not generally a critical factor in FETs. Typically, the real part of Y_{rs} (or g_{rs}) is zero over the useful frequency range. It is necessary, however, to know the values of Y_{rs} to calculate impedance-matching networks for FET *RF* amplifiers. For that reason, most FET datasheets list some values for Y_{rs}.

Figure 3-37 shows some typical Y_{rs} curves for a MOSFET. Note that the real part of g_{rs} remains at zero for all conditions and at all frequencies; that

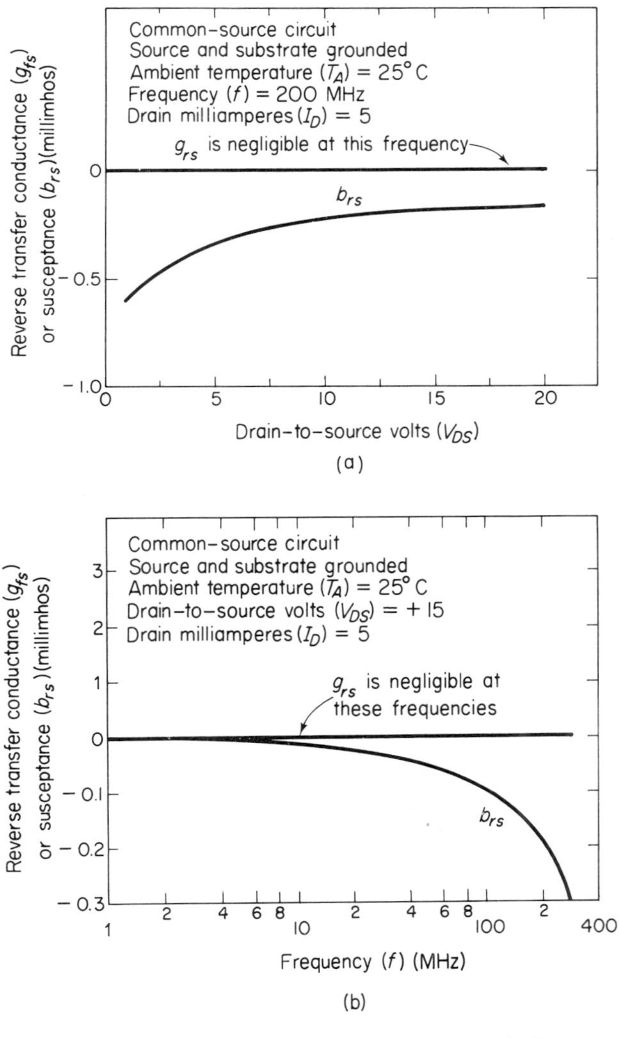

FIGURE 3-37 (a) Reverse transadmittance versus drain-to-source voltage (b) frequency (courtesy RCA).

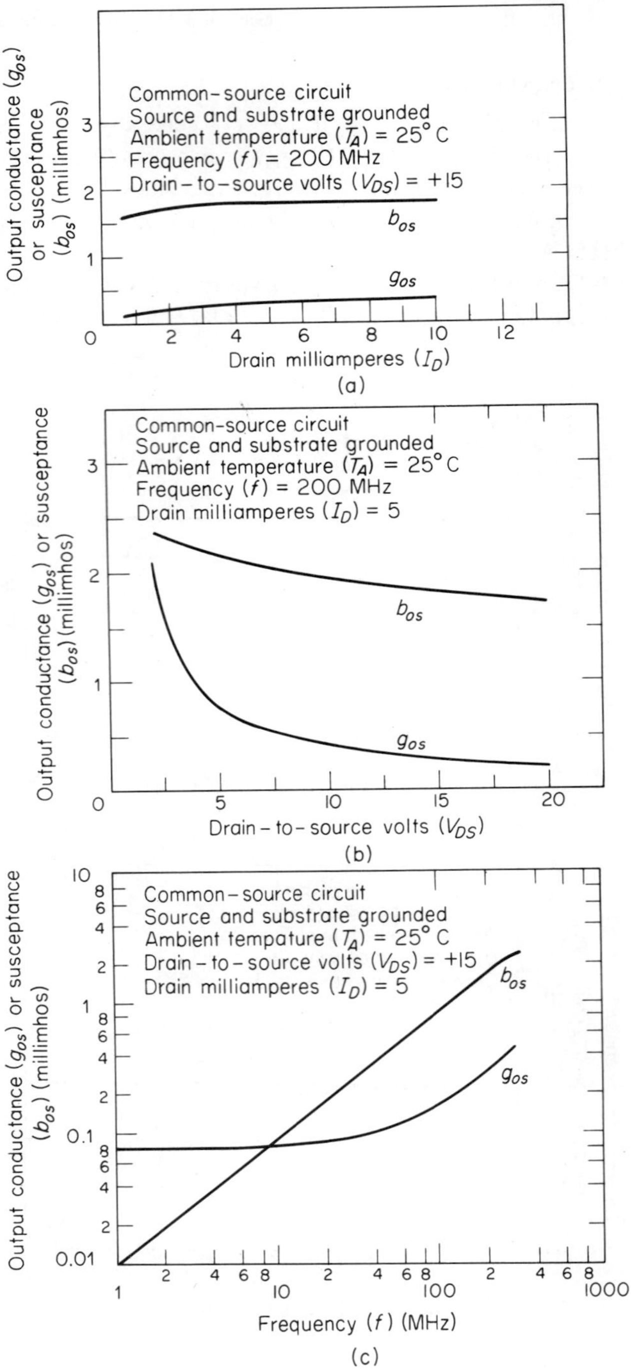

FIGURE 3-38 (a) Output admittance versus drain current (b) drain-to-source voltage (c) frequency for 3N152 (courtesy RCA).

is, there is no reverse conductance. The imaginary part of b_{rs}, however, does vary with voltage, current, and frequency; that is, the reverse susceptance (or reactance) does vary with conditions. Thus, under the right conditions, there can be undesired feedback from output to input (reverse transadmittance). This condition must be accounted for in the design of *RF* amplifier stages.

Output Admittance

Output admittance Y_{os} is another important dynamic characteristic for FETs. Y_{os} is also represented by various Y and g parameters, such as Y_{22}, g_{os}, g_{22}. Y_{os} is even specified in terms of drain resistance, or r_d, where $r_d = 1/Y_{os}$. This is similar to the vacuum-tube characteristic of output admittance, where Y_{output} is $1/r_p$ (or output admittance is equal to one divided by plate resistance).

No matter what symbol is used, Y_{os} defines the relationship between output signal current and output voltage, with the input voltage held constant, or

$$Y_{os} = \frac{\Delta I_D}{\Delta V_{DS}}$$

with V_{GS} held constant.

Y_{os} is expressed in mhos (current divided by voltage). Types A and B FETs are measured with the gates and source grounded. For Type C FETs, Y_{os} is measured at some specified value of V_{GS} that will permit a substantial drain current to flow.

Some datasheets give Y_{os} as a complex number with both the real (g_{os}) and imaginary (b_{os}) values shown by means of curves. Figure 3-38 illustrates some typical Y_{os} curves.

Input Admittance

Input admittance Y_{is} (or Y_{11}) is not generally a critical factor in FETs. The real part of Y_{is} (or g_{is}) is nearly zero at low frequencies. It is necessary, however, to know the values of Y_{is} to calculate impedance-matching networks for FET *RF* amplifiers. Figure 3-39 shows some typical Y_{is} curves.

Amplification Factor

Amplification factor μ does not usually appear on most FET datasheets. This is because it does not usually have a great significance in most small-signal applications; however, amplification factor is sometimes used as a figure of merit in isolated cases.

Amplification factor defines the relationship between output signal

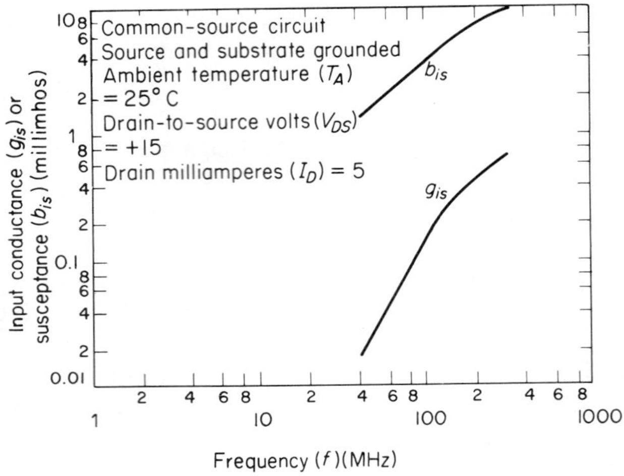

FIGURE 3-39 Input admittance versus frequency for 3N152 (courtesy RCA).

voltage and input signal voltage, with the output current held constant, or

$$\text{Amplification factor} = \frac{\Delta V_{DS}}{\Delta V_{GS}}$$

with I_D held constant.

Amplification factor can also be calculated by Y_{fs}/Y_{os}.

Input Capacitance

Input capacitance C_{iss} is the common-source input capacitance with the output shorted and is used as a low-frequency substitute for Y_{is}. This is because Y_{is} is entirely capacitive at low frequencies.

To find an approximate value for Y_{is} at low frequencies (below about 1 MHz), multiply C_{iss} by 6.28 F. The result will be b_{is}, or the imaginary part of Y_{is}. At these low frequencies, g_{is} can be considered as zero.

Note that C_{iss} is an important characteristic for FETs used in switching or chopper applications. This is because a large voltage swing at the gate must appear across the input capacitance C_{iss}.

Output Capacitance

Output capacitance C_{oss} is the common-source output capacitance with the input shorted. C_{oss} does not appear on all FET datasheets. Generally, C_{oss} is found on dual-gate datasheets. The output capacitance of a typical FET is usually quite low (in the order of 1 or 2 pF).

Reverse Transfer Capacitance

Reverse transfer capacitance C_{rss} is defined as the common-source reverse transfer capacitance with the input shorted. C_{rss} is often used in place of Y_{rs}, the short-circuit reverse transfer admittance, since Y_{rs} is almost entirely capacitive over the useful frequency range of most FETs and is of relatively constant capacity. Consequently, the low-frequency C_{rss} is an adequate specification.

C_{rss} is also of major importance in FETs used as switches. Similar to the C_{os} of two-junction transistors, C_{rss} must be charged and discharged during the switching interval.

For chopper applications, C_{rss} is the feed-through capacitance for the chopper drive.

C_{rss} is also known as the Miller-effect capacitance since the reverse capacitance can produce a condition similar to the Miller effect in vacuum tubes (such as constantly changing frequency-response curves).

Element Capacitances

The elements of an FET have some capacitance between them, as do two-junction transistors. Some of these capacitances have an effect on the dynamic characteristics of the device. Figure 3-10 shows the element capacitances. Of course, drain-substrate junction capacitance $C_{d(\text{sub})}$ is the most important.

$C_{d(\text{sub})}$ is usually found on datasheets for FETs used in switching applications. This is because $C_{d(\text{sub})}$ appears in parallel with the load in a switching circuit and must be charged and discharged between the two logic levels during the switching interval.

C_{ds} drain-to-source capacitance is another specification found on some switching FET datasheets. C_{ds} also appears in parallel with the load in switching and logic applications.

Channel Resistance

The channel resistance is an important characteristic for FETs used in switching applications. Channel resistance describes the bulk resistance of the channel *in series* with the drain and source. Channel resistance is described as $r_{d(\text{on})}$, r_{DS}, R_{DS}, r_{ds}, $r_{d(\text{off})}$, etc., depending on the datasheet. Some of these descriptions are static, some are dynamic, and some are mixed (again depending on the datasheet).

From a practical standpoint, there are two channel resistance specifications of concern in switching applications. These are the "on" and "off" specifications, such as $r_{ds(\text{on})}$ and $r_{ds(\text{off})}$, which can be either static or dynamic.

The "on" specification is the channel resistance with the FET biased "on." In a depletion mode device, the "on" condition can be produced by zero bias

($V_{GS} = 0$). In the enhancement mode, the "on" condition requires some forward bias. The opposite is true for the "off" condition. In a depletion device, the "off" condition requires some reverse bias. For enhancement, the "off" condition requires zero bias.

As a point of reference, a typical MOSFET "on" resistance is in the order of 220 Ω; whereas the "off" resistance is greater than 10^{10}.

Switching Time

Switching-time characteristics are of particular importance for FETs used in digital applications. The datasheets that describe those FETs specifically designed for switching include *timing characteristics*. Typically, these include t_d (delay time), t_s (storage time), t_r (rise time), and t_f (fall time). These characteristics are measured with a pulse source and a multiple-trace oscilloscope, as are similar characteristics for two-junction transistors.

Gain

The datasheets for FETs used as amplifiers usually list some gain characteristics. Typically, these are *power-gain* figures, at some specified frequency and under specified test conditions (such as V_{DS} and I_D). The gain figures are expressed in decibels, sometimes as maximum or minimum values. Such terms as *MAG* (maximum available gain), *MUG* (maximum usable gain), and G_{PS} (power gain with the device connected as common source) are used.

The datasheet notes concerning gain should always be consulted. For example, the term *MAG* usually means the maximum available gain under "ideal" or theoretical conditions. The term *MUG* usually means the maximum usable gain when the amplifier is neutralized or mismatched to produce the stability effects of neutralization.

On some datasheets, subscripts are added to identify the gain figures. For example, the subscript (*c*) can be added to *MUG* to indicate maximum usable gain when the FET is used as a converter.

Noise Figure

Noise figure for FETs is usually listed as *NF* on the datasheets and represents a common-source noise figure. *NF* represents a ratio between input signal-to-noise ratio and output signal-to-noise ratio and is measured in decibels. In most FET datasheets, *NF* includes the effects of e_n (equivalent short-circuit input noise expressed in volts-per-root cycle) and i_n (the equivalent open-circuit input noise current).

As shown in Fig. 3-40, *NF* attains its highest value for a small generator resistance and decreases for increasing generator resistance, indicating a large noise contribution from the noise-voltage generator. For this reason, some

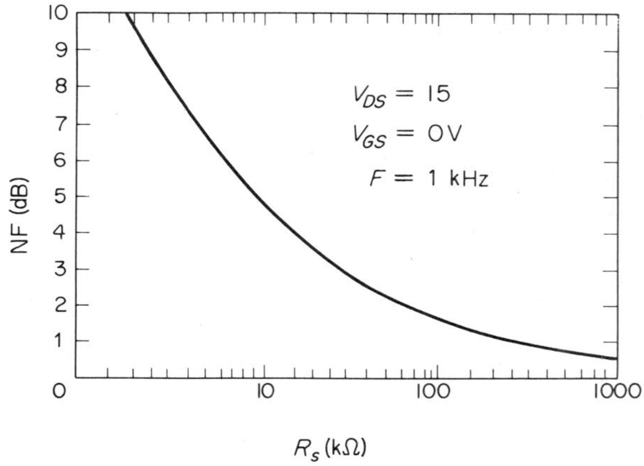

FIGURE 3-40 Noise figure versus source resistance (courtesy Motorola).

datasheets specify both *NF* and e_n but neglect i_n. In effect, *NF* is independent of operating current and proportional to voltage; however, the voltage effects are slight over the normal operating range of a typical FET.

Figure 3-41 is a nomograph for converting noise figure to equivalent input noise voltage for different generator source impedances R_S. This nomograph can be used with any FET. Since *NF* and e_n are frequency dependent, Fig. 3-41 must be used in conjunction with Fig. 3-42 (noise figure versus frequency at a specified source impedance) to determine e_n.

For example, to find the input noise at 50 Hz, with an R_S of 1 MΩ, the

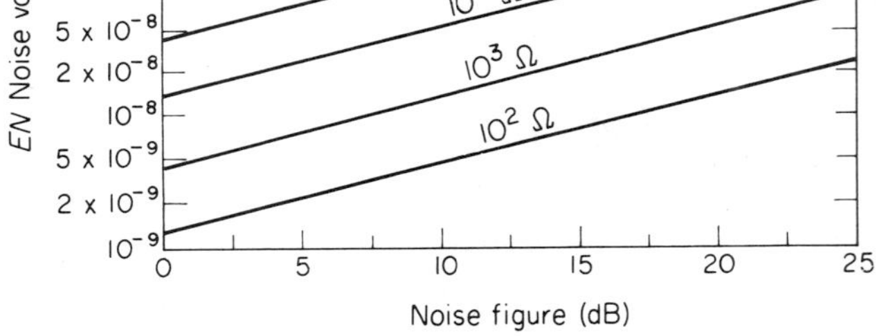

FIGURE 3-41 Noise figure conversion chart (courtesy Motorola).

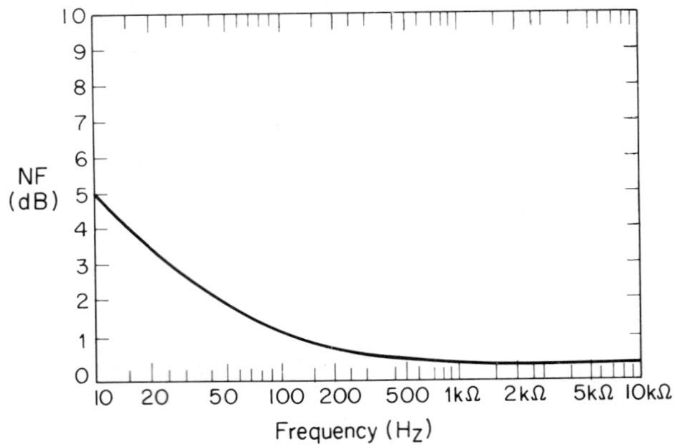

FIGURE 3-42 Noise figure versus frequency (courtesy Motorola).

NF (from Fig. 3-42) is about 1.5 dB. Next, from Fig. 3-41, noise voltage (e_n) is found to be 1.5×10^{-7} V (on the 10^6 or 1-MΩ curve), or about 150 nV. If the stage has a voltage gain of 10, the output noise is 10×150 nV or 1.5 μV of 50-Hz noise. As another example, the *NF* for an RCA 3N154 is given as 5 dB (maximum) at 200 MHz. With the same 1-MΩ source resistance R_S, the noise voltage (e_n) is approximately 2×10^{-7} V (200 nV).

3-13. ADVANTAGES AND LIMITATIONS OF FETS

From a study of the material presented thus far in this chapter, it is obvious that FETs have many advantages over two-junction transistors. Likewise, MOSFETs have certain advantages over JFETs and vice versa. In this section, we shall summarize the advantages and limitations as they apply to typical applications.

Many of the typical FET characteristics provide definite advantages for specific applications. For example, in digital work, the high input impedance of the FET results in high fan-out for digital circuits. Likewise, the capacitive input of a MOSFET permits direct-coupled circuits and a lower component count.

3-13.1 Basic FET device advantages

FETs are available in both single-gate and dual-gate types. Both types offer the advantages of extremely high input resistance, low input capacitance, very low feedback capacitance, high forward transconductance, and low noise at very high frequencies. Because of their insulated-gate characteristics, MOSFETs have extremely low leakage current and are relatively insensitive to temperature variations. In addition, the drain currents have a negative temperature coefficient that makes "thermal runaway" almost impossible.

The extremely high input resistance of FETs permits the use of simple, electron-tube-type biasing techniques. The high input resistance makes MOSFETs capable of handling relatively large positive and negative input signal excursions without degradation of input impedance due to diode-current loading.

Because of this capability, MOSFETs have considerably greater dynamic ranges than JFETs and two-junction transistors. In turn, because of the higher dynamic range, MOSFETs can provide superior performance in amplifier circuits with AGC (automatic gain control). For example, because of the high input resistances, MOSFETs impose virtually no loading on AGC voltage sources (whereas a two-junction transistor requires considerable AGC current).

In addition to these basic advantages, FETs are notably superior to other solid-state devices in cross-modulation characteristics and in their relative freedom from spurious responses. FETs are also characterized by zero offset voltage, a feature that makes them especially desirable for chopper applications.

3-13.2 Typical FET ranges

Low Feedback Capacitance

All FETs have low reverse or feedback capacitance C_{rss}. JFETs have C_{rss} values in the range of 2 pF. Single-gate MOSFETs have C_{rss} values in the range of 0.1 to 0.3 pF. The C_{rss} of a typical dual-gate MOSFET without diode protection is in the order of 0.02 to 0.03 pF. When protective diodes are added, the C_{rss} increases to the 0.03 to 0.05 range.

This low feedback capacitance often eliminates the need for neutralization, or mismatching, in *RF* amplifiers. A typical low-power two-junction transistor will have a reverse capacitance (generally listed as C_{ob}) of 5 to 10 pF and higher.

Low Gate Leakage Current

All FETs have low gate leakage current I_{GSS}. A typical JFET will have an I_{GSS} in the range of 100 pA at normal ambient temperature and 100 nA at elevated temperatures. Because of the insulated-gate construction, MOSFETs have even lower I_{GSS} values. For example, some single-gate MOSFETs have an I_{GSS} of less than 1 pA. Even at maximum operating temperatures, I_{GSS} remains in the order of picoamperes rather than nanoamperes. In turn, low gate leakage results in high static input resistance or impedance, even at maximum operating temperatures.

The very low gate leakge current of MOSFETs also eliminates the problem of diode-current loading of the input circuit under strong signal conditions, which is common to bipolar transistors and JFETs. With these other devices, a strong input signal will result in considerable input leakage current, thus loading the input signal source. MOSFET gate leakage remains low, even in the presence of maximum permitted input signal.

WIDE DYNAMIC RANGE

Very little gate current is drawn by a MOSFET in the presence of an input signal (either reverse- or forward-bias input). Since there is very little current loading, a MOSFET will handle both positive and negative input signal excursions over a wide range. A typical MOSFET will handle ac inputs of ± 15 V and continuous dc inputs of ± 10 V. Generally, a JFET is limited in input signal-handling capability (for signals of one polarity, either positive or negative). This is because of the diode input of a JFET.

WIDE-RANGE, LOW-POWER AGC

The wide input signal range also applies to AGC voltages, thus producing a wide AGC range. This is particularly useful in high-performance receiver applications. Because little gate current is drawn in either the forward or reverse input conditions, MOSFET AGC systems are voltage-operated rather than current-operated (as are most two-junction transistor AGC systems); that is, MOSFET AGC systems operate on a feedback of voltage changes rather than current changes. Thus, little power is required to operate MOSFET AGC systems.

Remote AGC is possible with dual-gate MOSFETs. Since a voltage applied to either gate affects the drain current, it is possible to incorporate remote AGC with dual-gate MOSFETs; that is, the signal can be applied to one gate and the AGC feedback voltage applied to the other gate. This results in increased gain reduction for a given AGC voltage.

HIGH FORWARD TRANSCONDUCTANCE AND POWER GAIIN

FETs have a high forward transconductance, typically in the order of 7500 μmhos. Of course, there is no comparable specification for two-junction transistors; however, forward transconductance (output current versus input voltage) is a measure of power gain. Single-gate MOSFETs have power gains in the order of 15 to 20 dB. Generally, JFETs have similar power gains.

The forward transconductance of a typical dual-gate MOSFET is in the order of 10,000 to 15,000 μmhos, with 20,000 μmhos possible. As the result of this high forward transconductance, the power gain for a dual-gate MOSFET is in the order of 18 to 20 dB, with 28 to 30 dB possible.

Negative Temperature Coefficient

The drain current of an FET generally operates on a negative temperature coefficient (NTC); that is, as temperature increases, drain current decreases. This prevents "thermal runaway" common in two-junction transistors, as described in Chapter 2. No such precautions are required with MOSFETs. It is only necessary to provide a heat sink sufficient to dissipate the maximum rated power of the FET. Generally, this is done quite simply by keeping the FET case in good contact with the metal chassis.

Zero Temperature Coefficient

An important advantage of all FETs is their ability to operate a zero-temperature-coefficient (0TC) point. This means that if the gate source is biased at a specific voltage and is held constant, the drain current will not vary with changes in temperature.

Typically, JFETs show the 0TC characteristic over a wide range of temperature, approximately 150°C. Likewise, it is not always practical to operate an FET at some fixed V_{GS}. For these reasons, 0TC is not always considered an important MOSFET characteristic, even though 0TC can be an advantage in certain applications. The 0TC characteristic is discussed thoroughly in Sec. 3-14.

Zero Offset Voltage

There is no inherent offset voltage associated with FETs, as there is with two-junction transistors. This is a particular advantage in chopper applications or in amplifiers where FETs are used as the input stage. In a typical circuit, input offset voltage can be defined as that voltage that must be applied at the input terminals to obtain zero output voltage. With an FET in a typical circuit, zero volts in produces zero volts out.

3-13.3 FET limitations

As is the case with any electronic component, all FETs have certain characteristic limitations. In some applications these limitations present no particular problem. In different applications, the characteristics of FETs must be compared with two-junction capacitors (or MOSFETs must be compared with JFETs) to make an intelligent selection. Of course, future design of FETs may be improved, thus removing the limitations. Until that time, the informed FET user should consider the following:

Power Output

The maximum power dissipation of a typical FET is in the range of 300 mW, with 400 mW generally the top limit. Drain current (I_D) is typically

50 mA or less. Thus the maximum output power of present-day FETs is a fraction of 1 W.

Operating Frequency

The maximum operating frequency of a typical FET is in the range of 200 MHz, with 500 MHz generally the top. This compares to 1200 MHz (and higher) for two-junction transistors.

Handling MOSFETs

As discussed in Sec. 3-9 and 3-10, MOSFETs require special handling when out of circuit. This means that extra care must be used during service as well as assembly of equipment with MOSFETs. Special problems are created with automatic assembly equipment. All of these problems are considered as limitations by some technicians and engineers.

Durability

In circuit, MOSFETs are about as rugged as JFETs and two-junction transistors, with one possible exception. When a MOSFET is subjected to an input voltage that causes the metal oxide gate to break down, the MOSFET is destroyed. Other junction-type transistors can often withstand momentary overloads (voltage and/or current) without permanent damage.

Protective Diode Limitations

As discussed in Sec. 3-10, the problem of damage due to momentary input overloads can be eliminated by means of protective diodes; however, the addition of protective diodes presents an operating limitation on the MOSFET. When the protective diodes are used, the maximum input signal (and dynamic range) is set by the diode rather than the MOSFET characteristics. Typically, the protective diodes restrict the input to a range of about 10 or 15 V.

The use of protective diodes also increases gate leakage current. The I_{GSS} of a typical MOSFET without protective diodes is 1 nA (or less). With protective diodes, the I_{GSS} is increased to 50 nA.

Chopper Limitations

Both MOSFETs and JFETs have a low drain-source resistance when switched on; however, JFETs that are specifically designed for switch or chopper use have a lower $r_{ds(\text{on})}$ than comparable MOSFETs. Typically, a chopper MOSFET will have an $r_{ds(\text{on})}$ of about 200 Ω; whereas a JFET for chopper use will have an $r_{ds(\text{on})}$ of about 25 Ω. Of course, this advantage is generally offset by the much higher capacitances and input leakage common to JFETs.

3-14. TEMPERATURE COEFFICIENT OF I_D OR V_{GS}

While there is no single specification dealing with the zero-temperature-coefficient (0TC) effect of FETs, its importance warrants discussion. Figure 3-43 shows the typical I_D temperature coefficient θI_D versus I_D for a JFET (2N5265). The FET shown in Fig. 3-43 has a temperature coefficient of zero at approximately $I_D = 0.5$ mA.

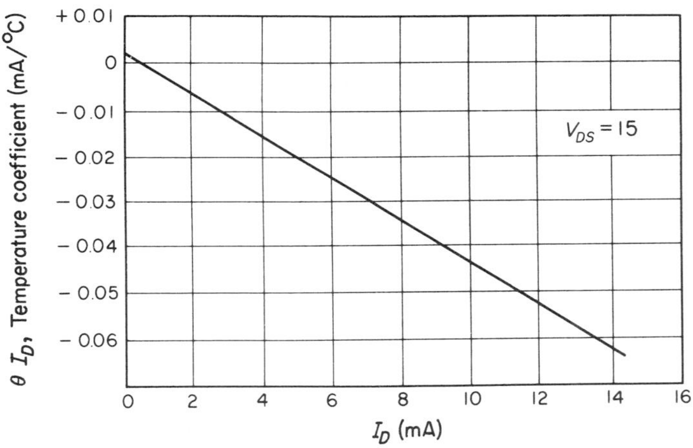

FIGURE 3-43 Drain current temperature coefficient versus drain current for 2N5265 through 2N5270 (courtesy Motorola).

The positive temperature coefficient occurring at small values of drain current is caused by a change in the width of the thermally generated depletion layer at the gate-channel junction of the JFET and changes at a rate equivalent to −2.2 mV/°C across the gate-source junction. This is the same value and phenomenon as the forward-biased base-emitter junction of a bipolar transistor.

The negative temperature coefficient occurs at high values of I_D due to a decrease in carrier mobility (increasing resistivity) in the channel. Because the FET usually operates with higher values of I_D (above 0.5 mA), the negative temperature coefficient predominates.

At some value of I_D (about 0.5 mA in this case) these two effects cancel each other, and zero temperature coefficient of I_D or V_{GS} occurs. The design equations for 0TC operation are

$$V_{GSZ} \approx V_P - 0.63 \tag{3-1}$$

$$I_{DZ} \approx I_{DSS}\left(\frac{0.63}{V_P}\right)^2 \tag{3-2}$$

$$\text{Drift in mV/°C} \approx 2.2\left(\frac{1\sqrt{I_D}}{I_{DZ}}\right) \tag{3-3}$$

where subscript Z indicates the 0TC point for the parameter.

Equation 3-3 is valid near, but not at, I_{DZ}. Equations 3-1 and 3-2 must both be satisfied for 0TC operation. Since V_P and I_{DSS} vary from device to device, an adjustment for V_{GS} or I_D is necessary. Note that resistor temperature coefficients and the effects of I_{GSS} with temperature must also be considered.

3-15. DESIGNING THE BASIC FET STAGE

The following example is presented to show how the transfer characteristic curves from a datasheet are used in design of a typical FET bias network. The circuit involved is shown in Fig. 3-44. Although a MOSFET is used in the example, the basic design principles also apply to JFETs.

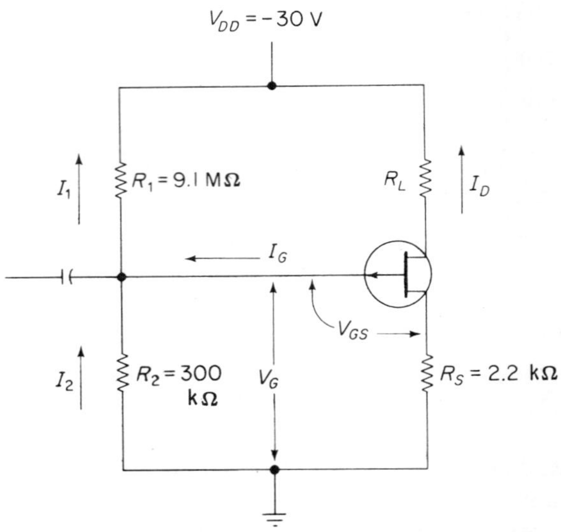

FIGURE 3-44 The basic FET stage (courtesy Motorola).

Note that both external bias (or fixed bias) and self-bias are used. As discussed in later chapters, the basic bias network can be modified to produce an FET amplifier stage with desired characteristics (stage gain, input/output impedance, etc.).

Keep one point in mind when studying the following bias scheme. The purpose of the basic bias circuit is to establish a given I_D and to maintain that I_D (plus or minus some given tolerance) over a given temperature range. Generally, this is to keep the transconductance at some given level. In the following discussion, no particular consideration is given to stage characteristics (operating point, gain, impedance, etc.). Instead, we concentrate on how to set the I_D and how to keep the I_D constant.

Of course, since FETs rarely operate under static conditions, the basic bias circuit is used as a reference or starting point for design. The actual circuit configuration, and especially the bias circuit values, should be selected on the basis of dynamic circuit conditions (desired output voltage swing, expected input signal level, etc.).

For example, if an FET is used as a linear amplifier, the output (drain terminal) should be at one-half of the supply voltage (for a typical stage). This permits maximum voltage swing. As a result, the value of R_L should be selected to provide the voltage drop equal to one-half the supply.

3-15.1 Design example

Assume that the circuit of Fig. 3-44 is to maintain I_D at 1 ± 0.025 mA, over a temperature range from -55 to $+125$°C, with a supply voltage V_{DD} of 30 V, using an FET with transfer characteristics similar to those of Fig. 3-45 (*N*-channel depletion type).

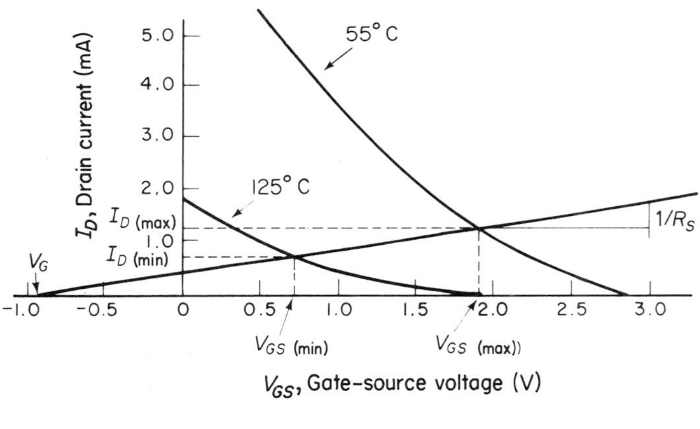

$$R_S = \frac{V_{GS\,(max)} - V_{GS\,(min)}}{I_{D\,(max)} - I_{D\,(min)}}$$

$$V_{R_L} = V_{DD} - (1.5 \times V_{GS\,(off)}) - (I_{DC\,(max)} \times R_S)$$

$$R_2 = \frac{V_G R_1}{V_{DD} - V_G} \qquad R_{L\,(max)} = \frac{V_{R_L}}{I_{DC\,(max)}}$$

$$V_G = \frac{I_{D\,(min)} \cdot V_{GS\,(max)} - I_{D\,(max)} V_{GS\,(min)}}{I_{D\,(max)} - I_{D\,(min)}}$$

FIGURE 3-45 $1/R_S$ load line for basic FET stage (courtesy Motorola).

The first step is to draw a $1/R_S$ load line on the transfer characteristics as shown in Fig. 3-45. As noted by the equations, the value of R_S is set by the limits of V_{GS} and I_D. The value of $V_{GS(\min)}$ is the point where $I_{D(\min)}$ of 0.75 mA crosses the high temperature limit curve of 125°C, or approximately 0.8 V. The value of $V_{GS(\max)}$ is the point where the $I_{D(\max)}$ of 1.25 mA crosses the low temperature limit curve of −55°C, or approximately 1.9 V.

Using these values, the first trial value for R_S is

$$R_S = \frac{1.9 - 0.8}{1.25 - 0.75} = 2.2\ \text{k}\Omega$$

The fixed-bias voltage V_G is determined from the intercept of the $1/R_S$ load line with the V_{GS} axis and is computed by using the same set of values as shown in the Fig. 3-45 equations:

$$V_G = \frac{(0.75 \times 1.9) - (1.25 \times 0.8)}{0.5} = 0.85\ \text{V}$$

The maximum value of R_1 is determined by the maximum value of I_{GSS}; that is, the value of I_{GSS} at the highest temperature involved. Assume that the FET has a maximum I_{GSS} of 5 nA at +125°C (typical for a MOSFET).

The variation in V_G versus temperature will not be too great if a value for R_1 is chosen such that I_1 *is at least 6 times greater* than the maximum I_{GSS}. For extra stability, let I_1 equal 10 times maximum I_{GSS}, or 10×5 nA = 50 nA.

The drop across R_1 is equal to $V_{DD} - V_G$, or $30 - 0.85 = 29.15$ V. With an I_1 of 50 nA and a drop across R_1 of 29.15 V, the approximate value of R_1 is 29.15 V/50 nA = 583 MΩ.

With the value of R_1 established at 583 MΩ, the value of R_2 is found from a simple voltage divider relationship as shown in the Fig. 3-45 equations.

The value of R_L is determined by desired circuit performance or output characteristics or both. For example, an increase in the value of R_L increases gain but decreases stability. Similarly, the value of R_L sets the *approximate output impedance of the circuit* (when the FET is connected as a common-source amplifier). These factors are discussed in Sec. 3-16.

With all the bias resistance values established, the circuit of Fig. 3-44 can be converted to a single FET amplifier stage by the addition of input and output coupling capacitors. This conversion is described in Chapter 5.

3-16. SMALL-SIGNAL ANALYSIS OF BASIC FET STAGE

Small-signal analysis of an FET amplifier stage is easily accomplished with reasonable accuracy by using a few simple equations. The FET model for the analysis is shown in Fig. 3-46. Note that this model omits all capacitance, including C_{rss}. By omitting all capacitance and using only the real parts of

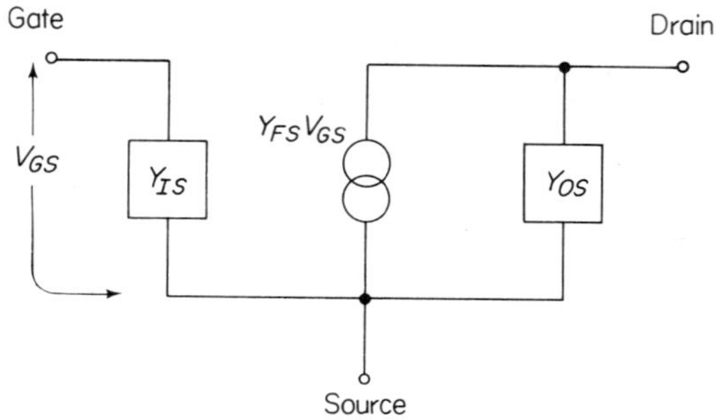

FIGURE 3-46 Small-signal analysis of basic FET stage (courtesy Motorola).

Y_{is}, Y_{fs} (or g_{fs}), and Y_{os}, the model and the accompanying equations are useful up to about 100 kHz. If capacitance effects are included, the equations are useful up to several megahertz. The effects of capacitance on *RF* amplifiers are discussed in Chapter 6.

Figures 3-47, 3-48, and 3-49 show schematics for common-source, common-drain (source-follower), and common-gate circuits. In addition, the *approximate* equations for voltage gain, input impedance, and output impedance are included. The equations for finding the resistance values of R_1, R_2, R_S, and R_L are discussed in Sec. 3-15. Note that the approximate equations are based on certain assumptions. The assumptions are given with the corresponding equations.

The circuits of Fig. 3-47 through 3-49 are for *N*-channel JFETs. Simply reverse the polarity for *P*-channel FETs. All MOSFETs may not conform exactly to the relations shown but are sufficiently close for first trial values.

For the common-source circuit (Fig. 3-47), the omission of C_{rss} will affect the equations for input impedance and voltage gain. For low frequencies (below about 100 kHz), however, the error is minimal. If *only dc feedback* is required, then the source resistor R_S is bypassed. With a capacitor across R_S, the effects on the design equations are to set R_S at zero. Under these conditions, voltage gain is the product of R_L and Y_{fs}. (Note that g_{fs} can be substituted in any equation of Fig. 3-47 through 3-49.)

With R_S unbypassed, the circuit characteristics are virtually independent of FET parameters (with the exception of Y_{fs}). Instead, the circuit characteristics (impedances, gain, etc.) are dependent on R_L and R_S. By using precision resistors with close temperature coefficients, the common-source circuit can be made very stable over a wide temperature range.

The common-drain (source-follower) configuration (Fig. 3-48) is a very useful basic circuit. Some of its properties are a voltage gain always less than unity with no phase inversion, low output impedance (essentially set by the

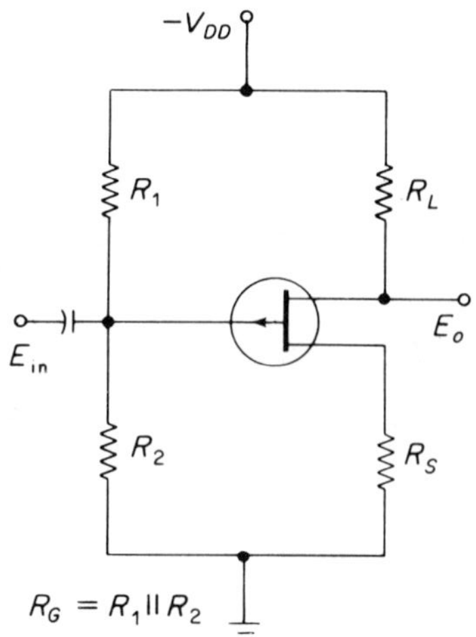

	Exact	Assuming	Approximate
Voltage gain	$A_V = \dfrac{-Y_{fs} R_L}{1 + Y_{os}(R_L + R_S) + Y_{fs} R_S}$	$Y_{os}(R_L + R_S) \ll 1$	$A_V \approx \dfrac{R_S}{\dfrac{1}{Y_{fs}} + R_S}$
Input impedance	$Z_{in} = R_g \,\|\, \dfrac{\dfrac{1}{Y_{is}}}{1 - \dfrac{Y_{fs} R_S}{1 + Y_{os}(R_L + R_S) + Y_{fs} R_S}}$	$Y_{os}(R_L + R_S) \ll 1$; $\dfrac{1}{Y_{is}} \gg R_G$	$Z_{in} \approx R_1 \,\|\, R_2$
Output impedance	$Z_o = \dfrac{R_L}{\dfrac{Y_{os}(R_L + R_S)}{1 + Y_{fs} R_S} + 1}$	$Y_{os}(R_L + R_S) \ll 1$	$Z_o \approx R_L$

FIGURE 3-47 Basic common-source FET stage circuit and characteristics (courtesy Motorola).

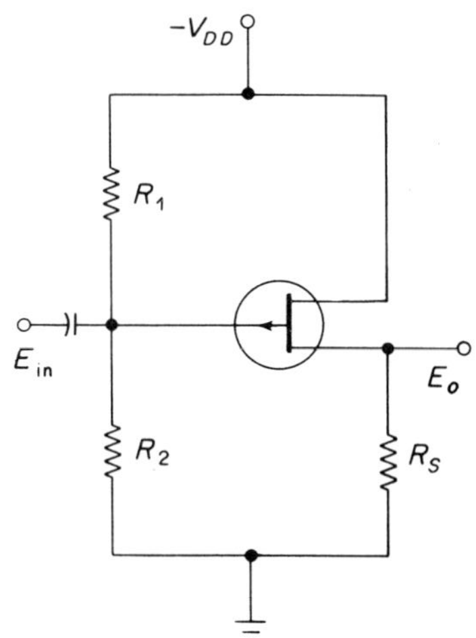

	Exact	Assuming	Approximate
Voltage gain	$A_V = \dfrac{Y_{fs} R_S}{1 + R_S (Y_{fs} + Y_{os})}$	$R_S (Y_{fs} + Y_{os}) \ll 1$	$A_V \approx \dfrac{R_S}{\dfrac{1}{Y_{fs}} + R_S}$
Input impedance	$Z_{in} = R_g \parallel \dfrac{\dfrac{1}{Y_{is}}}{1 - \dfrac{Y_{fs} R_S}{1 + R_S (Y_{fs} + Y_{os})}}$	$R_S (Y_{fs} + Y_{os}) \ll 1$ $R_G \gg \dfrac{\dfrac{1}{Y_{is}}}{1 - Y_{fs} R_S}$	$Z_{in} \approx R_1 \parallel R_2$
Output impedance	$Z_o = \dfrac{R_S + Y_{os} R_S^2}{1 + R_S (Y_{os} + Y_{fs})}$	$Y_{os} \rightarrow 0$	$Z_o \approx R_S \parallel \dfrac{1}{Y_{fs}}$

FIGURE 3-48 Basic common-drain (source follower) FET stage circuit and characteristics (courtesy Motorola).

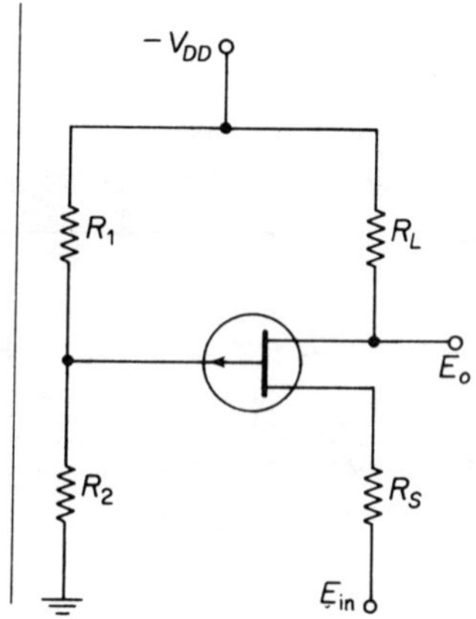

	Exact	Assuming	Approximate
Voltage gain	$A_V = \dfrac{Y_{fs}R_L}{1 + Y_{os}\ (R_L + R_S) + Y_{fs}R_S}$	$Y_{os}\ (R_L + R_S) << 1$	$A_V \approx \dfrac{R_L}{\frac{1}{Y_{fs}} + R_S}$
Input impedance	$Z_{in} = \dfrac{1}{Y_{os} + Y_{fs}} + (R_S + R_L)\ \dfrac{Y_{os}}{Y_{os} + Y_{fs}} + \dfrac{R_S Y_{fs}}{Y_{os} + Y_{fs}}$	$Y_{os} << Y_{fs}$ $Y_{os}\ (R_L + R_S) << 1$	$Z_{in} \approx R_S + \dfrac{1}{Y_{fs}}$
Output impedance	$Z_o = \dfrac{R_L}{\dfrac{Y_{os}\ (R_L + R_S)}{1 + Y_{fs}R_S} + 1}$	$Y_{os}\ (R_L + R_S) << 1$	$Z_o \approx R_L$

FIGURE 3-49 Basic common-gate FET stage circuit and characteristics (courtesy Motorola).

value of R_S), high input impedance, large signal swing, and active impedance transformation.

The common-gate stage (Fig. 3-49) offers impedance transformation opposite that of the source follower. Common gate produces low input impedance and high output impedance. The voltage gain is less than unity, with no phase inversion.

3-17. TESTING FETS

As in the case of two-junction transistors, FETs must be tested on both a static and a dynamic basis. Static tests indicate the FET response to direct-current variations. Dynamic tests show the response of an FET to alternating currents or signals. All the characteristics covered in Sec. 3-12 are discussed. Keep in mind that FETs operate in three modes (Type A, or depletion only, Type B or depletion/enhancement, and Type C or enhancement only).

3-17.1 $V_{GS(off)}$ *test*

The basic circuit for $V_{GS(\text{off})}$ test is shown in Fig. 3-50. This test applies to depletion-only and depletion/enhancement mode devices. As shown, V_{DS} is set at some fixed value, and reverse-bias V_{GS} is adjusted until I_D is at some specific negligible value. This is essentially a *cutoff value* test. As an example, the *N*-channel 3N128 specifies a V_{DS} of 15 V and an I_D of 50 μA. The V_{GS} should be between −0.5 and −8 V for the 50 μA of I_D. As a guideline, V_{DS} should be a minimum of 1.5 times $V_{GS(\text{off})}$ to provide proper circuit operation.

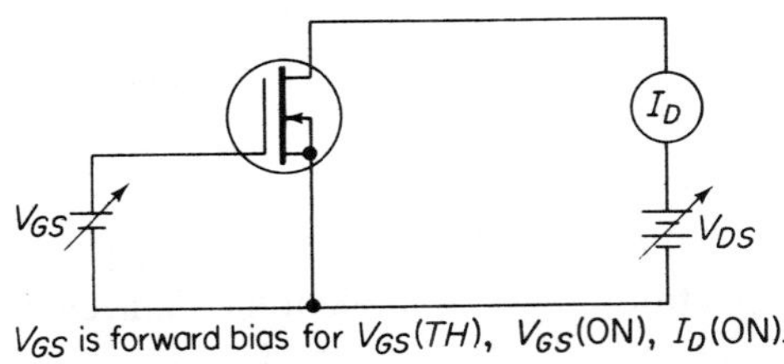

Polarities shown are for *N*-channel MOS devices. Reverse all polarities for *P* channel.

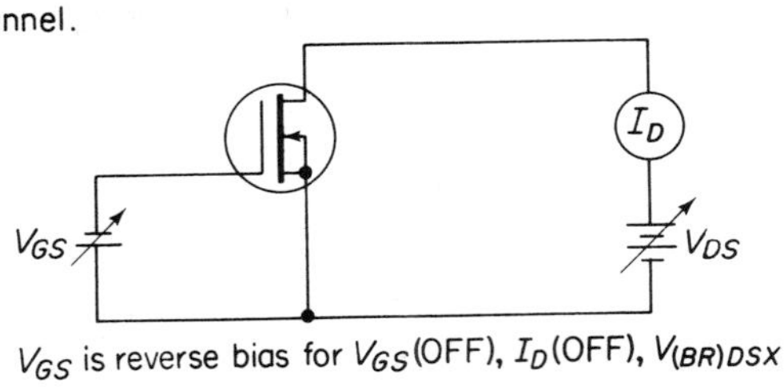

FIGURE 3-50 Test circuit for $V_{GS(\text{off})}$, $V_{GS(\text{th})}$, $V_{GS(\text{on})}$, $I_{D(\text{on})}$, $I_{D(\text{off})}$, and $V_{(BR)DSX}$.

3-17.2 $V_{GS(th)}$ or $V_{GS(on)}$ test

$V_{GS(\text{th})}$ and $V_{GS(\text{on})}$ are essentially the same specification as $V_{GS(\text{off})}$, except that $V_{GS(\text{th})}$ and $V_{GS(\text{on})}$ are generally applied to enhancement-only mode devices. The basic circuit of Fig. 3-50 is used, except that V_{GS} is connected for forward bias (gate positive for N channel). A forward bias is required since I_D flow is an enhancement mode device only when the gate is forward biased. During test, V_{GS} is increased from zero volts until the I_D is at a specified value.

3-17.3 $V_{DS(off)}$ and $V_{DS(on)}$ saturation test

The basic circuit for saturation tests is shown in Fig. 3-51. Saturation tests are generally of concern only when the FET is used for switch or chopper applications. During either test, V_{GS} is held at some fixed value, and V_{DS} is increased until I_D is at maximum or increases very little for further increases in V_{DS}. Take care not to exceed the maximum V_{DS} when making this test. Generally, $V_{DS(\text{on})}$ is made with some specific value of V_{GS}; whereas $V_{DS(\text{off})}$ is made with V_{GS} at zero (usually shorted to the source).

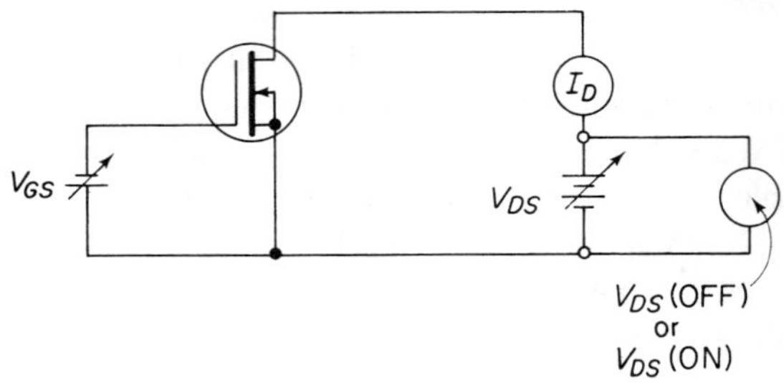

FIGURE 3-51 Test circuit for $V_{DS(\text{off})}$ and $V_{DS(\text{on})}$.

3-17.4 I_{DSS} test

The basic circuit for I_{DSS} test is shown in Fig. 3-52. This test is applied to depletion-only and depletion/enhancement mode devices. As shown, V_{DS} is set at some fixed value, and V_{GS} is zero (gate shorted to source). I_{DSS} is a zero-bias current flow test. For example, the 3N128 specifies a V_{DS} of 15 V and V_{GS} of zero. Under these conditions, the I_D should be between 5 and 25 mA.

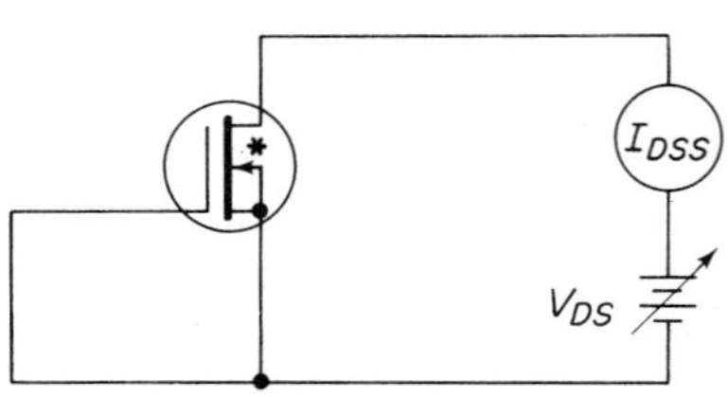

* Disconnect substrate from source (where possible) for $V_{(BR)DSS}$.

FIGURE 3-52 Test circuit for I_{DSS} and $V_{(BR)DSS}$.

3-17.5 $I_{D(on)}$ test

$I_{D(on)}$ is essentially the same specification as I_{DSS}, except that $I_{D(on)}$ is generally applied to enhancement mode devices (Type B or C). The basic circuit of Fig. 3-50 is used, except that V_{GS} is connected for forward bias (gate positive for N channel). Forward bias is required since I_D flows in an enhancement mode device only when the gate is forward biased. Some I_D will flow in a class B device without forward bias; however, $I_{D(on)}$ is the current near the maximum (or saturation current). During test, V_{GS} and V_{DS} are set to some fixed value, and the resulting value of $I_{D(on)}$ is measured.

3-17.6 $I_{D(off)}$ test

The test circuit for $I_{D(off)}$ is the same as for $V_{GS(off)}$, Fig. 3-50. This is because $I_{D(off)}$ is the current that flows when $V_{GS(off)}$ is applied. V_{DS} and V_{GS} are set to some fixed value, and the resulting value of $I_{D(off)}$ is measured. For example, the 3N128 specifies a V_{DS} of 20 V and a V_{GS} of -8 V. Under these conditions, the I_D should be 50 μA maximum. On enhancement-only mode devices (Type C), the value of $I_{D(off)}$ is the I_D that flows when $V_{GS(th)}$ is applied. Thus, the fixed value of V_{GS} is always some forward-bias voltage.

3-17.7 Voltage breakdown tests

For JFETs, $V_{(BR)GSS}$ is the maximum voltage that may be applied between any two terminals and is the lowest voltage that will lead to breakdown of the gate junction. As shown in Fig. 3-53, an increasingly high reverse voltage is applied between the common gates and the source. Junction breakdown can be determined by gate current (beyond normal I_{GSS}), which indicates the beginning of an avalanche (V_A) condition.

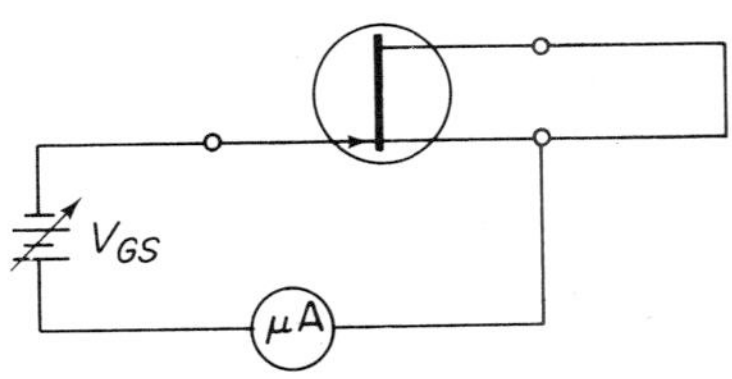

FIGURE 3-53 Test circuit for breakdown voltage.

For MOSFETs, $V_{(BE)GSS}$ is the breakdown voltage from gate to source and is a *maximum voltage, not a test voltage*, that is, for MOSFETs, $V_{(BR)GSS}$ is the voltage at which the gate oxide layer breaks down. If this full voltage is applied during test, the oxide layer can (and probably will) be damaged. If the test is performed on a MOSFET with gate protection, the resultant breakdown voltage is that of the protective diode, not the MOSFET. Thus, $V_{(BR)GSS}$ tests should not be made except during manufacture (as part of destructive testing to establish breakdown for a particular type of device). To perform a $V_{(BR)GSS}$ test, the drain and source are shorted and an increasing V_{GS} is applied until gate current I_G suddenly increases, indicating puncture of the gate oxide layer.

$V_{(BR)DGO}$ is essentially the same specification as $V_{(BR)GSS}$, except that breakdown is from gate to drain, and the same precautions apply.

$V_{(BR)DSX}$ applies to Types A and B MOSFETs. The basic circuit of Fig. 3-50 can be used. During test, reverse-bias V_{GS} is applied until the device is cut off. Usually, the $V_{GS(\text{off})}$ value is used. Then, V_{DS} is increased until there is some heavy flow of I_D.

$V_{(BR)DSS}$ applies to Type C MOSFETs. The basic circuit of Fig. 3-52 can be used, *except that the substrate must be disconnected from the source* (floating substrate). Of course, the substrate cannot be disconnected on those MOSFETs with internal connections between substrate and source or other element. For Type C devices, the gate and source are shorted since no reverse-bias V_{GS} is necessary for cutoff. With V_{GS} at zero, V_{DS} is increased until there is some heavy flow of I_D.

3-17.8 I_{GSS} test

The basic circuit for I_{GSS} test is shown in Fig. 3-54. This test applies to all modes of operation. As shown, V_{DS} is zero (drain shorted to source), V_{GS} is set to some given (reverse bias) values, and the resultant value of leakage current I_G is measured. For example, the 3N128 specifies a V_{DS} of zero and a V_{GS} of −8 V. Under these conditions, I_{GSS} should not exceed 50 pA (at 25°C) or 5 nA (at 125°C).

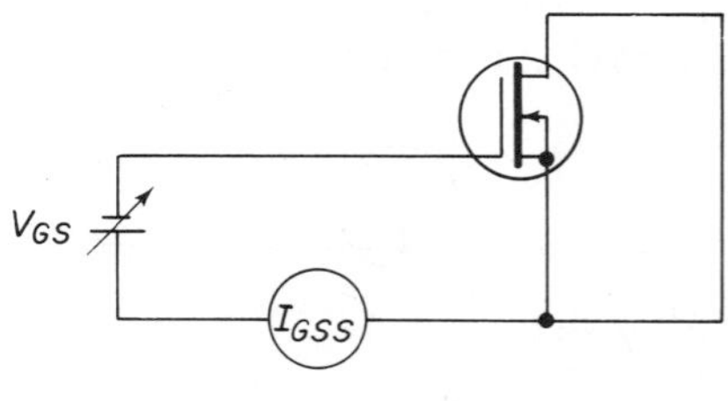

FIGURE 3-54 Test circuit for I_{GSS}.

3-17.9 $V_{G1S(off)}$ and $V_{G2S(off)}$ test

These tests apply to dual-gate devices, operating in the depletion and depletion/enhancement modes. Dual-gate devices can also be tested for $V_{GS(\text{off})}$ as described in Sec. 3-17.1, when both gates are tied together. The basic circuit for $V_{G1S(\text{off})}$ and $V_{G2S(\text{off})}$ is shown in Fig. 3-55. As shown, V_{DS} is set at some specific value. One gate is forward biased to a specific value and the other gate is reverse biased. The reverse-bias V_{G1S} or V_{G2S} is adjusted until I_D is at some negligible value, indicating a cutoff condition.

For example, to measure $V_{G1S(\text{off})}$, the 40841 specifies a V_{DS} of 15 V, a V_{G2S} of +4 V, and an I_D of 50 μA. V_{G1S} should be −2 V for the 50 μA of

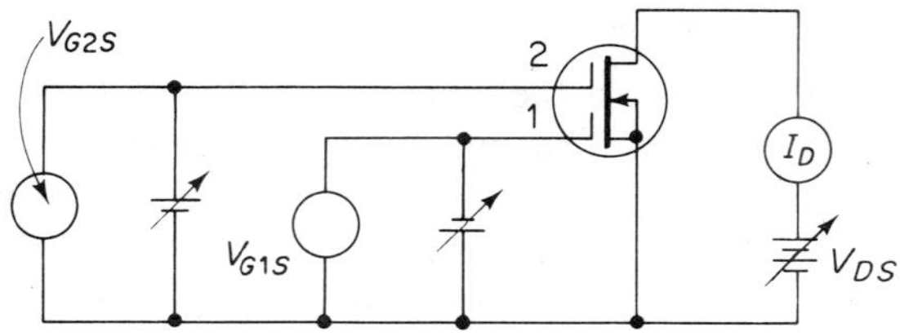

FIGURE 3-55 Test circuit for $V_{G1S(\text{off})}$ and $V_{G2S(\text{off})}$.

I_D. To measure $V_{G2S(\text{off})}$, the 40841 specifies a V_{DS} of 15 V, a V_{G1S} of 0 V, and an I_D of 50 μA. V_{G2S} should also be -2 V for the 50 μA of I_D.

3-17.10 Dual-gate voltage breakdown tests

Dual-gate MOSFETs are often tested for forward and reverse gate-to-source breakdown voltage. The basic test circuit is shown in Fig. 3-56. As shown, V_{DS} and one gate are at 0 V (both shorted to source). A variable voltage is applied to the opposite gate and the gate current is measured. For example, to measure $V_{(BR)G1SSF}$, the 40841 specifies a V_{DS} and V_{G2S} of 0 V and an I_{G1SSF} of 100 μA, with a typical 9 V of V_{G1} applied. $V_{(BR)G1SSR}$ is measured in the same way, except that the gate is reverse biased. The results should be the same (100 μA of I_{G1SSR} for the 40841). $V_{(BR)G2SSF}$ and $V_{(BR)G2SSR}$ are measured in the same way, except that gate 1 is connected to the source and the voltage is applied to gate 2.

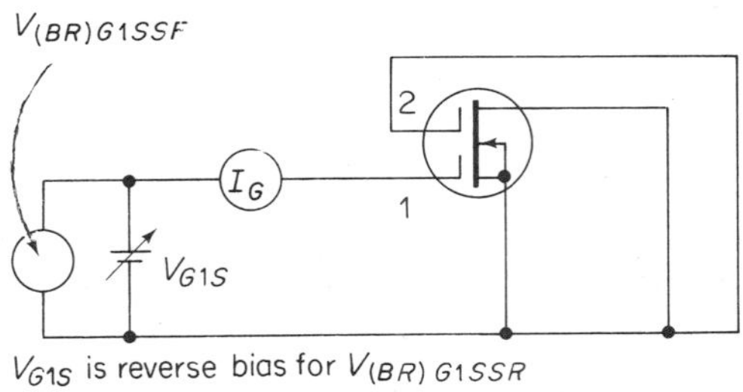

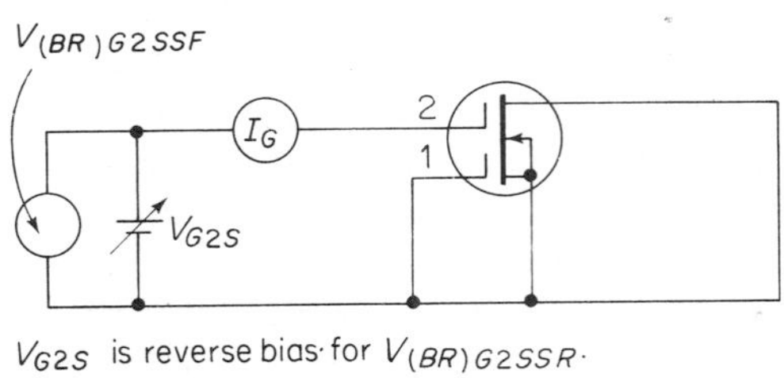

FIGURE 3-56 Test circuits for dual-gate voltage breakdown and current.

3-17.11 Dual-gate current tests

Dual-gate current tests use the same basic test circuit as for gate-voltage breakdown tests, Fig. 3-56. The difference in procedure is that the gate voltage is set to a specific value and the resultant current is measured. For example, to measure I_{G1SSF}, the 40841 specifies a V_{DS} and V_{G2S} of 0 V and a V_{G1S} of 6 V. A *maximum* I_{G1SSF} of 60 nA should flow under these conditions. I_{G1SSR} is measured the same way, except that V_{G1S} is -6 V. I_{G2SSF} and I_{G2SSR} are measured in the same way, except that gate 1 is connected to the source and the voltage is applied to gate 2.

3-17.12 I_{DS} *test*

This test applies to dual-gate devices, operating in the depletion and depletion/enhancement modes. Dual-gate devices can also be tested for I_{DSS} as described in Sec. 3-17.4, when both gates are tied together.

The circuit for I_{DS} is shown in Fig. 3-57. As shown, V_{DS} is set at some specific value, gate 1 is shorted to the source, and gate 2 is set to a specific value. I_{DS} is considered as a zero-bias current flow test, even though one gate has a forward bias. For example, the 40841 specifies a V_{DS} of 15 V, a V_{G1S} of 0 V, and a V_{G2S} of +4 V. Under these conditions, the I_D should be a typical 10 mA.

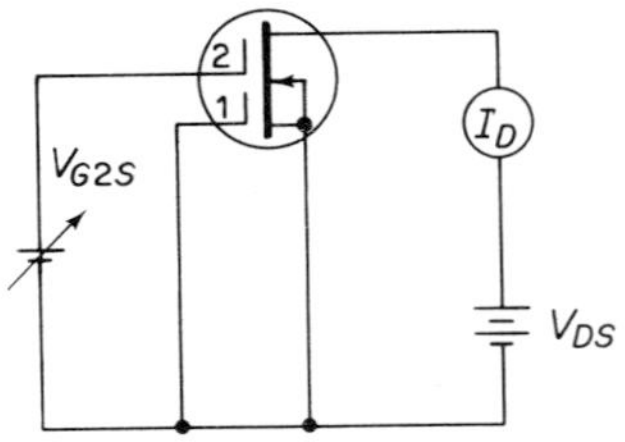

FIGURE 3-57 Test circuit for I_{DS} (dual-gate zero-bias current test).

3-17.13 y*-parameter tests*

The *y*-parameter tests establish the four basic admittances (forward transadmittance, reverse transadmittance, input admittance, and output admittance).

Admittance is the reciprocal of impedance and is composed of conductance (*g*, the real part of admittance) and susceptance (*b*, the imaginary part of admittance). The *y* parameters can be expressed as complex numbers, where both the *g* and *b* values are given, or as a simple number in mhos. This is the same as expressing impedance (*Z*) in simple terms of ohms or a complex number composed of resistance (*R*) and reactance (*X*).

When it is necessary to test a device to find a complex number (both the conductance and susceptance), it is necessary to use an *admittance meter* or possibly an R_X meter. An admittance meter consists essentially of a signal source and a bridge circuit. The y parameter under test is connected to form one leg of the bridge. The signal source is adjusted to the frequency of interest and applied to the bridge. The bridge conductance and susceptance leg elements (usually a resistor and a capacitor) are adjusted until the bridge is balanced. The conductance and susceptance values required to produce balance are equal to the y-parameter conductance and susceptance, at that frequency.

The R_X meter functions in essentially the same way, except that the readout is in resistance and reactance. The reciprocal of these R and X values are taken to find g and b. As discussed in Chapter 6, the design of some *RF* networks is based on impedance (Z, R, and X) values rather than on admittance (y, g, and b) values.

Operation of admittance meters and R_X meters is discussed in *Handbook of Electronic Test Equipment*, by John D. Lenk, Prentice-Hall, Inc., Englewood Cliffs, N.J., 1971, and will not be repeated here. Instead, we shall present test procedures that can be accomplished with basic test equipment to find y parameters, expressed in *simple number terms.*

3-17.14 Forward transadmittance (transconductance) test

Forward transadmittance Y_{fs} is often referred to as transconductance and may be given as y_{21}, g_m, or even g_{fs}.

The basic circuit for Y_{fs} test is shown in Fig. 3-58. This circuit applies to single-input devices, such as two-junction transistors, and single-gate FETs. The circuit for dual-gate is the same, except that a forward bias is generally applied to gate 2. For example, the 40841 specifies a V_{G2S} of +4 V for test of forward transconductance (which the datasheet lists as g_{fs}). This forward bias produces an I_D of 10 mA when the V_{DS} is 15 V.

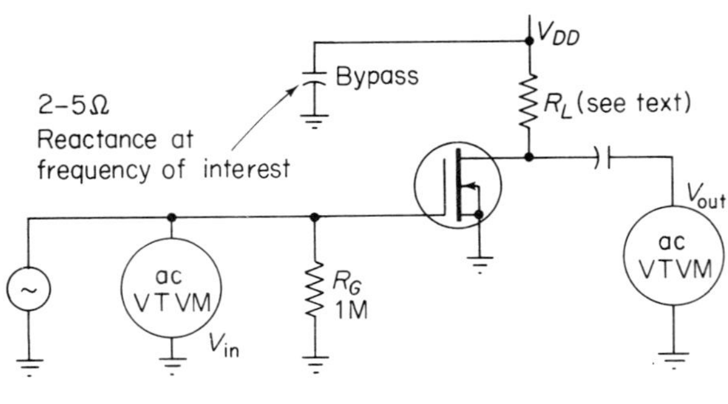

$$Y_{FS} \approx \frac{V_{out}}{V_{in} \times R_L}$$

FIGURE 3-58 Test circuit for Y_{fs} (as a simple number).

As indicated in Fig. 3-58, the value of R_L must be such that the drop is negligible at I_{DSS} (for FETs). Just as important, the value of R_L must be such that the operating point voltage (V_{DS} in the case of an FET) will be correct for a given supply voltage (V_{DD}) and operating current (I_D).

For example, if I_D is 10 mA, V_{DD} is 20 V, and V_{DS} must be 15 V, R_L must drop 5 V at 10 mA. Thus, the value for R_L is 5 V/0.01 A = 500 Ω. For single-gate devices, the value of I_{DSS} should be used to set the value of R_L. The value of R_G is not critical and is typically 1 MΩ.

During test, the signal source is adjusted to the frequency of interest. The amplitude of the signal source V_{in} is set to some convenient number such as 1 V or 100 mV. The value of Y_{fs} is calculated from the equation of Fig. 3-58 and is expressed in mhos (or millimhos or μmhos).

As an example, assume that the value of R_L is 1000 Ω, V_{in} is 1 V and V_{out} is 8 V. The value of Y_{fs} is

$$\frac{8}{1 \times 1000} = 0.008 \text{ mho} = 8 \text{ mmhos} = 8000\ \mu\text{mhos}$$

3-17.15 Re ($\mathbf{Y_{fs}}$) *tests*

For certain high-frequency applications, it is necessary to know the real part of Y_{fs}. At low frequencies, the Y_{fs} value found as described in Sec. 3-17.14 is sufficiently accurate. As frequency increases, Y_{fs} will *appear to be exceptionally high* if tested by the circuit of Fig. 3-58. To obtain a true picture of the forward transconductance characteristics, it may be necessary to measure Y_{fs} as a complex number, using an admittance meter or an R_X meter. This is almost always true at 30 MHz or higher.

3-17.16 Reverse transadmittance test

Reverse transadmittance Y_{rs} (also known as reverse transconductance, or Y_{12}) is generally not a critical value. For practical test purposes, the real value of Y_{rs} (or g_{rs}) can be considered as nearly zero at all useful frequencies. If it is necessary to establish the imaginary part of Y_{rs} (b_{rs}), such as for calculations involved in *RF* matching networks, an admittance meter or R_X meter must be used.

3-17.17 Output admittance test

Output admittance Y_{os} may be given as Y_{22}, g_{os}, or g_{22}. The basic circuit for Y_{os} test is shown in Fig. 3-59. The circuit for dual gate is the same, except that a forward bias is generally applied to gate 2. As indicated, the value of R_S must be such as to cause a negligible drop (so that V_{DS} can be maintained at a desired level, with a given V_{DD} and I_D).

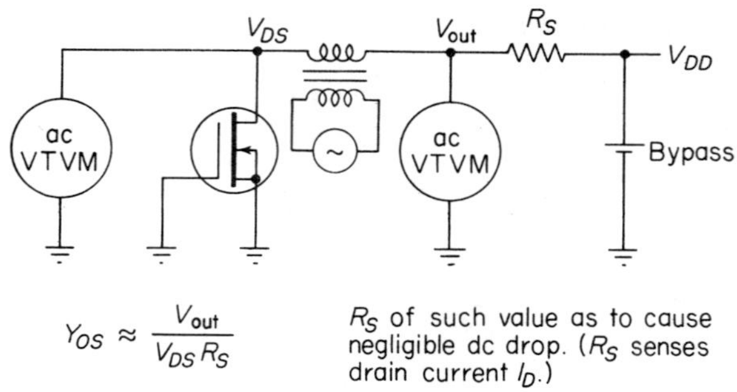

FIGURE 3-59 Test circuit for Y_{os} (as a simple number).

During test, the signal source is adjusted to the frequency of interest. Both V_{out} and V_{DS} are measured, and the value of Y_{os} is calculated from the equation of Fig. 3-59.

3-17.18 Input admittance test

Input admittance Y_{is} (or Y_{11}) is generally not a critical value. For practical test purposes, the real value of Y_{is} (or g_{is}) can be considered as nearly zero at all useful frequencies. If it is necessary to establish the imaginary part of Y_{is} (or b_{is}), such as for calculations involved in *RF* matching networks, an admittance meter or R_X meter must be used.

3-17.19 Input capacitance test

The basic circuits for C_{iss} tests are shown in Fig. 3-60 and 3-61. Figure 3-60 is for JFETs and Fig. 3-61 is for MOSFETs. For dual-gate MOSFETs, the capacitance measurement is between gate 1 and all other terminals. For single-gate MOSFETs, the capacitance measurement is between the gate and all other terminals.

The circuit of Fig. 3-61(a) is for MOSFETs where a specific V_{GS} must be applied but V_{DS} is of no concern. The circuit of Fig. 3-61(b) is used where the capacitance is measured by means of a bridge with three terminals (high, low, and guard or ground). The reason for the two circuits is that some MOSFET datasheets specify a given V_{DS}, I_D, and V_{GS} when C_{iss} is measured. Other datasheets specify zero V_{DS} and sometimes zero V_{GS}.

3-17.20 Output capacitance test

The basic circuit for C_{oss} is shown in Fig. 3-62. Generally, C_{oss} is used only in dual-gate MOSFET specifications. In a typical specification, gate 1 is shorted to the source and gate 2 has a specific voltage value applied to produce a given I_D.

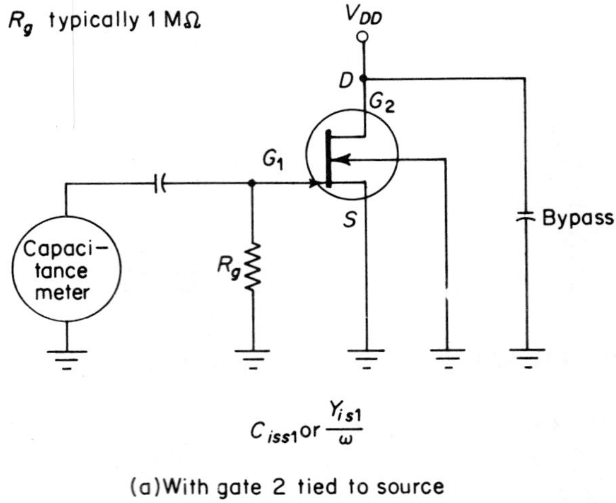

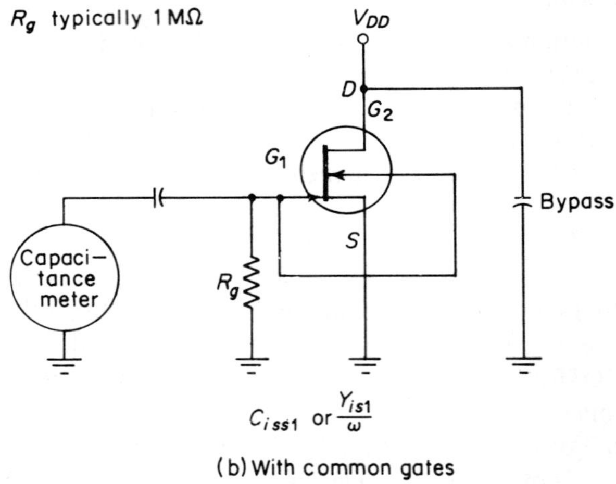

FIGURE 3-60 Test circuit for C_{iss} in tetrode JFETs (courtesy Motorola).

3-17.21 Reverse transfer capacitance test

The basic circuits for C_{rss} tests are shown in Fig. 3-63 and 3-64. Figure 3-63 is for JFETs and Fig. 3-64 is for MOSFETs. Three-terminal measurements are used in all cases. For dual-gate MOSFETs, gate 2 and the source are returned to the guard terminal and the capacitance measurement is made between gate 1 and the drain.

Note that the capacitors in series between the meter or bridge and the

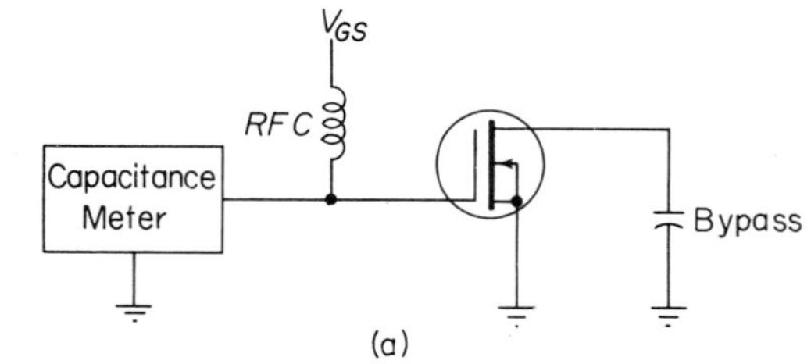

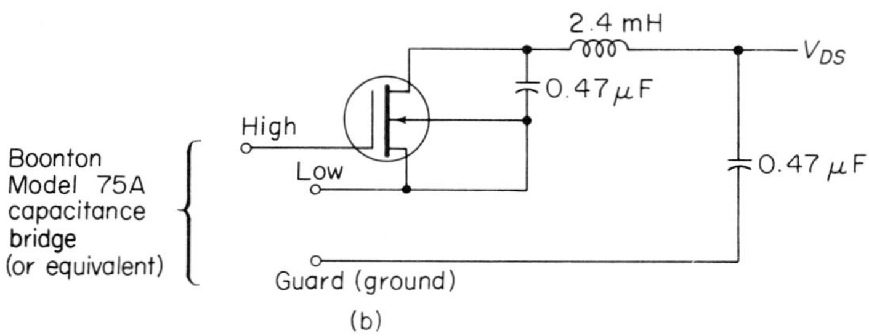

FIGURE 3-61 Test circuit for C_{iss} in MOSFETs.

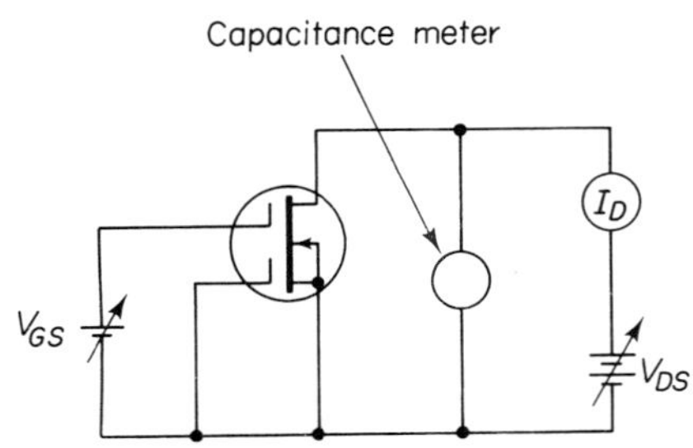

FIGURE 3-62 Test circuit for C_{oss}.

device under test will be included in the measurement. The value of C_{rss} is so low, however, that the series capacitance value indicated by the meter will be that of the device under test.

3-17.22 Element capacitance tests

The capacitance between elements of an FET can be measured with a capacitance meter. No special test connections are required. Some datasheets specify certain test connections or conditions, however, such as all remaining elements connected to source or gate connected to source. Generally, the only element capacitances of any consequence are $C_{d(sub)}$ drain-substrate junction capacitance and C_{ds} drain-to-source capacitance. Similarly, these capacitances are of little concern except for switching applications.

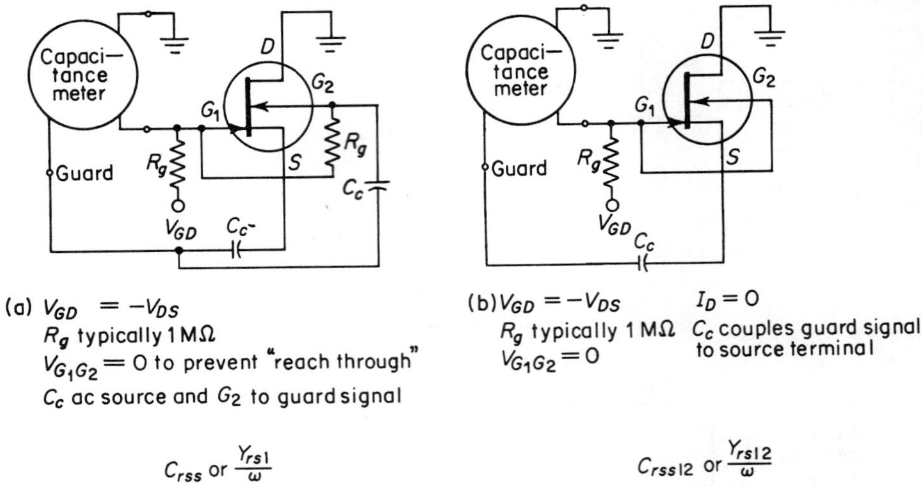

FIGURE 3-63 Test circuits for C_{rss} in JFETS. (Lenk, *Practical Semiconductor Databook for Electronic Engineers and Technicians*, Prentice-Hall.)

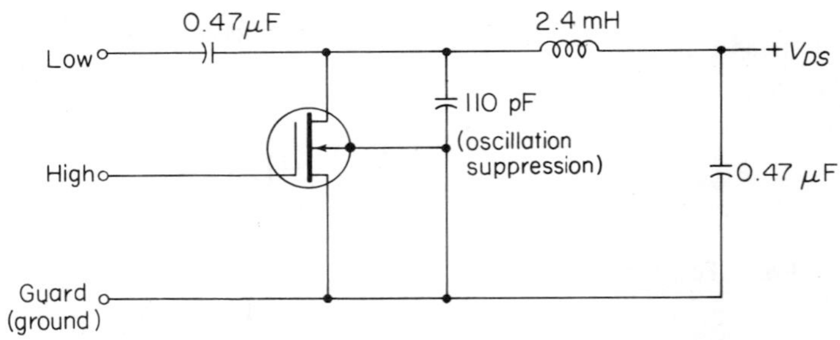

FIGURE 3-64 Test circuit for C_{rss} in MOSFETS.

3-17.23 Channel resistance tests

The basic circuit for channel resistance tests is shown in Fig. 3-65. Both $r_{ds(\text{on})}$ and $r_{ds(\text{off})}$ can be measured with the basic circuit; however, $r_{ds(\text{on})}$ is generally the test of interest.

If the FET operates in the depletion or depletion/enhancement modes, $r_{ds(\text{on})}$ is measured by adjusting V_{GS} to 0 V (or simply connecting the gate to the source). For example, the 3N128 specifies a V_{DS} and V_{GS} of 0 V for an $r_{ds(\text{on})}$ of 200 Ω (at a test frequency of 1 kHz).

During test, the ac voltage source V is adjusted to some convenient value (1 V, 10 V, etc.), and the channel current I is measured. The value of r_{ds} is

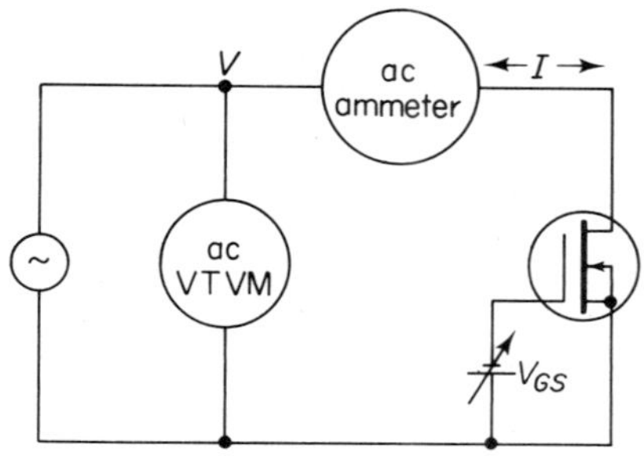

FIGURE 3-65 Test circuit for $r_{ds(\text{on})}$ and $r_{ds(\text{off})}$.

found when V is divided by I. For example, if V is adjusted to 10 V and 50 mA channel current is measured, the "on" channel resistance, or $r_{ds(\text{on})}$, is 200 Ω (10/0.05 = 200).

If the FET operates in the enhancement-only mode (Type C), it is necessary to forward bias the gate by adjusting V_{GS} to some specific value.

Figure 3-65 can also be used to measure the "off" channel resistance $r_{ds(\text{off})}$; however, the bias conditions are opposite those for $r_{ds(\text{on})}$. In the depletion and depletion/enhancement modes, the gate must be reverse biased by adjusting V_{GS} to some specific value. In the enhancement-only mode, the gate can be connected directly to the source.

Figure 3-65 can also be used to measure channel resistance of dual-gate FETs. Generally, the simplest way is to connect both gates together; however, some datasheets specify a fixed bias on one or both gates.

3-17.24 Switching-time tests

Switching-time tests for FETs are the same as for two-junction transistors. The procedures for finding delay, storage, rise, and fall times are fully discussed in Chapter 2. A typical test circuit for measurement of FET switching characteristics is shown in Fig. 3-66.

3-17.25 Gain tests

The Y_{fs} test described in Sec. 3-17.14 can be used to establish the gain of an FET; however, the test will not establish the gain in a working circuit. The only true test of circuit gain is to operate the device in a working circuit and to measure actual gain. The most practical method is to operate the device in the circuit with which the device is to be used; however, it may be convenient to have a "standard" or "universal" circuit for test of FET gain.

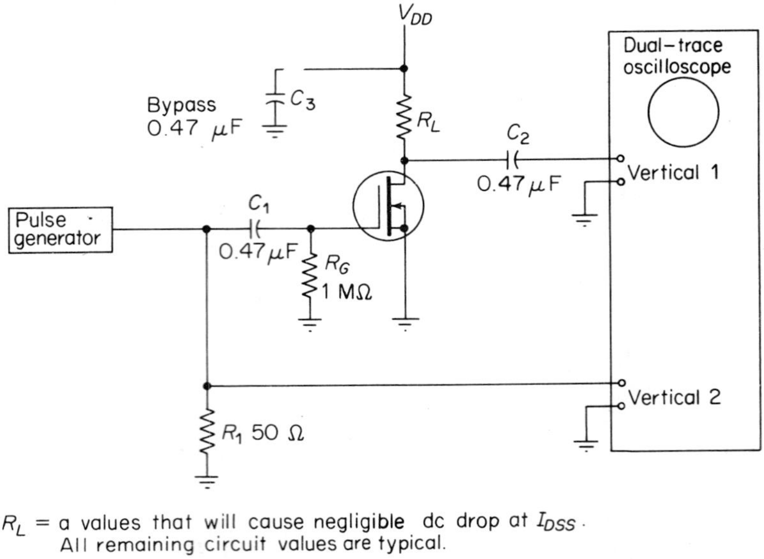

FIGURE 3-66 Test circuit for switching characteristics.

Generally, the main concern is with power gain in *RF* circuits. Most FET datasheets include a power-gain test circuit diagram. Figures 3-67 and 3-68 are typical examples of such power-gain test circuits. The circuit of Fig. 3-68 is for dual gate, where gate 2 is connected to a variable AGC voltage. This permits the power gain to be measured under various AGC conditions. For example, the 40821 specifies power gain between 14 and 17 dB, with V_{DS} of 15 V, I_D of 10 mA, and V_{G2S} of +4 V (applied at the AGC line).

Keep in mind that these test circuits are for specific FETs, operating with specific loads (50 Ω) and at specific frequencies (200 MHz); however, similar test circuits can be fabricated for other loads and frequencies.

Once the test circuit has been fabricated, the procedures for measuring power gain are relatively simple. In brief, an *RF* signal voltage is applied across the input load, and the input power is calculated from voltage and load resistance. The amplified output voltage is measured across a similar load and the power calculated. The ratio of output power to input power is the power gain and is expressed in decibels.

For those FETs used as converters, conversion gain is of importance. Figures 3-69 and 3-70 are typical examples of conversion gain test circuits (supplied as part of the datasheet). The circuit of Fig. 3-69 is for a single-gate device where both the *RF* input and local oscillator input are applied to the gate. Figure 3-70 is for dual gate where *RF* input is applied to gate 1

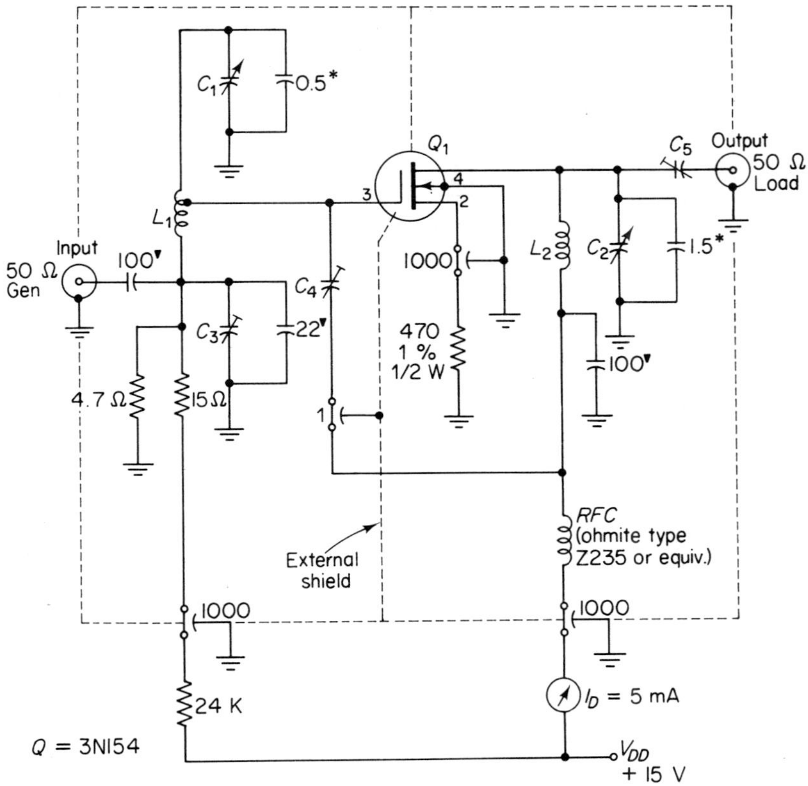

All resistors in ohms and 1/4 W unless otherwise specified.
All capacitors in pF.

* Tubular ceramic
▼ Disc ceramic

C_1, C_2 : 1.5–5 pF variable air capacitor: E.F. Johnson Type 160-102 or equivalent.

C_3 : 1–10 pF piston-type variable air capacitor: JFD Type VAM-010, Johanson Type 4335, or equivalent.

C_4, C_5 : 0.3–3 pF piston-type variable air capacitor: Roanwell Type MH-13 or equivalent.

L_1 : 5 turns silver-plated 0.02" thick, 0.07"–0.08" wide copper ribbon. Internal diameter of winding = 0.25"; winding length approx. 0.65". Tapped at $1\frac{1}{2}$ turns from C_1 end of winding.

L_2 : Same as L_1 except winding length approx. 0.7"; no tap.

FIGURE 3-67 Power gain and noise figure (NF) test circuit for 3N128 (courtesy RCA).

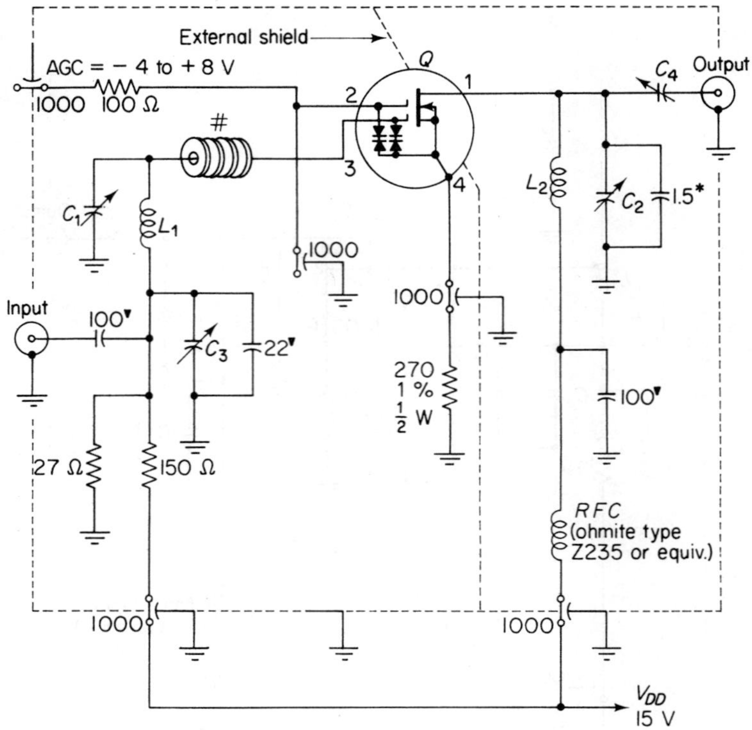

#Ferrite bead (4); Pyroferric Co. "Carbonyl J" 0.09 in OD; 0.03 in ID; 0.063 in thickness.

Q = 40820
▾ Disc ceramic.
* Tubular ceramic.

All resistors in ohms.

All capacitors in picofarads.

C_1: 1.8 − 8.7 pF variable air capacitor: E. F. Johnson Type 160-104, or equivalent.

C_2: 1.5 − 5 pF variable air capacitor: E. F. Johnson Type 160-102, or equivalent.

C_3: 1 − 10 pF piston-type variable air capacitor: JFD Type VAM-010; Johanson Type 4335, or equivalent.

C_4: 0.8 − 4.5 pF piston-type variable air capacitor: Erie 560-013, or equivalent.

L_1: 4 turns silver-plated 0.02 in thick, 0.075–0.085 in wide, copper ribbon. Internal diameter of winding = 0.25 in, winding length approx. 0.80 in.

L_2: $4\frac{1}{2}$ turns silver-plated 0.02 in thick, 0.085-0.095 in wide; $\frac{5}{16}$ in; ID Coil ≈ 0.90 in. long.

FIGURE 3-68 Power-gain and noise figure (NF) test circuit for 40820 (courtesy RCA).

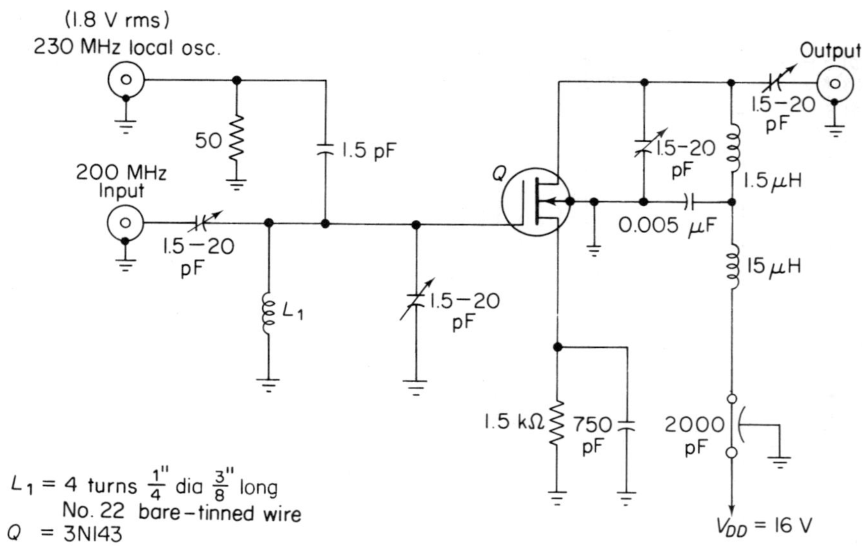

FIGURE 3-69 Conversion power-gain test for 3N143 (courtesy RCA).

and the local oscillator input is applied to gate 2. Again, these circuits are for specific devices and frequencies. However, the circuits will serve as a design "starting point" for standard test circuits.

3-17.26 Noise figure tests

The circuits of Fig. 3-67 and 3-68 can also be used to measure noise figure. Such noise figure tests are best accomplished using specialized test equipment, such as a *VHF* noise source and *VHF* noise meter. The instructions supplied with the noise measurement test equipment are generally complete and provide full descriptions of noise tests. In any event, the test procedures will not be duplicated here. The test equipment instructions generally omit the amplifier test circuit, however, thus creating a need for circuits as shown in Fig. 3-67 and 3-68.

3-17.27 Cross-modulation tests

Figure 3-71 shows the basic block diagram and amplifier circuits for cross-modulation tests. These circuits and values are for a 3N128 being tested at frequencies of 200 MHz (desired frequency) and 150 MHz (interfering frequency). The circuits provide for unneutralized, neutralized, and cascode amplifier operation.

During test, signals of both frequencies are applied to the input. The test

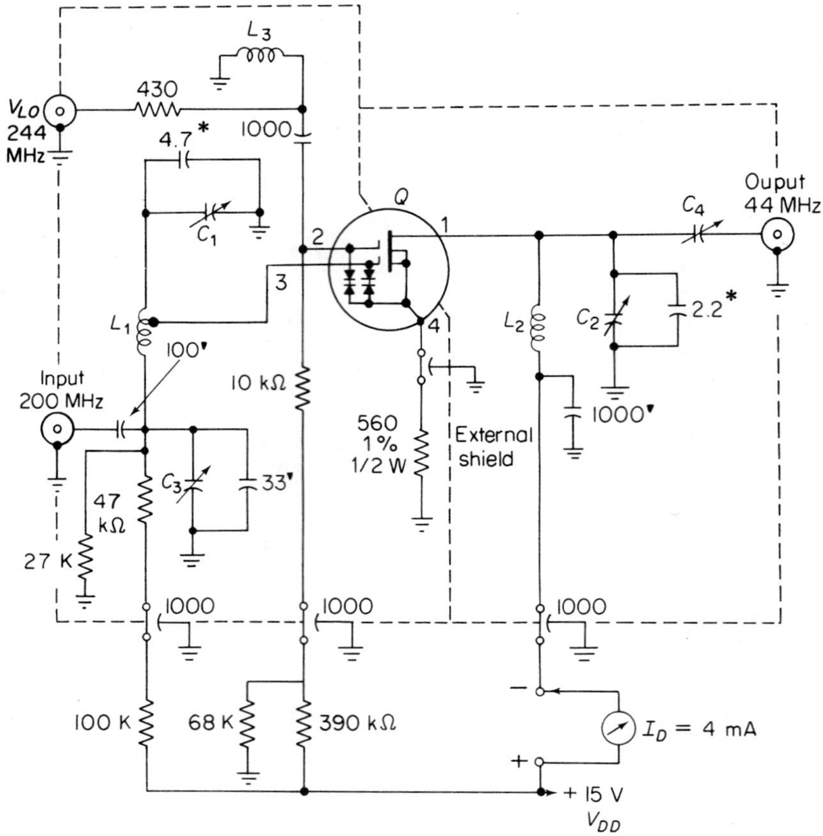

Q = 40821

▾ Disc ceramic.

* Tubular ceramic

All resistors in ohms.

All capacitors in pF.

C_1,C_2: 1.5 – 5 pF variable air capacitor: E.F. Johnson Type 160–102 or equivalent.

C_3: 1–10 pF piston-type variable air capacitor: JFD Type VAM-010, Johanson Type 4335, or equivalent.

C_4: 0.9 – 7 pF compression-type capacitor: ARCO 400 or equivalent.

L_1: 5 turns silver-plated 0.02" thick, 0.07"–0.08" wide copper ribbon. Internal diameter of winding = 0.25", winding length approx. 0.65. Tapped at $1\frac{1}{2}$ turns from C_1 end of winding.

L_2: Ohmite Z-235 RF choke or equivalent.

L_3: J.W. Miller Co. #4580 μH *RF* choke or equivalent.

Note: If 50-Ω meter is used in place of sweep detector, a low-pass filter must be provided to eliminate local oscillator voltage from load.

FIGURE 3-70 Conversion power-gain test for 40821 (courtesy RCA).

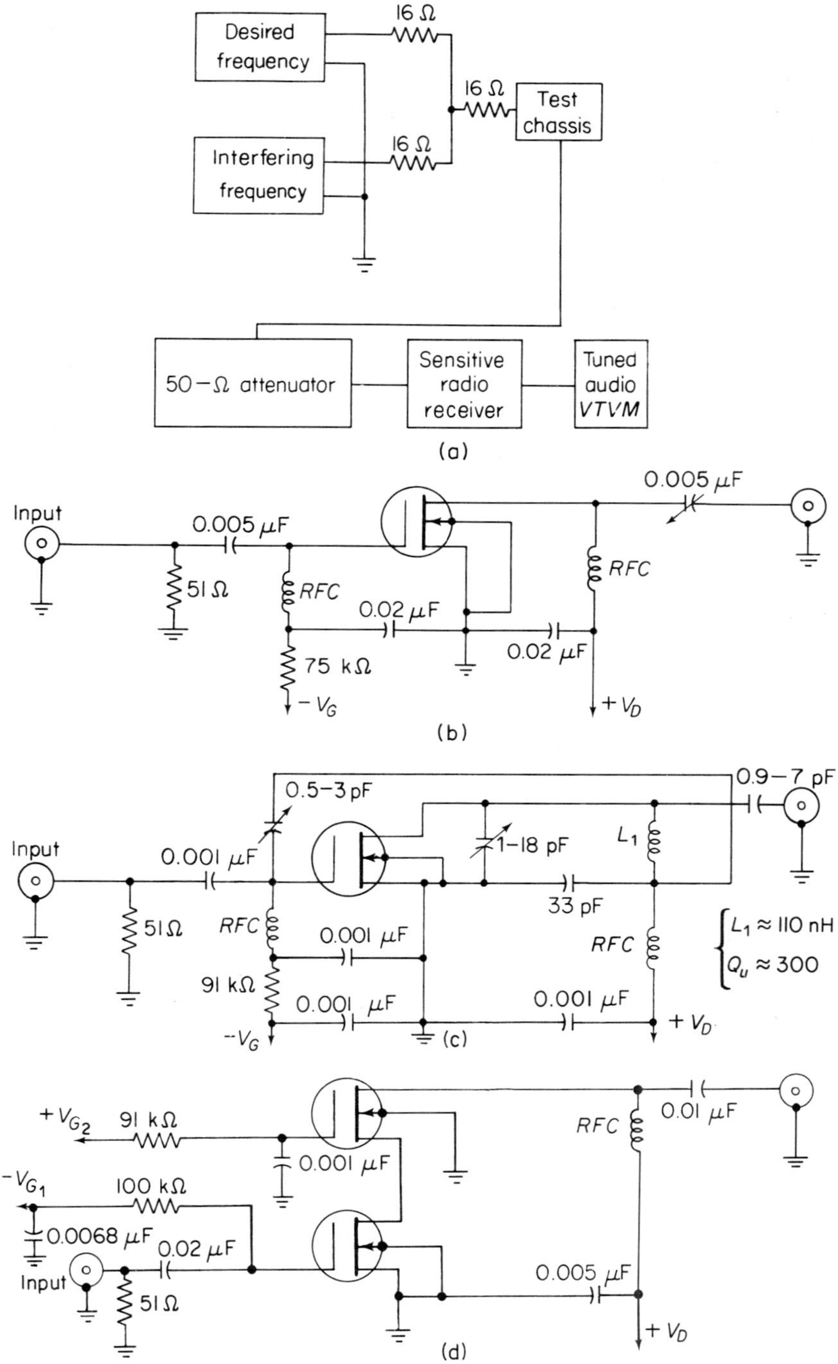

FIGURE 3-71 Test circuits used to measure cross-modulation distortion in MOSFETs: (a) block diagram, (b) unneutralized-stage test circuit, (c) neutralized-stage test circuit, (d) cascode-stage test circuit (courtesy RCA).

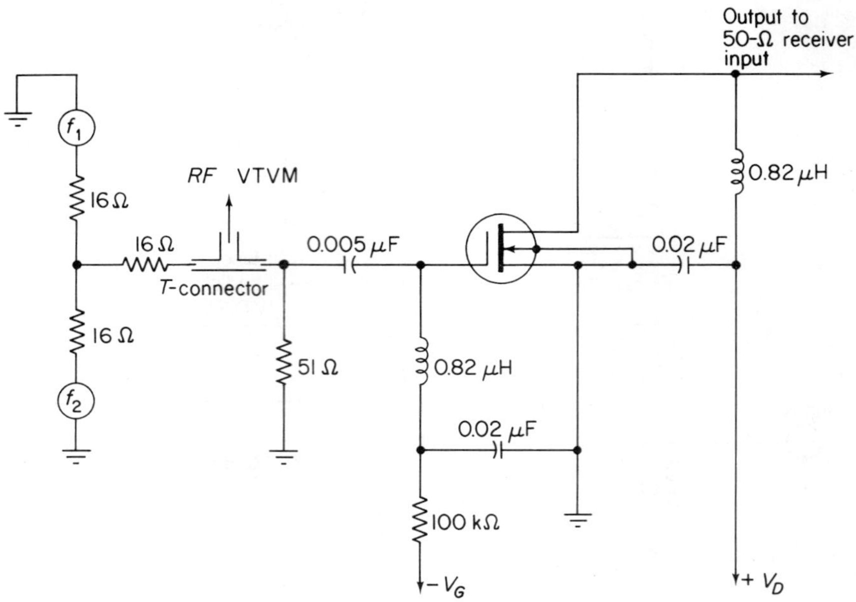

FIGURE 3-72 Test circuit used to measure intermodulation distortion in MOSFETs (courtesy RCA).

circuit output is monitored by the receiver. The amount of attenuation produced on the interfering signal is measured at various signal levels.

3-17.28 Intermodulation tests

Figure 3-72 shows the basic circuit for intermodulation tests. This circuit and values are for a 3N128 being tested at frequencies of 175 MHz (f_1) and 200 MHz (f_2). If intermodulation is present, there will be an intermodulation signal of 150 MHz.

During test, f_1 and f_2 are set to zero (amplitude) and the background noise level is measured (on the *RF* indicator of the receiver at the output of the test circuit). The amplitudes of f_1 and f_2 are increased until the reading on the *RF* indicator is 1 mV above the noise level. The voltage levels required to produce this output indication are measured on the *RF* voltmeter at the test circuit input.

4

Unijunction Transistors

The unijunction transistor (or UJT) operates on an entirely different principle from that of the two-junction transistor or the field effect transistor. The UJT is a *negative resistance* unit. That is, under the proper conditions, the input voltage or signal can be decreased; yet the output or load current will increase. Once the UJT is "turned on," it will not "turn off" until the circuit is broken or the input voltage is removed. For this reason, the UJT makes an excellent trigger source. The UJT can be biased just below the "firing" point. When a small trigger voltage (either intermittent or constant) is applied, the UJT will "fire" and produce a large output voltage pulse or signal that will remain on until the circuit is broken (by switching off the base voltage).

The *programmable* unijunction transistor (or PUT) is a four-layer device similar to an SCR (semiconductor or silicon-controlled rectifier), except that the anode gate (rather than the cathode gate) is brought out. The PUT is normally used in conventional unijunction transistor (UJT) circuits.

The characteristics of both UJT and PUT devices are similar, but the triggering voltage of the PUT is programmable and can be set by an external resistive voltage divider network. The PUT is faster and more sensitive than the UJT. The PUT finds limited application as a phase control element and is most often used in *long duration timer* circuits. In general, the PUT is more versatile and is a more economical device than the UJT and can replace the UJT in many applications.

Although the basic function of the UJT is that of a switch, a *relaxation oscillator* is the primary building block in most UJT circuits. This basic relaxation oscillator circuit provides phase control of thyristors such as SCRs, Triacs, etc. For that reason, much UJT theory is based on the relationship to relaxation oscillator principles.

4-1. OPERATING THEORY OF UNIJUNCTION TRANSISTORS

Figure 4-1 shows the standard unijunction transistor symbol with appropriate terms for current and voltage. As shown, the UJT is a three-terminal device. The three terminals are emitter (*E*), base 1 (*B*1), and base 2 (*B*2). Figure 4-2 shows the commonly used UJT symbols and their proper definitions in accordance with JEDEC (Joint Electron Device Engineering Council) standards.

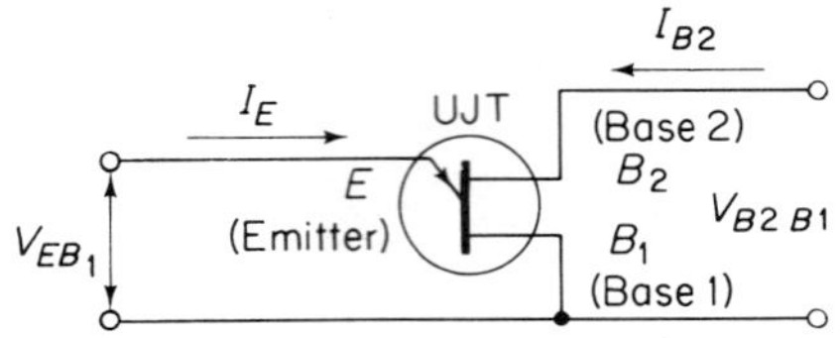

FIGURE 4-1 UJT symbols and nomenclature.

Note that the base 1 and base 2 leads are shown (in Fig. 4-1) at right angles to the base because they are *nonrectifying* (not a junction, as two-junction transistors and JFETs). The emitter connection is represented by an arrow, however, because it is a rectifying *PN* junction and the arrow is slanted to indicate the emitting properties of the junction. The arrowhead is pointing toward the base, showing a *P*-type emitter and an *N*-type base.

Figure 4-3 shows a simplified bar structure and equivalent circuit for a UJT. This bar structure was typical of the structures used in early UJTs and is still in use today. The bar structure is used here (to explain operation of the UJT) for the sake of convenience. Many different structures are used in the manufacture of UJTs, however, as discussed in later sections.

When voltage V_{B2B1} is applied, a current will flow in the silicon bar from base 2 to base 1. Since the bar is essentially a resistor (r_{BB}), the current that flows into base 2 is determined by

$$I_{B2} = \frac{V_{B2B1}}{r_{BB}} \tag{4-1}$$

A fraction of the applied voltage V_{B2B1} will appear at point *A* (Fig. 4-3), where the emitter is alloyed into the silicon bar. This fraction is the *intrinsic standoff ratio* η. The voltage at point *A* is therefore $\eta \times V_{B2B1}$. The *PN* junction formed by the emitter and the silicon bar is reverse biased, and only a small reverse leakage current flows in the emitter lead.

As the voltage V_E at the emitter is increased, a point will be reached where V_E equals the voltage at point *A*, plus the forward voltage drop of the *PN* junction V_D. The emitter voltage at this point is called the *peak-point emitter voltage* V_P. The peak-point voltage can be written

$$V_P = V_D + \eta V_{B2B1} \tag{4-2}$$

Unijunction Transistor Nomenclature

Symbol	Definition
I_E	*Emitter current.*
I_{EO}	*Emitter reverse current.* Measured between emitter and base 2 at a specified voltage; base 1 is open circuited.
I_p	*Peak-point emitter current.* The maximum emitter current that can flow without allowing the UJT to go into the negative resistance region. Peak point is the lowest current on the emitter characteristic where $$\frac{\Delta V_{EB1}}{\Delta I_E} = 0$$
I_V	*Valley-point emitter current.* The current flowing in the emitter when the device is biased to the valley point. Valley point is the second lowest current on the emitter characteristic where $$\frac{\Delta V_{EB1}}{\Delta I_E} = 0$$
r_{BB}	*Interbase resistance.* Resistance between base 2 and base 1, measured at a specified interbase voltage.
V_{B1B2}	*Voltage between base* 2 *and base* 1. Positive at base 2.
V_p	*Peak-point emitter voltage.* The maximum voltage seen at the emitter before the UJT goes into the negaive resistance region.
V_D	*Forward voltage drop of the emitter junction.*
V_{EB1}	*Emitter-to-base* 1 *voltage.*
$V_{EB1(sat)}$	*Emitter saturation voltage.* Forward voltage drop from emitter to base 1 at a specified emitter current (larger than I_V) and specified interbase voltage.
V_v	*Valley-point emitter voltage.* The voltage at which the valley point occurs with a specified V_{B2B1}.
V_{OB1}	*Base* 1 *peak-pulse voltage.* The peak voltage measure across a resistor in series with base 1 when the UJT is operated as a relaxation oscillator in a specified circuit.
η	*Intrinsic standoff ratio.* Defined by the relationship $$\eta = \frac{V_p - V_D}{V_{B2B1}}$$
αr_{BB}	*Interbase resistance temperature coefficient.* Variations of resistance between *B*2 and *B*1 over the specified temperature range and measured at the specified interbase voltage and temperature, with emitter open circuited.
$I_{B2(mod)}$	*Interbase modulation current. B*2 current modulation due to firing. Measured at a specified interbase voltage, emitter voltage, and temperature.

FIGURE 4-2 Commonly used UJT symbols and definitions.

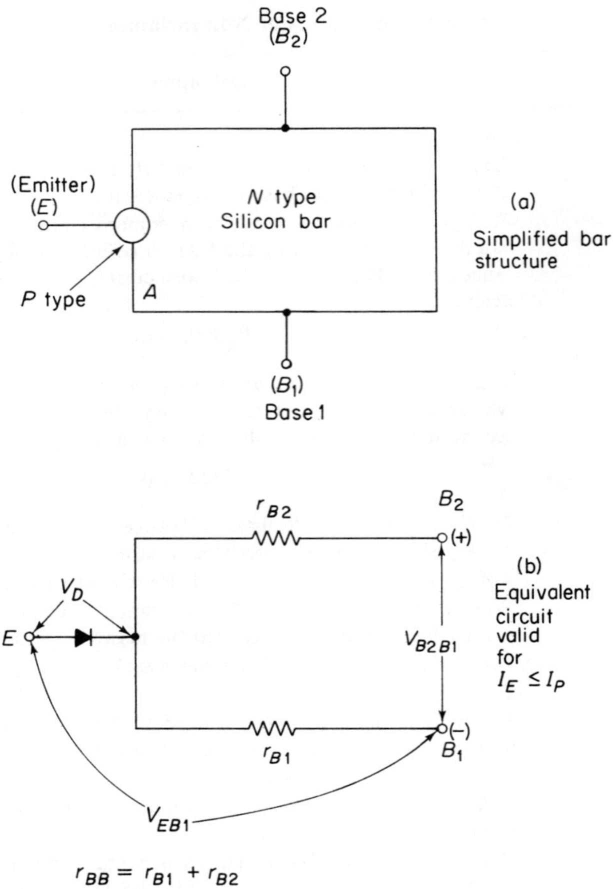

FIGURE 4-3 Simplified bar structure and equivalent circuit for a UJT.

The *PN* junction is now forward biased, and holes will be injected from the emitter into the silicon bar. The electric field inside the bar set up by V_{B2B1} is of such a direction that the injected holes will be moved toward the base 1 terminal.

When the holes are injected into the bar from the emitter, an equal amount of electrons will be injected from base 1 to maintain charge neutrality. Since both the electron and hole concentrations increase in the silicon bar between the emitter and base 1, the conductivity will also increase and the resistance will decrease. This process is called *conductivity modulation.*

The decrease in resistance will cause a decrease in the voltage drop from emitter to base 1, which in turn allows more holes to be injected from the emitter. This regenerative action continues until the UJT is in the so-called

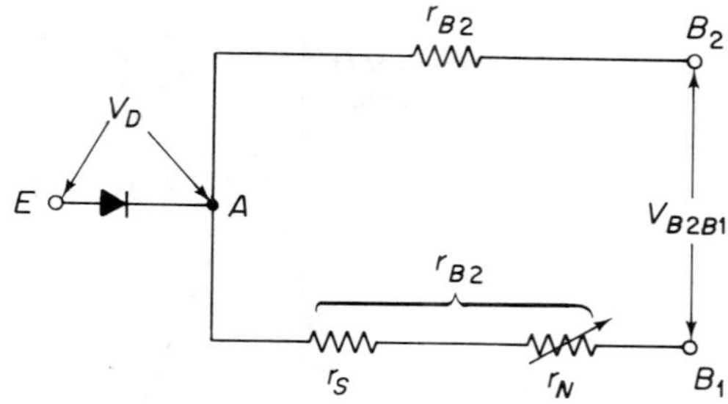

FIGURE 4-4 Equivalent circuit for UJT in negative resistance region.

negative resistance region. An equivalent circuit for this region is shown in Fig. 4-4. At emitter currents equal to or less than I_P, the resistance r_{BB} can be divided into two parts: r_{B1} and r_{B2} according to the relations

$$r_{B1} = \eta r_{BB} \quad \text{and} \quad r_{B2} = r_{BB} - r_{B1} \tag{4-3}$$

In the negative resistance region, resistor r_{B1} can be thought of as consisting of a fixed portion r_S and a variable portion r_N, where r_S is the saturation resistance and r_N is the negative resistance (the magnitude of which decreases with increasing emitter current). When the negative resistance r_N decreases to zero and r_{B1} is equal to r_S only, the UJT is no longer in the negative resistance region.

The point on the emitter characteristic where r_{B1} just reaches its minimum value is called the *valley point*. The emitter current and voltage at this point are the valley-point emitter current I_V and the valley-point emitter voltage, respectively.

When the emitter current is increased beyond I_V, the UJT enters the so-called *saturation region* where the emitter current is essentially a linear function of the emitter voltage. The equivalent circuit for the saturation region is shown in Fig. 4-5.

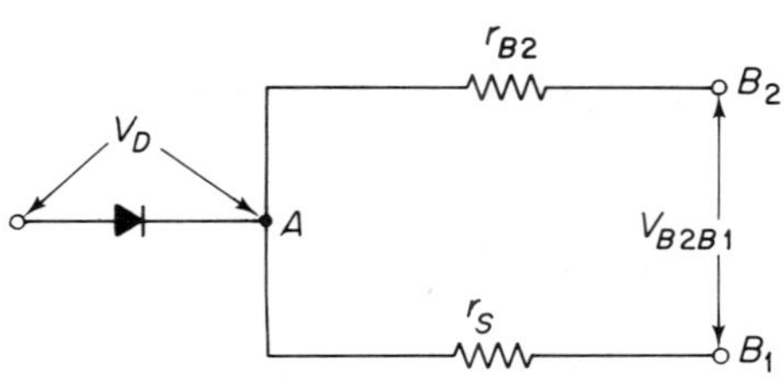

FIGURE 4-5 Equivalent circuit for UJT in saturation region.

A static emitter-characteristic curve for a single value of V_{B2B1} is shown in Fig. 4-6. The emitter curve is not drawn to scale; this shows the different operating regions in more detail. The region to the left of the peak point is called the *cutoff region*. The emitter junction is reversed biased in most of the cutoff region but is slightly forward biased at the peak point.

The region between the peak point and the valley point, where the emitter junction is forward biased and conductivity modulation takes place, is called the *negative resistance* region.

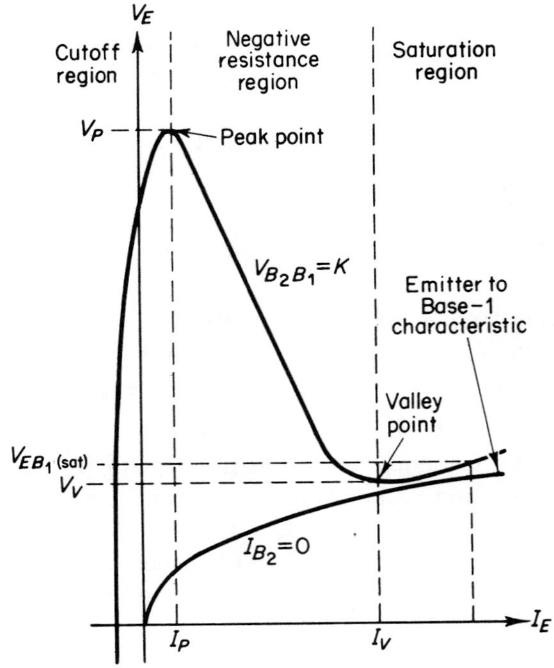

FIGURE 4-6 Static UJT emitter characteristic curves (courtesy Motorola).

The region to the right of the valley point, where the emitter current is limited only by r_S, is called the *saturation region*.

The curve for base 2 current (I_{B2}) equal to zero is essentially the forward characteristic of a conventional silicon diode.

4-2. OPERATION OF THE PUT

Figure 4-7 shows the symbol and a transistor equivalent circuit for a PUT. As shown, the PUT has three terminals: an anode (A), a gate (G), and a cathode (K). As seen from the equivalent circuit, the device is actually an anode-gated SCR. This means that if the gate is made negative with respect to the anode, the device will switch from a blocking state to its "on" state.

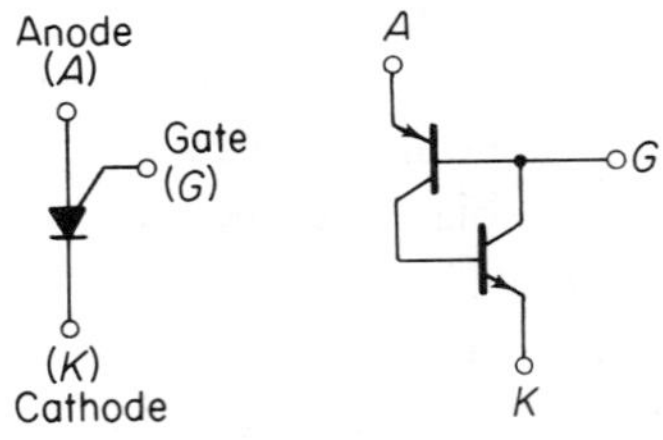

FIGURE 4-7 PUT symbol and equivalent circuit (courtesy Motorola).

Since the PUT is normally used as a unijunction transistor, the UJT terminology is used to describe its parameters. In order to operate a PUT as a UJT, an external reference voltage must be maintained at the gate terminal.

A typical relaxation-type oscillator circuit is shown in Fig. 4-8. The voltage divider shown is a typical way of obtaining the gate reference. In this circuit, the characteristic curve (looking into the anode–cathode terminals) would appear as shown in Fig. 4-9. The peak and valley points are stable operating points at either end of a negative resistance region.

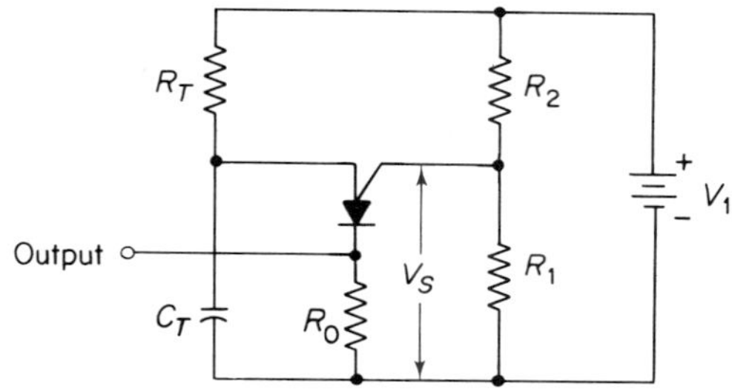

FIGURE 4-8 Typical PUT relaxation oscillator circuit.

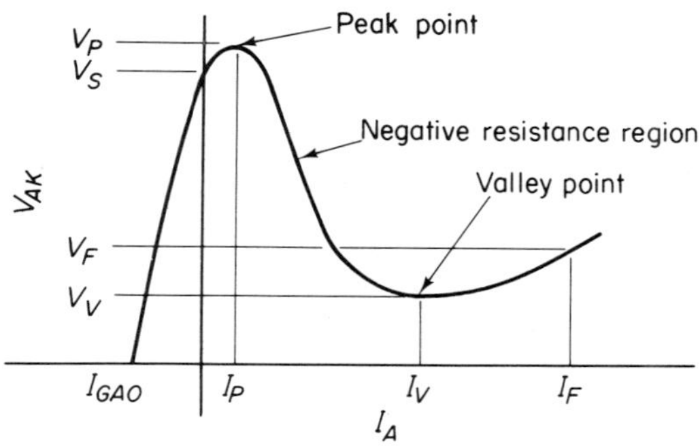

FIGURE 4-9 Static characteristics of PUT.

The peak-point voltage (V_P) is essentially the same as the external gate reference, the only difference being the gate diode drop. Since the reference is circuit dependent (rather than device dependent), the reference may be varied. Thus, V_P is made programmable by varying the reference. This feature is the most significant difference between the UJT and the PUT.

In characterizing the PUT, it is convenient to speak of the Thevenin equivalent circuit for the external gate voltage (V_S) and the equivalent gate resistance (R_G). The parameters are defined in terms of the divider resistors R_1 and R_2 and supply voltage as follows:

$$V_S = \frac{R_1 V_1}{R_1 + R_2} \tag{4-4}$$

$$R_G = \frac{R_1 R_2}{R_1 + R_2} \tag{4-5}$$

Most device parameters are sensitive to changes in V_S and R_G. For example, decreasing R_G will cause peak and valley currents to increase. This is easy to see since R_G actually shunts the device and will cause its sensitivity to decrease.

4-3. STRUCTURE OF UNIJUNCTION TRANSISTORS

The bar structure (discussed in Sec. 4-1) was used frequently during the early development of the UJT. The *cube structure* was also used to some extent. Figure 4-10 is a cross-section diagram of the bar and cube structures.

The bar structure in Fig. 4-10(a) is formed by mounting a high-resistivity *N*-type silicon bar on a ceramic platform having an air gap in the center and gold–antimony film deposited on each side of the gap. Base 1 and base 2 are ohmic contacts that are formed between the silicon bar and the gold. A single *P*-type emitter is formed by alloying an aluminum wire onto the bar opposite from the base contacts.

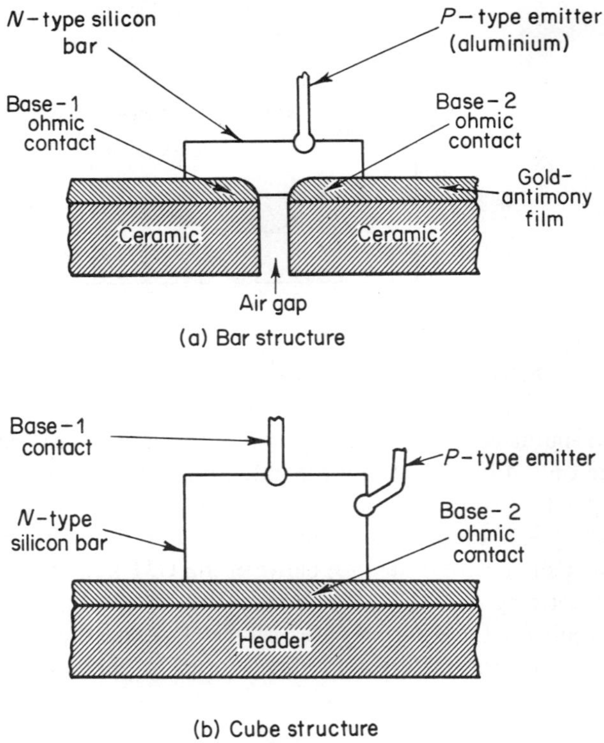

FIGURE 4-10 Cross-section diagram of UJT bar and cube structure (courtesy Motorola).

The cube structure in Fig. 4-10(b) uses a high-resistance *N*-type silicon cube. The cube is mounted on a header with a gold–antimony alloy contact between the bottom of the cube and the header. The base 2 ohmic contact is made to the gold–antimony area. Base 1 is formed by alloying a gold wire to the top of the silicon bar. The emitter is formed in the same way by alloying an aluminum wire to the side of the cube.

Figure 4-11 shows the production steps of an *annular* UJT developed

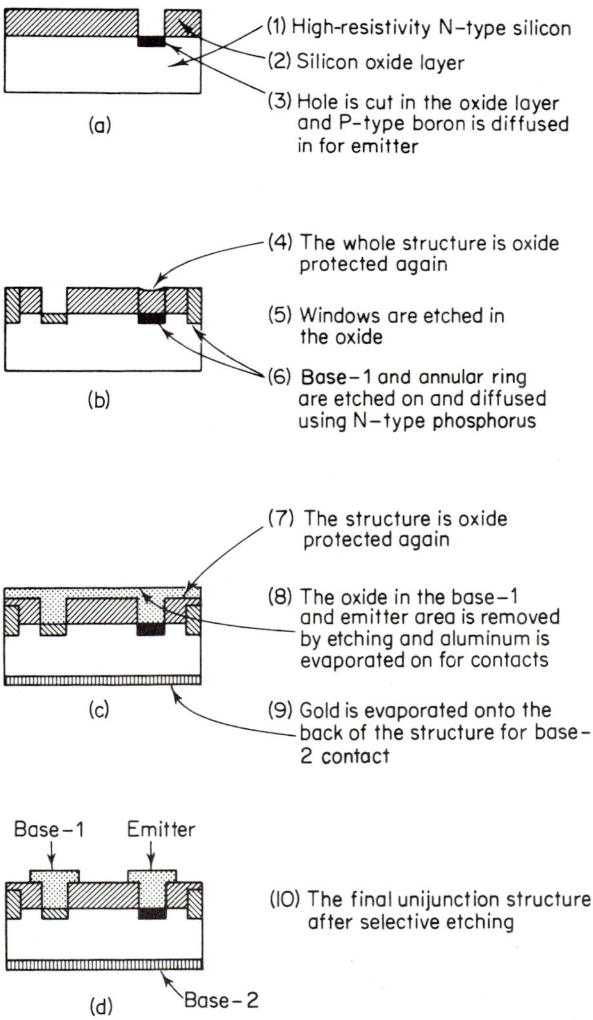

FIGURE 4-11 Motorola annular UJT construction (courtesy Motorola).

by Motorola. This process lends itself to modern automatic production methods, while remaining quality construction.

4-4. CHARACTERISTICS OF UNIJUNCTION TRANSISTORS

The static emitter characteristics shown in Fig. 4-6 were not drawn to scale, in order to show all three regions (cutoff, negative resistance, saturation) of a UJT. Figure 4-12 and 4-13 show a typical UJT emitter curve for $V_{B2B1} = 20$ V, drawn to scale.

Figure 4-12 shows part of the cutoff region plotted on a linear scale. When the emitter voltage is zero, the emitter current is negative. Peak-point voltage is reached at a forward emitter current of about 10 μA. As seen in Fig. 4-13, the voltage remains constant until the emitter current reaches approximately 0.1 μA, at which point the voltage starts to decrease.

Peak current I_p for the UJT graphed in Fig. 4-12 and 4-13 therefore equals 0.1 μA, while peak voltage V_P equals 16 V. Referring to Fig. 4-13, valley voltage V_V is seen to be about 1.6 V and the valley current I_V is approximately 8 mA.

The saturation resistance r_S can be found from the slope of the emitter characteristic in the saturation region (above 8 mA) and is approximately 5 Ω.

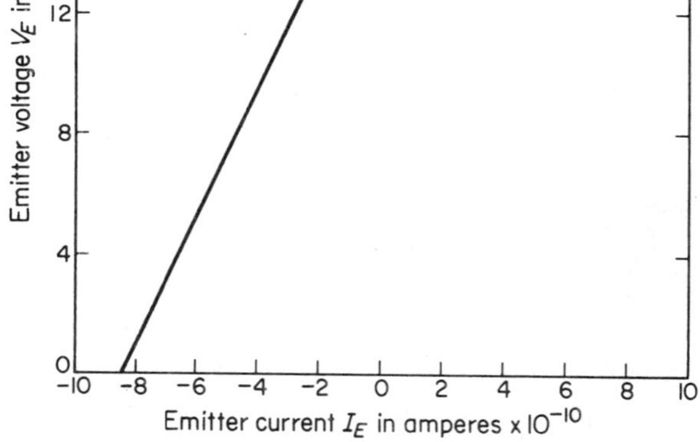

FIGURE 4-12 UJT static emitter characteristics in cutoff region (courtesy Motorola).

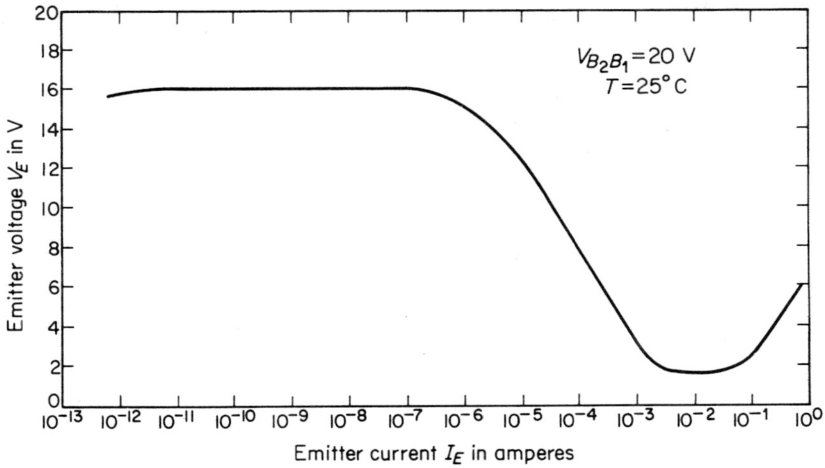

FIGURE 4-13 UJT static emitter characteristics in negative resistance region (courtesy Motorola).

The diode voltage drop V_D is defined as the forward voltage drop of the emitter junction. Since V_D is essentially equivalent to the forward voltage drop of a silicon diode, the value of V_D is dependent on both forward current and temperature. V_D is particularly important since it is one of the major factors in peak voltage V_P of a UJT. The equation for V_P is

$$V_P = V_D + \eta V_{B2B1} \tag{4-6}$$

In applications such as timers and oscillators, any changes in V_P would result in inaccuracy since oscillator and timer circuit accuracy would be dependent on the repeatability of V_P.

There are several ways to measure V_D; however, it is important to hold the emitter current near peak I_P when the measurement is made since it really is V_D at I_P that is required in Eq. 4-6.

One simple way to measure V_D is shown in Fig. 4-14. A constant-current signal equal to I_P is applied between the emitter and base 1, and a potentiometric voltmeter is used to measure the voltage from emitter to base 2. A potentiometric voltmeter has essentially infinite input impedance when the meter is nulled, and there is no current flow in the base 2 circuit. Therefore, the voltage measured is equal to V_D.

Figure 4-15 shows V_D as a function of temperature for an emitter current of 1 μA. The variation of V_D is essentially linear over the temperature range considered and is equal to -2.7 mA/°C. The diode voltage drop therefore decreases with increasing temperature, and V_D can be written as

$$V_D = V_{DN} - (T - 25)°\text{C}\, K_D \tag{4-7}$$

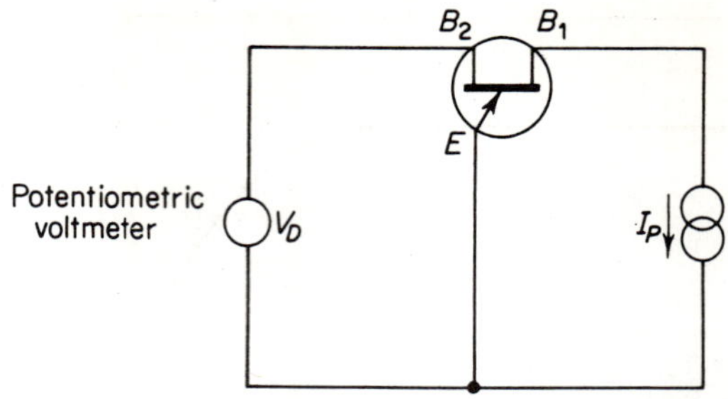

FIGURE 4-14 Basic diode voltage drop V_D test circuit (courtesy Motorola).

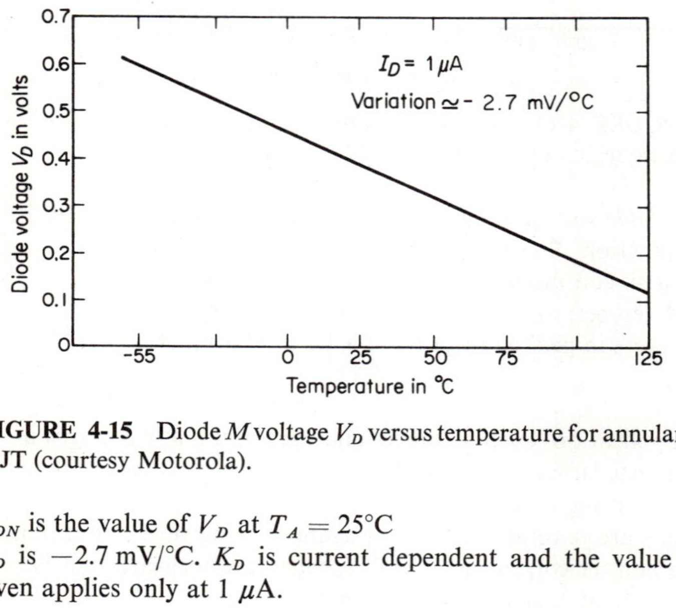

FIGURE 4-15 Diode *M* voltage V_D versus temperature for annular UJT (courtesy Motorola).

where V_{DN} is the value of V_D at $T_A = 25°C$
K_D is -2.7 mV/°C. K_D is current dependent and the value for K_D given applies only at 1 μA.

The *intrinsic standoff ratio* is defined by

$$\eta = \frac{V_P - V_D}{V_{B2B1}} = \frac{r_{B1}}{r_{BB}} \tag{4-8}$$

The intrinsic standoff ratio is somewhat temperature dependent. Figure 4-16 shows typical variations of standoff ratio with temperature for UJTs from three different manufacturers.

The intrinsic standoff ratio can be calculated if the V_{B2B1}, V_P, and V_D are known, using Eq. 4-8. The ratio can also be measured directly with a sensitive ohmmeter (base 1-to-emitter resistance divided by interbase resistance), using Eq. 4-8. As in the case of any transistor, the voltages and currents used to make the resistance measurements should be kept well within the limits of the UJT.

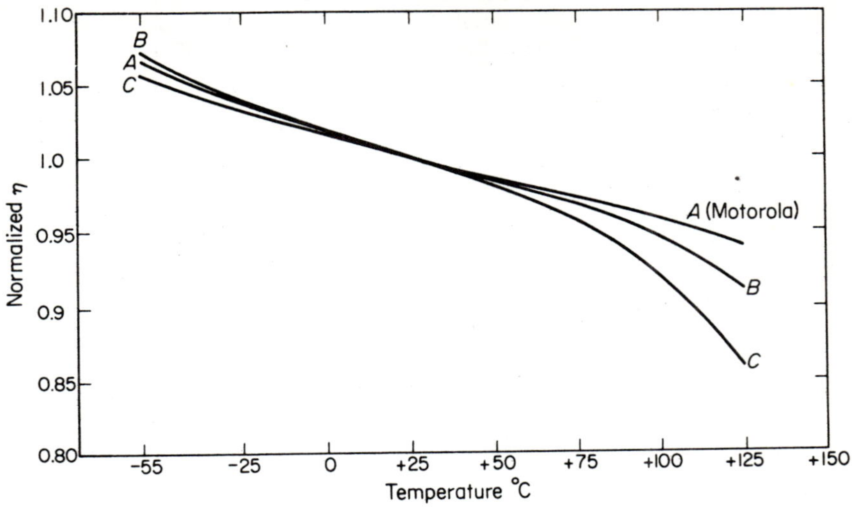

FIGURE 4-16 Typical variations of standoff ratios versus temperature (courtesy Motorola).

The intrinsic standoff ratio is also slightly dependent on V_{B2B1} but the variation is so small that it can be ignored for practical purposes.

The interbase resistance r_{BB} is highly temperature dependent. A typical r_{BB} versus temperature characteristic curve for $V_{B2B1} = 3$ V is shown in Fig. 4-17.

Interbase resistance r_{BB} is also found to vary with interbase voltage V_{B2B1}. A typical curve is shown in Fig. 4-18.

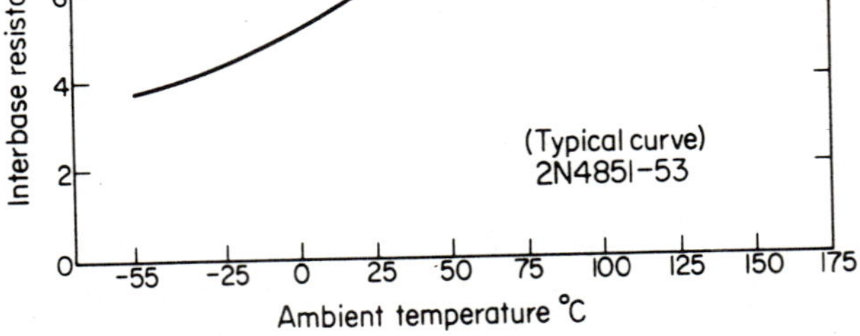

FIGURE 4-17 Interbase resistance r_{BB} versus temperature (courtesy Motorola).

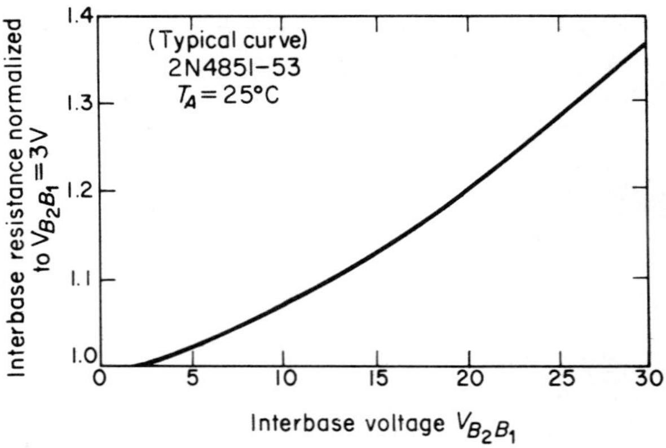

FIGURE 4-18 Typical variation of r_{BB} as a function of interbase voltage V_{B2B1} (courtesy Motorola).

The peak-point characteristics V_P and I_P decrease as temperature is increased. The variation of V_P with temperature can be minimized by the addition of an external resistor in the base 2 circuit, as shown in Fig. 4-19. With such a circuit (as the ambient temperature increases) the interbase resistance r_{BB} will increase and the interbase voltage V_{B2B1} will also increase due to the voltage divider action of resistor R_2 (Fig. 4-19), r_{BB}, and R_1.

If R_2 is chosen correctly, the increase in interbase voltage will compensate for the decrease in V_D. The *approximate* value of R_2 is

$$R_2 \approx \frac{0.7r_{BB}}{\eta V_1} + \frac{(1-\eta)R_1}{\eta} \tag{4-9}$$

If R_2 satisfies Eq. 4-9, the peak-point voltage V_P will be given by

$$V_P = \eta V_1 \tag{4-10}$$

The circuit of Fig. 4-19 is most effective if the temperature-compensating resistor R_2 is in close physical contact with the emitter diode. This will

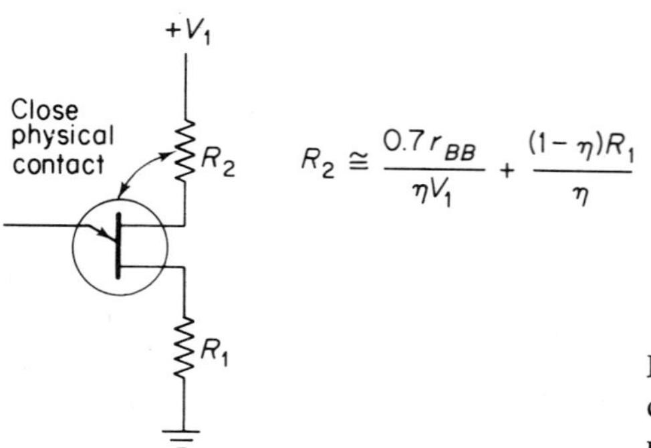

FIGURE 4-19 Temperature compensation circuit for UJT using external bias resistor R_2.

compensate for changes in power dissipation or sudden changes in ambient temperature.

The valley-point characteristics V_V and I_V decrease as ambient temperature is increased. A curve showing typical temperature behavior for an annular (Motorola) UJT is shown in Fig. 4-20, where the curves are normalized to the value of 25°C.

V_V and I_V are also dependent on the interbase voltage. When V_{B2B1} increases, both V_V and I_V will increase only. Typical curves showing this behavior are given in Fig. 4-21, where the curves are normalized to the value of 10 V (V_{B2B1}).

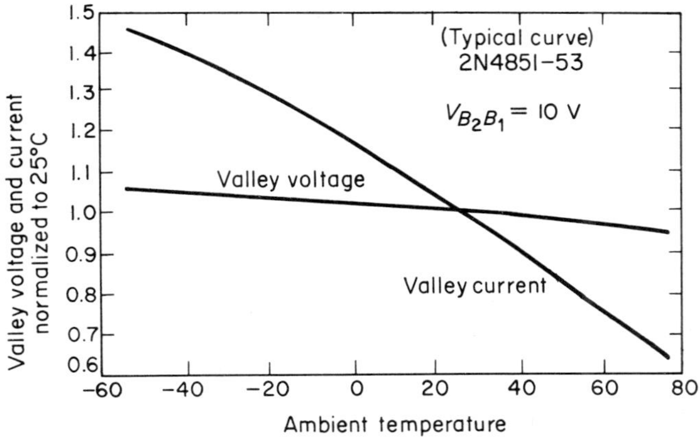

FIGURE 4-20 Valley voltage and valley current versus temperature (courtesy Motorola).

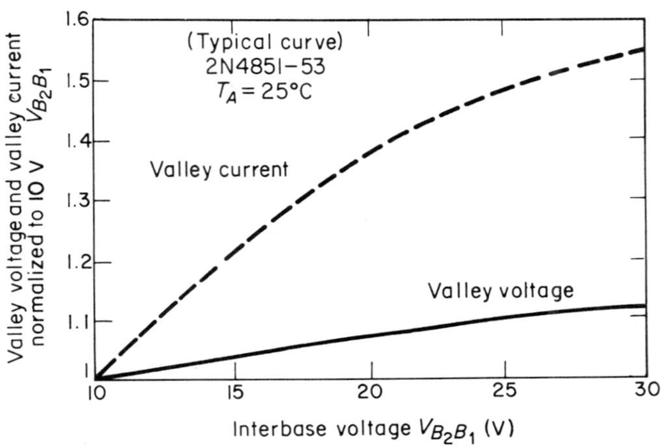

FIGURE 4-21 Valley voltage and valley current versus interbase voltage V_{B2B1} (courtesy Motorola).

Valley current I_V is a relatively difficult characteristic to measure since, as seen in Fig. 4-13, the voltage is relatively constant for large variations in emitter-current around the valley point.

The emitter reverse current I_{EO} is generally specified as the current flowing from base 2 to the emitter with 30 V applied between base 2 and the emitter (positive at base 2) and the base 1 terminal open circuited.

I_{EO} is highly temperature dependent since it is (in effect) the leakage current of a silicon diode. A curve showing variation of I_{EO} with temperature is shown in Fig. 4-22 with I_{EO} being approximately 1.5 nA at 25°C.

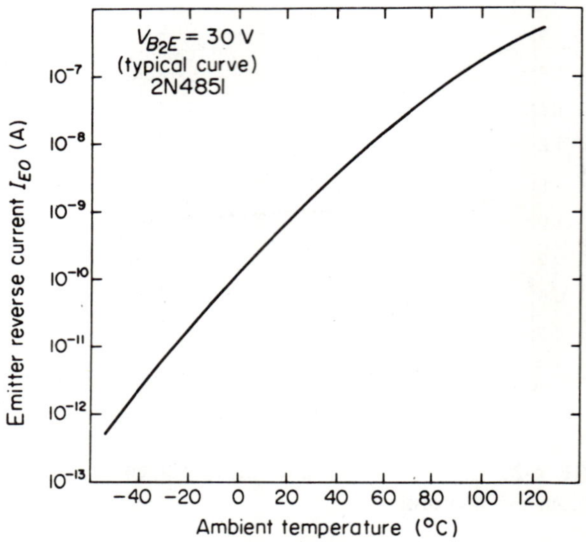

FIGURE 4-22 Emitter reverse current I_{EO} versus ambient temperature (courtesy Motorola).

Interbase characteristics are usually indicated by measurement of base 2 current I_{B2} as a function of interbase voltage V_{B2B1} and emitter-current I_E. Usually, interbase characteristics are measured on a *sweep basis* rather than with constant voltages and currents, to avoid heating effects due to power dissipation.

A circuit for sweep test of interbase characteristics is shown in Fig. 4-23. A constant current is applied to the emitter from time t_0 to t_1. Simultaneously, a voltage ramp going from 0 to 30 V is applied to base 2. Base 1 is grounded to complete the circuit. The current I_{B2} is measured with a *current probe* and applied to the vertical input of an oscilloscope. The voltage ramp, applied to base 2, is also applied to the oscilloscope horizontal input.

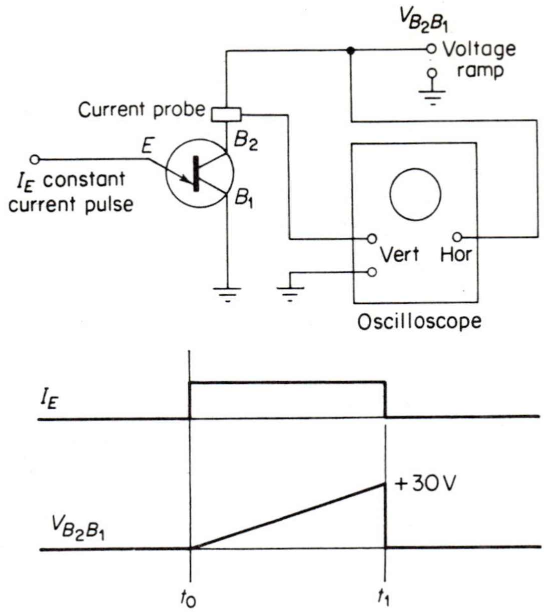

FIGURE 4-23 Interbase characteristic test circuit (courtesy Motorola).

Figure 4-24 shows typical interbase characteristics at ambient temperatures of −55, +25, and +125°C. As shown, the percentage increase in I_{B2} decreases with increasing emitter current and temperature.

The modulated interbase current $I_{B2(\text{mod})}$ is usually measured at $I_E = 50$ mA and $V_{B2B1} = 10$ V. As shown in Fig. 4-24, the approximate modulated interbase current is 11 mA (at +25°C).

The transient characteristics of a UJT are usually not specified in the same way as for a conventional two-junction transistor or FET. For example, switching times are usually not specified on a UJT datasheet. Instead, a parameter of $f_{\max}$ is given which indicates the maximum frequency of oscillation which can be obtained using the UJT in a specified relaxation oscillator circuit.

In some applications such as critical timers, however, it may be of interest to determine *turn-on* and *turn-off* times associated with the UJT. The following describes procedures for these measurements.

The circuit of Fig. 4-25 can be used to measure t_{on} and t_{off} for the case where the UJT emitter circuit is *purely resistive*. Typical switching-time values are $t_{\text{on}} = 1\ \mu$s; $t_{\text{off}} = 2.5\ \mu$s. The waveform observed at the base 1 terminal when the UJT turns off is shown in Fig. 4-26.

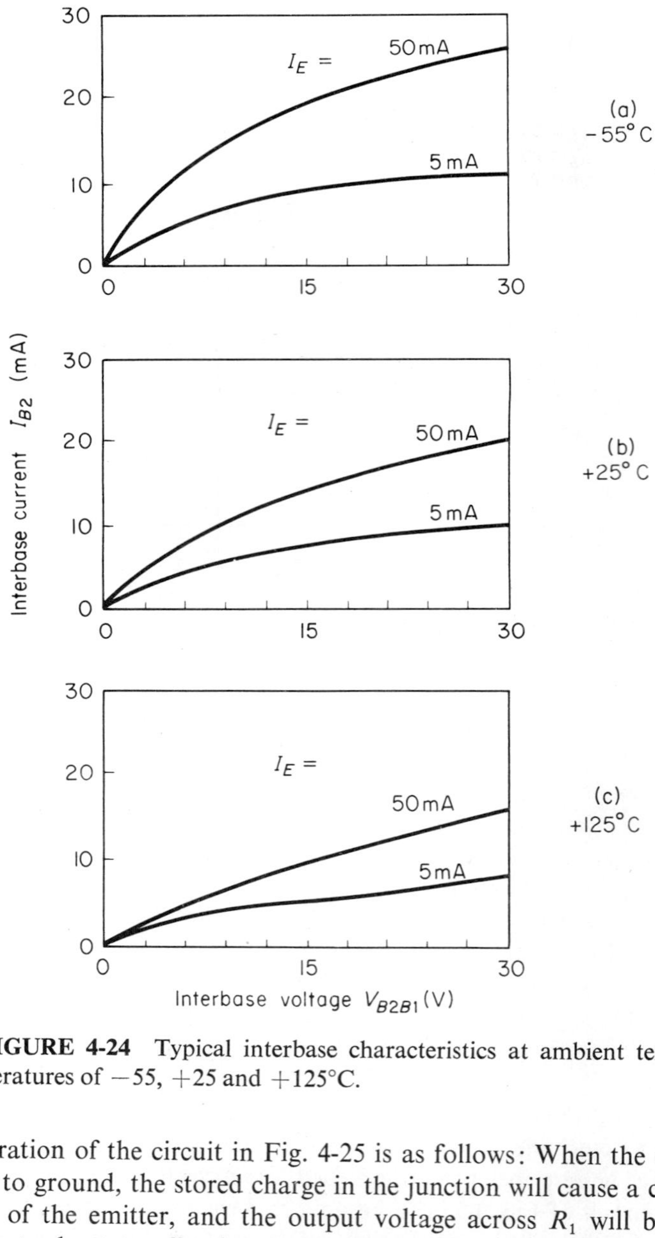

FIGURE 4-24 Typical interbase characteristics at ambient temperatures of −55, +25 and +125°C.

Operation of the circuit in Fig. 4-25 is as follows: When the emitter is returned to ground, the stored charge in the junction will cause a current to flow out of the emitter, and the output voltage across R_1 will be smaller than the steady-state off value. Immediately following the removal of the excess charge, the voltage across R_1 will go higher than the steady-state off value because r_{B1} has still not returned to normal, following the conductivity

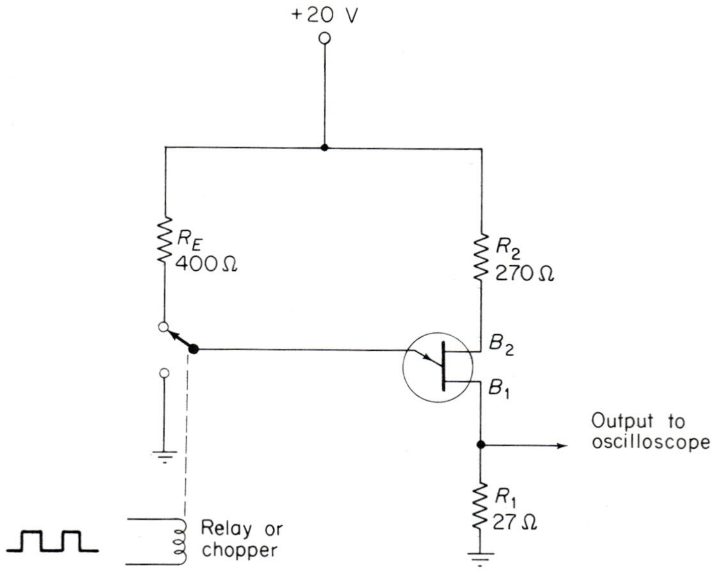

FIGURE 4-25 Circuit for measurement of t_{on} and t_{off} where UJT emitter load is purely resistive.

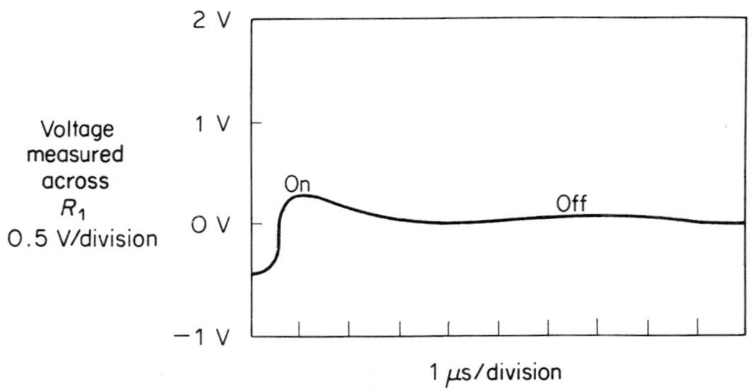

FIGURE 4-26 Turn-off waveform for typical UJT.

modulation in the "on" state, and I_{B2} will be larger than the steady-state off value.

The circuit of Fig. 4-27 can be used to measure t_{on} and t_{off} for the case where the UJT emitter circuit has *both capacitance and resistance* (as usually is the case). The test circuit of Fig. 4-27 is a *relaxation oscillator*, with turn-on and turn-off time being measured at the base 1 terminal. The turn-on and turn-off waveforms are shown in Fig. 4-28.

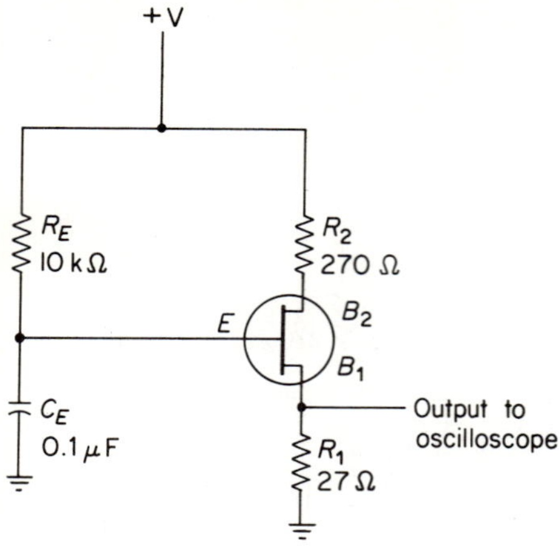

FIGURE 4-27 Circuit (relaxation oscillator) for measurement of t_{on} and t_{off} where UJT emitter load is both resistive and capacitive.

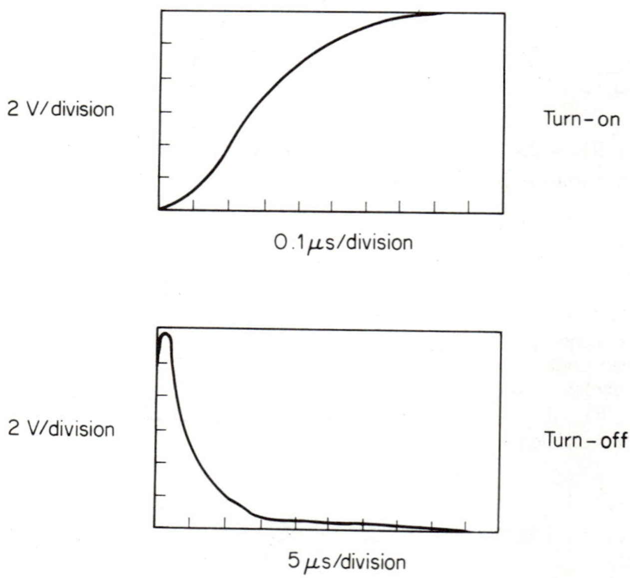

FIGURE 4-28 Turn-on and turn-off waveforms for UJT using relaxation oscillator circuit.

Turn-on time is measured from the start of the turn-on to the 90% point. A typical turn-on time is 0.5 μs with the capacitance of C_E shown in Fig. 4-27. An increase in C_E capacitance will cause an increase in turn-on time. Turn-off time is measured from the start of turn-off to the 90% point and is about 12 μs (due to the long discharge time of the capacitor). Turn-off time also increases with an increase in C_E capacitance.

The effect of C_E capacitance on switching time is shown in Fig. 4-29.

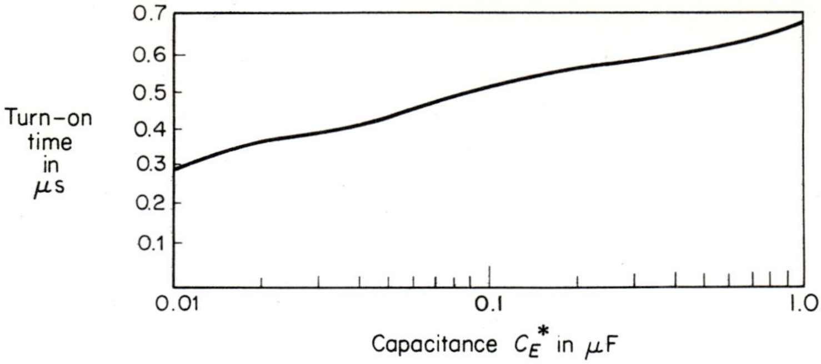

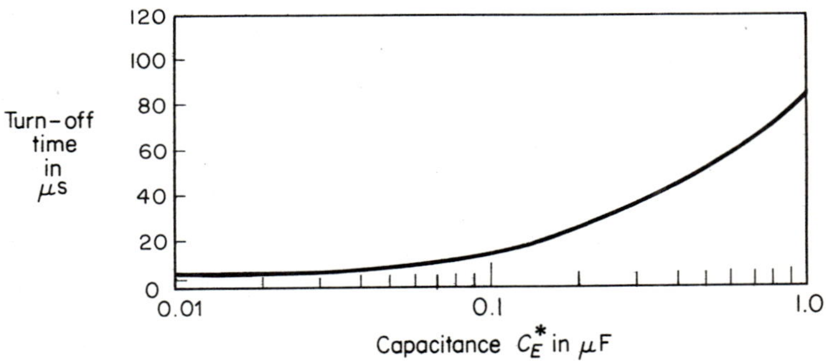

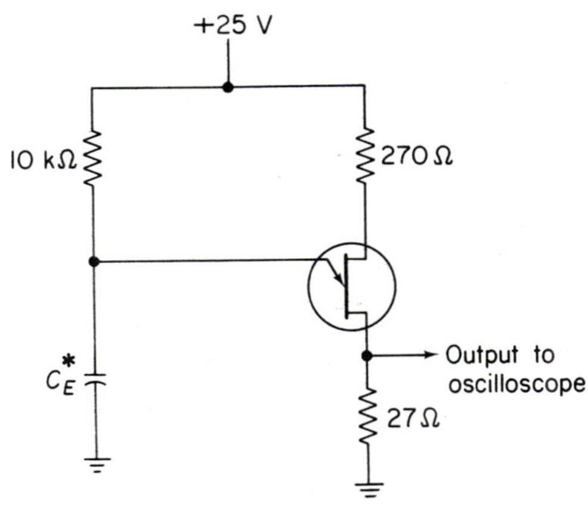

FIGURE 4-29 Turn-on and turn-off times versus emitter capacitance in relaxation oscillator test circuit.

Symbol	Test Circuit Figure	Test Conditions	MPU 131	MPU 132	MPU 133	Unit
I_P	4–31	$R_G = 1$ MΩ	1.25	0.19	0.08	μA
		$R_G = 10$ kΩ	4.0	1.20	0.70	μA
I_V	4–31	$R_G = 1$ MΩ	18.0	18.0	18.0	μA
		$R_G = 10$ kΩ	270	270	270	μA
V_{AG}		(See Fig. 4-32)				
I_{GAO}		$V_S = 40$ V	(see Fig. 4-33)			
I_{GKS}		$V_S = 40$ V	5.0	5.0	5.0	nA
V_F	Curve tracer used	$I_F = 50$ mA	0.8	0.8	0.8	V
V_O	4–34		16	16	16	V
t_r	4–35		40	40	40	ns

FIGURE 4-30 Typical PUT characteristics (courtesy Motorola).

4-5. CHARACTERISTICS OF PUT

Figure 4-30 shows some typical characteristics of selected PUTs. These characteristics are usually given on most UJT datasheets, together with the necessary test circuits and conditions.

The information in Fig. 4-30 defines the static PUT curve shown in Fig. 4-9, for a 10-V gate reference V_S with various gate resistances R_G. The information also indicates the leakage currents of the devices and describes the output pulse. Values given in Fig. 4-30 are for 25°C unless otherwise noted. The following paragraphs define these parameters and indicate how they are measured and used.

PEAK-POINT CURRENT I_P

The peak point is indicated graphically by the static curve in Fig. 4-9. Reverse anode current flows with anode voltages less than the gate voltage V_S because of leakage from the bias network to the charging network. With currents less than I_P, the device is in a blocking state. With currents above I_P, the device goes through a negative resistance region to its "on" state.

The charging current, or the current through a timing resistor, must be greater than I_P at V_P to ensure that a device will switch from a "blocking" to an "on" state in an oscillator circuit. For this reason, maximum values of I_P are given on the datasheet. These values are dependent on V_S, temperature, and R_G. Typical curves on the datasheet indicate this dependence and must be consulted for most applications.

The test circuit of Fig. 4-31 is a sawtooth oscillator that uses a 0.01-μF

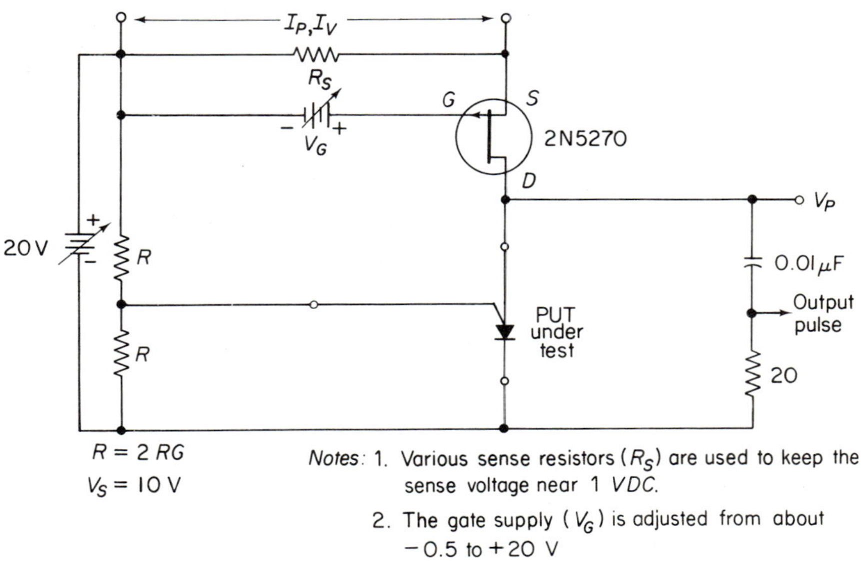

FIGURE 4-31 Test circuit for I_P, V_P, and I_V (courtesy Motorola).

timing capacitor, a 20-V supply, an adjustable charging current, and equal biasing resistors R. The two biasing resistors are chosen to give an equivalent R_G of 1 MΩ and 10 kΩ. The peak-point current is measured with the device "off" just prior to oscillation as detected by the absence of an output voltage pulse. The 2N5270 FET circuit is used as a current source. A variable gate-voltage supply is used to control this current.

VALLEY-POINT CURRENT I_V

The valley point is indicated graphically in Fig. 4-9. With currents slightly less than I_V, the device is in an unstable negative resistance state. A voltage minimum occurs at I_V, and with higher currents the device is in a stable "on" state.

When the device is used as an oscillator, the charging current or the current through a timing resistor must be less than I_V at the valley-point voltage V_V. For this reason, minimum values for I_V are given on the datasheet for $R_G = 10\ \text{k}\Omega$. With $R_G = 1\ \text{M}\Omega$, a reasonable "low" is 2 μA for all devices.

When the PUT is used as an SCR in the latching mode, the anode current must be greater than I_V. Maximum values for I_V are given with $R_G = 1\ \text{M}\Omega$. All devices have a reasonable "high" of 400-μA I_V with R_G at 10 kΩ.

LATCHING AND HOLDING CURRENT I_L AND I_H

Using the test circuit of Fig. 4-31, an attempt was made to differentiate among latching current I_L, holding current I_H, and valley current. With the PUT latched "on," reducing the current causes a voltage minimum that is the valley point. The device does remain on at lower currents until holding current I_H is reached. The holding current is measured as detected by the absence of an output voltage pulse just before oscillation occurs.

The latching current is generally higher than I_H and is measured by increasing the current while the PUT is oscillating and recording the value at which oscillations stop. With the PUT, there is no measurable difference between these parameters. This condition is not necessarily true with a UJT.

The valley current does not vary with circuit parameters and temperature as was true of I_P. Typical datasheet curves identify this dependence and are frequently used to approximate actual variations of I_V.

PEAK-POINT VOLTAGE V_P

The unique feature of the PUT is that the peak-point voltage can be determined externally. This programmable feature gives this device the ability to function in voltage-controlled oscillators or similar applications.

The triggering or peak-point voltage is approximated by

$$V_P \approx V_T + V_S \tag{4-11}$$

where V_S is the unloaded divider voltage and V_T is the offset voltage. The actual offset voltage will always be higher than the anode-gate voltage V_{AG} because I_P flows out of the gate just prior to triggering. This makes $V_T = V_{AG} + I_P R_G$. A change in R_G will affect both V_{AG} and $I_P R_G$ but in opposite ways. First, as R_G increases, I_P decreases and causes V_{AG} to decrease. Second, since I_P does not decrease so fast as R_G increases, the $I_P R_G$ product will increase and the actual V_T will increase. These second-order effects are difficult to predict and to measure. Allowing V_T to be 0.5 V as a first-order approximation gives sufficiently accurate results for most applications.

The peak-point voltage was tested using the circuit of Fig. 4-31, and a scope with 10-MΩ input impedance across the PUT. A Tektronix Type W plug-in is used to determine this parameter.

FORWARD ANODE-GATE VOLTAGE V_{AG}

The forward anode-to-gate voltage drop affects the peak-point voltage as was discussed. The drop is essentially the same as a small-signal silicon diode and is plotted in Fig. 4-32. The voltage decreases as current decreases, and the change in voltage with temperature is greater at low currents. At 10 nA, the temperature coefficient is about −2.4V/°C and it drops to about −1.6 mV/°C at 10 mA. This information is useful in applications where it is desirable to temperature compensate the effect of this diode.

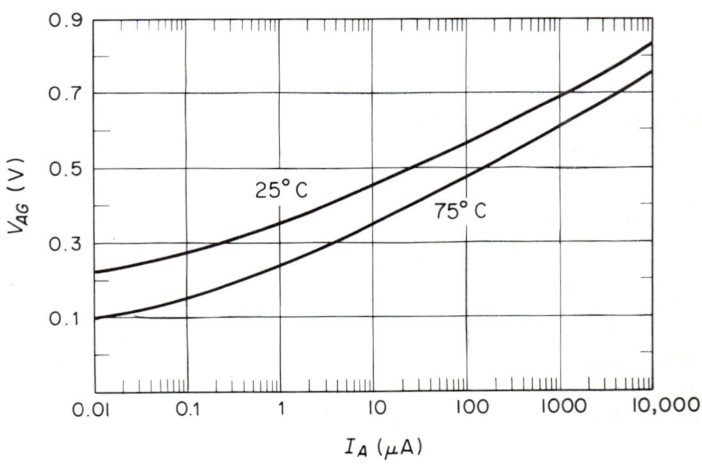

FIGURE 4-32 Gate voltage drop of MPU131, and 231 series (courtesy Motorola).

Gate-Cathode Leakage Current I_{GKS}

The gate-to-cathode leakage current is the current that flows from the gate to the cathode with the anode shorted to the cathode. I_{GKS} is the sum of the open-circuit gate-anode and gate-cathode leakage currents. The shorted leakage represents current that is shunted away from the voltage divider.

Gate-Anode Leakage Current I_{GAO}

The gate-to-anode leakage current is the current that flows from the gate to the anode with the cathode open. I_{GAO} is important in long duration timers since I_{GAO} adds to the charging current flowing into the timing capacitor. The typical leakage currents measured at 40 V are shown in Fig. 4-33. Leakage at 25°C is approximately 1 nA and the current appears to double for about every 10°C rise in temperature.

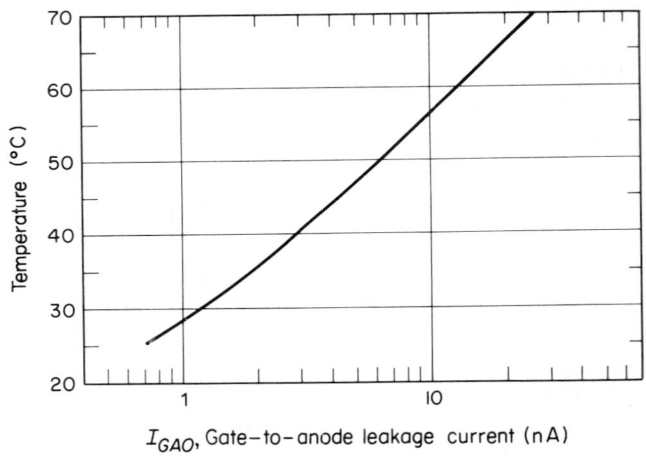

FIGURE 4-33 Typical leakage currents of the MPU131, 132, and 133 (reverse voltage equals 40V) (courtesy Motorola).

Forward Voltage V_F

The forward voltage V_F is the voltage drop between the anode and cathode when the PUT is biased on. V_F is the sum of an offset voltage and the drop across some internal dynamic impedance, which both tend to reduce the output pulse. The typical datasheet curve shows this impedance to be less than 1 Ω for up to 2 A of forward current. This is essentially an order of magnitude better than the UJT, which is closer to 20 Ω.

Peak Output Voltage V_O

The peak output voltage is not only a function of V_P, V_F, and dynamic impedance but is also affected by switching speed. This is particularly true when small capacitors (less than 0.01 μF) are used for timing since such capacitors lose part of their charge during the turn-on interval. The use of a relatively large capacitor (0.2 μF) in the test circuit of Fig. 4-34 tends to minimize this last effect. The output voltage is measured by placing a scope across the 20-Ω resistor, which is in series with the cathode lead.

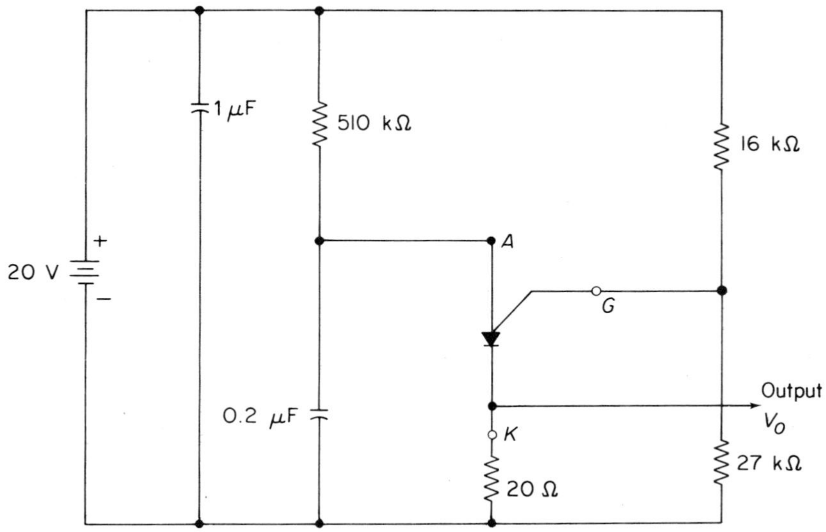

FIGURE 4-34 PUT test circuit for output voltage (V_o) (courtesy Motorola).

Rise Time t_r

Rise time is a useful parameter in pulse circuits that use capacitive coupling. It can be used to predict the amount of current that will flow between these circuits. Rise time is specified using a real-time scope and measuring between 0.6 and 6 V on the leading edge of the output pulse. Even fast scopes (100-MHz bandpass) can degrade rise-time measurement. For best results, rise-time readings should be corrected by calculations. It is preferable to use a 1000-pF capacitor and a sampling oscilloscope as shown in Fig. 4-35 to read the 10 and 90% points directly. The resulting typical rise times of 40 nA are significantly better than those of the UJT, which are about 100 ns.

Minimum and Maximum Frequency

In actual tests with PUTs whose parameters are known, it is possible to establish minimum and maximum values of timing resistors that will guarantee oscillation. The circuit under discussion is a conventional *RC* relaxation-type oscillator (such as shown in Fig. 4-27 and 4-29).

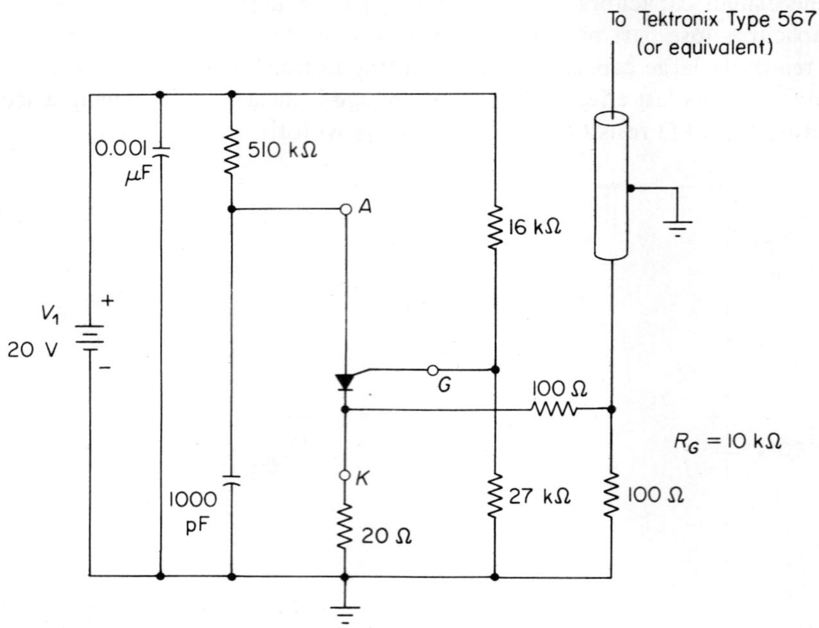

FIGURE 4-35 t_r test circuit for PUTs (courtesy Motorola).

To obtain maximum frequency, it is desirable to use low values of capacitance (1000 pF) and to select devices and bias conditions to obtain high I_V. It is possible to use stray capacitance but the results are generally unpredictable. The minimum value of timing resistance is obtained using the following rule-of-thumb:

$$R_{(\min)} = \frac{2(V_I - V_V)}{I_V} \tag{4-12}$$

where the valley voltage (V_V) is often negligible.

To obtain minimum frequency, it is desirable to use high values of capacitance (10 μF) and to select devices and bias conditions to obtain low I_P. It is important that the capacitor leakage be quite low. Glass and Mylar dielectrics are often used for these applications. The maximum timing resistor is as found by

$$R_{(\max)} = \frac{V_I - V_P}{2I_p} \tag{4-13}$$

In a circuit with a fixed value of timing capacitance, a sensitive PUT (the Motorola MPU133/233) offers the largest dynamic frequency range.

Allowing for capacitance and bias changes, the approximate frequency range of a PUT is from 0.003 Hz to 2.5 kHz.

Temperature Compensation

The PUT with its external bias network exhibits a relatively small frequency change with temperature. The uncompensated *RC* oscillator shown in Fig. 4-36 was tested at various frequencies by changing the timing resistor R_T. At discrete frequencies of 100, 200, 1000, and 2000 Hz, the ambient temperature was increased from 25 to 60°C. At these low frequencies, the negative temperature coefficient of V_{AG} predominated and caused a constant

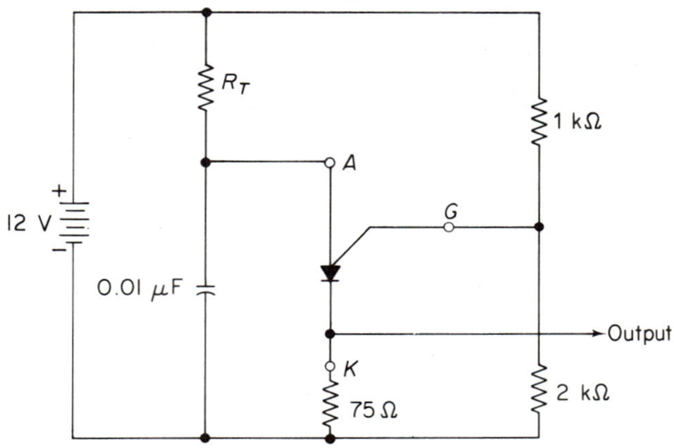

FIGURE 4-36 Uncompensated *RC* oscillator for PUT temperature compensation tests (courtesy Motorola).

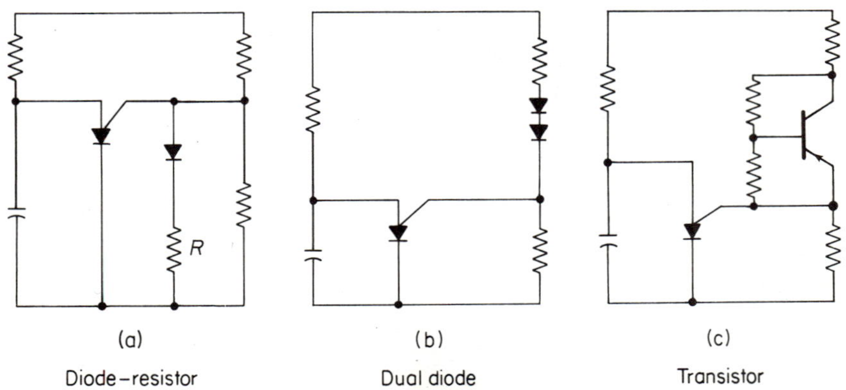

FIGURE 4-37 Temperature compensation techniques for PUTs (courtesy Motorola).

Symbol	Test Circuit Figure	Test Conditions	2N4870	2N4871	Unit
I_P	4–40(a)		1.0	1.0	μA
I_V	4–40(a)		5.0	7.0	μA
V_P	4–40(b)	$V_1 = 10$ V	6.6	8.0	V
I_{EB2O}		$V = 30$ V	See Fig. 4-39		
$V_{EB1(sat)}$	Curve tracer used	$I_E = 50$ mA	2.5	2.5	V
		$V_1 = 20$ V			
V_{OB1}	4–40b	$V_1 = 20$ V	6.0	8.0	V
t_r	4–40c		110	110	ns
R_{BB}	Curve tracer used	$V = 3$ V	6.0	6.0	kΩ

FIGURE 4-38 Typical UJT characteristics (courtesy Motorola).

2% increase in frequency. At 10 kHz, the frequency remained within 1% over the same temperature range. The storage time phenomenon that increases the length of the output pulse as temperature increases is responsible for this result. Since this parameter has not been characterized, it is obvious that temperature compensation is more practical with relatively low-frequency oscillation.

Various methods of compensation are shown in Fig. 4-37. In the low-cost diode-resistor combination of Fig. 4-37(a), the diode current is kept small to cause its temperature coefficient to increase. In Fig. 4-37(b), the bias current through the two diodes must be large enough so that their total coefficient compensates for V_{AG}. The transistor approach shown in Fig. 4-37(c) can be the most accurate since its temperature coefficient can be varied independently of bias current.

4-6. COMPARISON OF PUT AND UJT CHARACTERISTICS

Typical UJT characteristics are shown in Fig. 4-38 and 4-39. The related UJT test circuits are shown in Fig. 4-40. The UJT values can be compared against the corresponding PUT characteristics shown in Fig. 4-30 through 4-36. A study of the two sets of characteristics can be summarized as follows:

The UJT is not so sensitive as the PUT. The PUT is faster than the UJT. The UJT forward voltage drop is higher and the UJT intrinsic standoff ratio is not programmable. One distinct advantage of the UJT is that the range of frequencies of the UJT is consistently better than 100 times, whereas the PUT is not quite that good. This difference is primarily due to I_V.

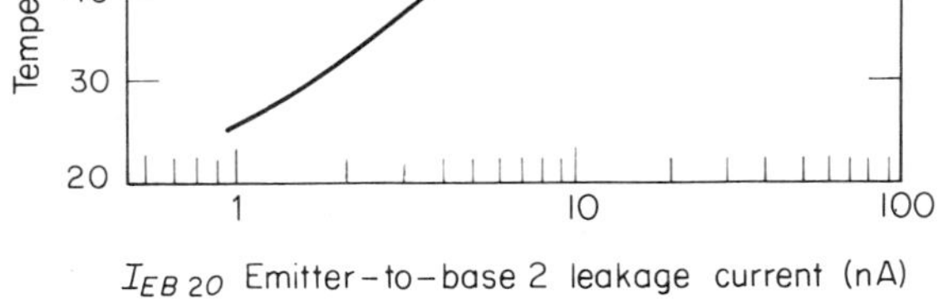

FIGURE 4-39 Typical leakage current of the 2N4870 and 71 (UJT) (reverse voltage equals 30V) (courtesy Motorola).

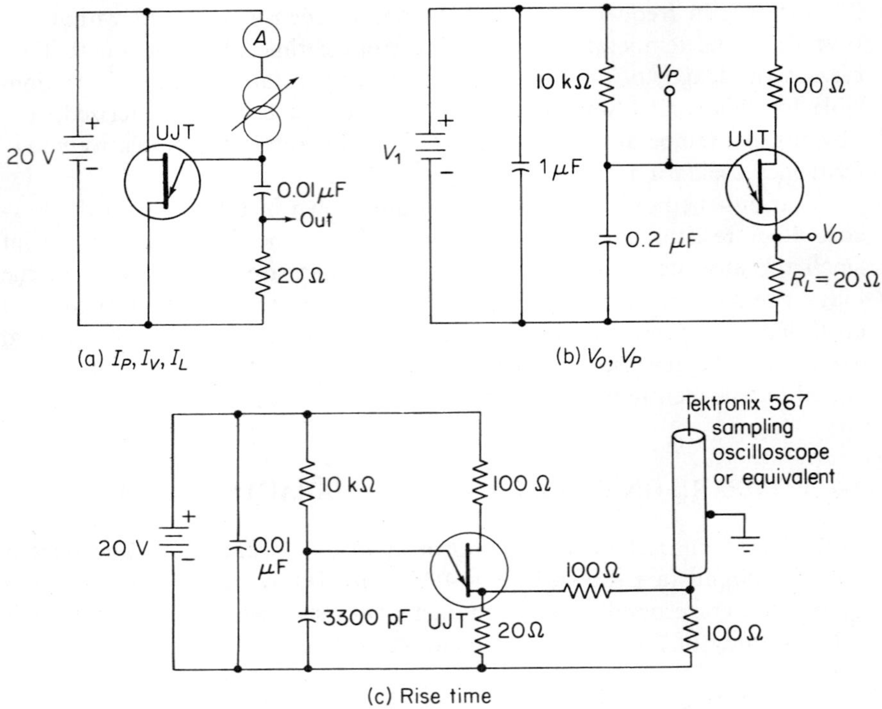

FIGURE 4-40 UJT test circuits (courtesy Motorola).

4-7. BASIC UJT RELAXATION OSCILLATOR

The relaxation oscillator is the basic building block in most UJT timer and oscillator circuits. Figure 4-41 shows the basic circuit, together with some typical waveforms. Operation of the circuit is as follows:

When power is applied, the capacitor C_E charges exponentially through the resistor R_E until voltage on the capacitor equals the emitter firing voltage V_P. At this voltage, the emitter base 1 junction becomes forward biased and the emitter goes into the negative resistance region of its characteristic curve. The capacitor C_E discharges through the emitter, and a positive-going pulse will be available at base 1. This pulse is shown in Fig. 4-41(b). The circuit values used to produce the pulses are also shown in Fig. 4-41.

Prior to firing, a current I_{B2} is flowing from base 2 to base 1. When emitter current starts to flow, this current will increase since the resistance from base 2 to ground is decreasing. A negative-going voltage pulse will appear at base 2. This waveform is shown in Fig. 4-41(c).

When voltage at the emitter has decreased to peak emitter voltage V_P (a voltage approximately equal to the valley voltage V_V when R_1 is purely

$$R_E < \frac{V_1 - V_p}{I_p} = R_E \text{ (max)}$$

$$R_E > \frac{V_1 - V_v}{I_v} = R_E \text{ (min)}$$

$$C_E \approx \frac{\text{period}}{R_E \times \ln \dfrac{1}{1 - \text{standoff ratio}}}$$

R_1 = see text

R_2 = see text

$$\text{Frequency} = \frac{1}{\text{period}}$$

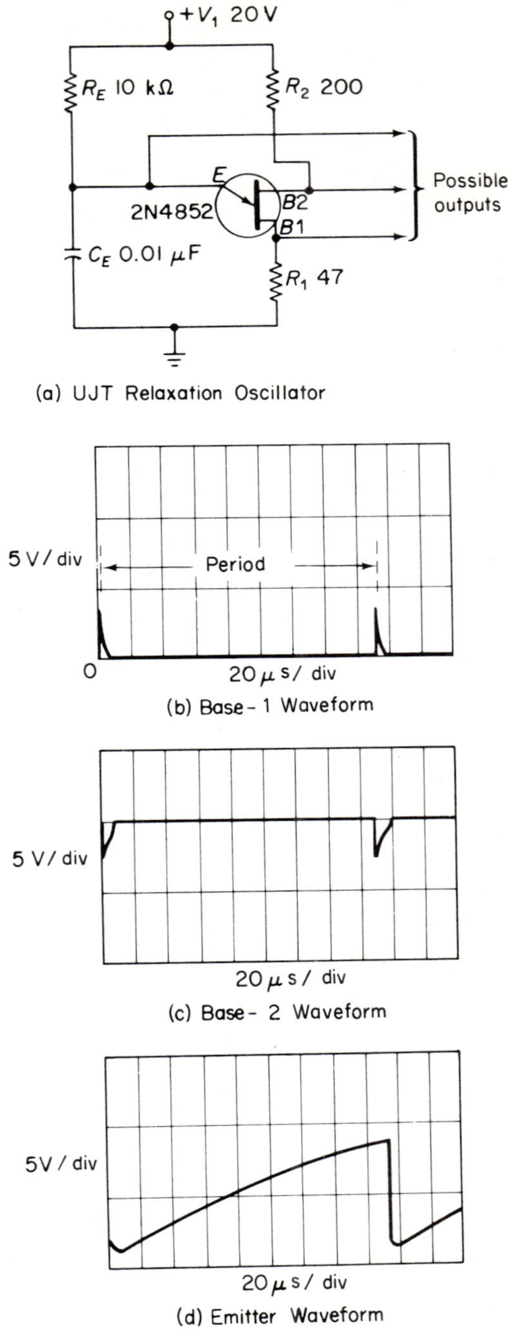

FIGURE 4-41 Basic UJT relaxation oscillator (courtesy Motorola).

resistive), the UJT will turn off if R_E meets certain conditions. C_E will start to charge up again, and the cycle repeats. The waveform that appears at the emitter is shown in Fig. 4-41(d).

4-7.1 Design considerations

The period of oscillation (and thus the frequency) is determined primarily by the values of R_E and C_E, as shown by the equations of Fig. 4-41. It is most practical to start design by selecting a trial value for R_E rather than for C_E because R_E must meet certain conditions for the oscillator to operate. If R_E is too large, the UJT will never fire; if R_E is too small, the UJT will not turn off.

These conditions can best be explained by means of the emitter-characteristic curve in Fig. 4-42. (This curve is not drawn to scale in order to show more detail.)

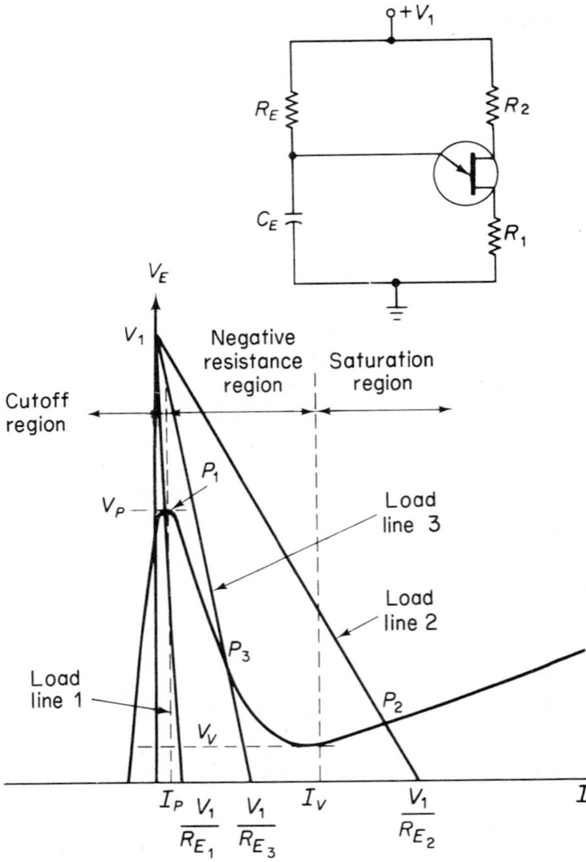

FIGURE 4-42 UJT emitter characteristic load lines (courtesy Motorola).

Selecting Emitter Resistor R_E

The emitter capacitor C_E will charge until the emitter voltage is equal to V_P. At this point on the characteristic curve, peak-point emitter current I_P will be flowing and in order to fire the UJT the value of R_E must be small enough to allow a current somewhat larger than I_P to flow. R_E must, therefore, meet the following requirement:

$$R_E < \frac{V_1 - V_P}{I_P} = R_{E(\max)} \tag{4-14}$$

where V_1 is the applied bias voltage.

Referring to Fig. 4-42, a load line intersecting the characteristic curve in the cutoff region, as illustrated by load line 1, would keep the UJT from ever firing. Therefore, R_E must be small enough to keep the UJT in the negative resistance region.

By keeping R_E smaller than $R_{E(\max)}$, the UJT will turn on and C_E will discharge through the emitter. If R_E is too small, however, an emitter current larger than the valley current I_V flows but the UJT will not turn off. Under these conditions, the UJT will operate in the saturation region, as illustrated by load line 2 in Fig. 4-42. The minimum R_E that can be used in order to assure oscillation is set by the following condition:

$$R_E > \frac{V_1 - V_V}{I_V} = R_{E(\min)} \tag{4-15}$$

where V_1 is the applied bias voltage.

An emitter resistance R_E selected to meet the requirements in Eq. 4-14 and 4-15 will result in a load line that intersects the characteristic curve somewhere in the negative resistance region. This is illustrated by load line 3 in Fig. 4-42. In a practical UJT circuit, however, the emitter-voltage variation in the neighborhood of the valley point is small. Thus, in order to assure turn-off, the value of R_E should be *two or three times larger than* $R_{E(\min)}$. This is best understood by a study of the following paragraphs, which summarize the dynamic operating conditions of a typical UJT oscillator.

Dynamic Operating Paths of UJT Oscillator

Figure 4-43 shows the basic UJT oscillator circuit with typical component values and the corresponding emitter voltage–current operating curves, both static and dynamic.

As shown on the curves, C_E will start to charge from point A. At point B, where the voltage of C_E equals V_P, the UJT will fire, and the characteristic curve goes into the negative resistance region. The voltage on C_E cannot change instantaneously, however, so the dynamic operating path will move

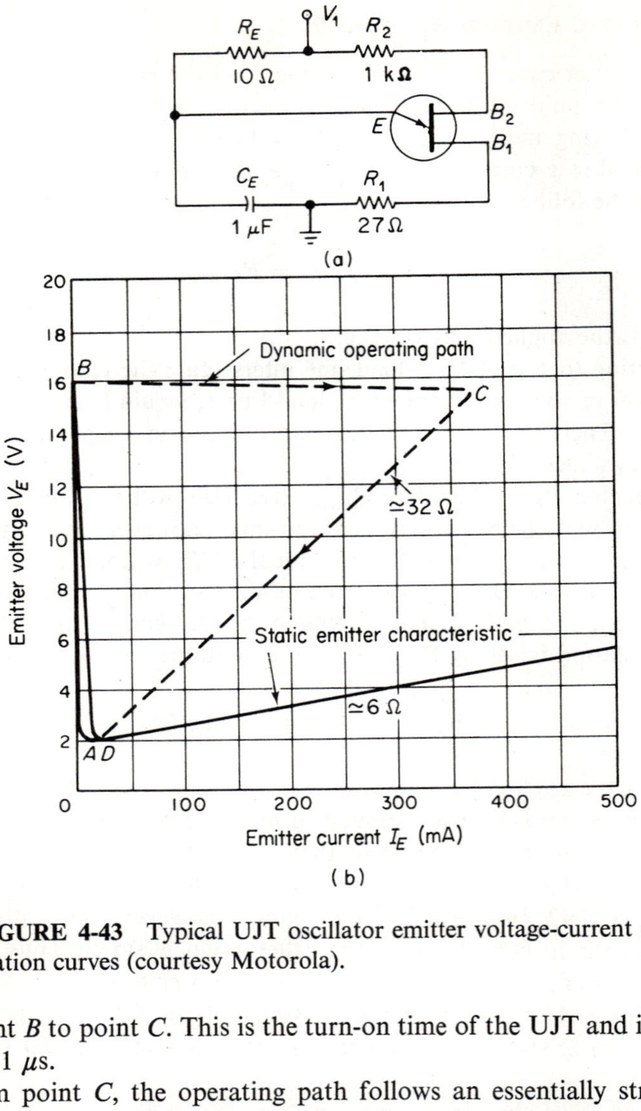

FIGURE 4-43 Typical UJT oscillator emitter voltage-current operation curves (courtesy Motorola).

from point *B* to point *C*. This is the turn-on time of the UJT and is typically less than 1 μs.

From point *C*, the operating path follows an essentially straight line to point *D*, which is about equal to the valley point. The straight line has a slope of approximately 32 V/A, or 32 **Ω**, and is composed of R_1 and the UJT emitter to base 1 saturation resistance. There is not enough emitter current available to sustain operation at point *D*, and the operating path tries to follow the characteristic curve to the point where it is intersected by the load line determined by R_E (similar to the load line 3 in Fig. 4-42).

If the emitter circuit had no capacitor C_E (an impractical circuit), this intersection point would be a stable operating point. To reach this point, however, the emitter voltage must increase. The resistance from emitter to base 1 will also increase and the emitter current will decrease somewhat. When the emitter voltage increases, however, current starts to flow into the capacitor and the emitter current is reduced more than that required by the characteristic curve.

In a practical circuit (with C_E) there are no stable operating points in the negative resistance region. Thus, from point D, the operating path goes to point A, and the cycle repeats. The total period of oscillation (and thus the frequency) is set by turn-on and turn-off characteristics of the UJT, plus the charge time of C_E through R_E (from point A to B).

The shape of the dynamic operating path is determined by the capacitor C_E, the interbase voltage, and the value of resistor R. Figure 4-44 shows operating paths for different values of C_E. Figure 4-45 shows operating paths for fixed C_E but varying interbase voltage. (Note that interbase voltage is shown as V_{B2B1}; however, V_{BB} is used for interbase voltage on some UJT datasheets.) No operating-path curves are given for various R_1 values. This

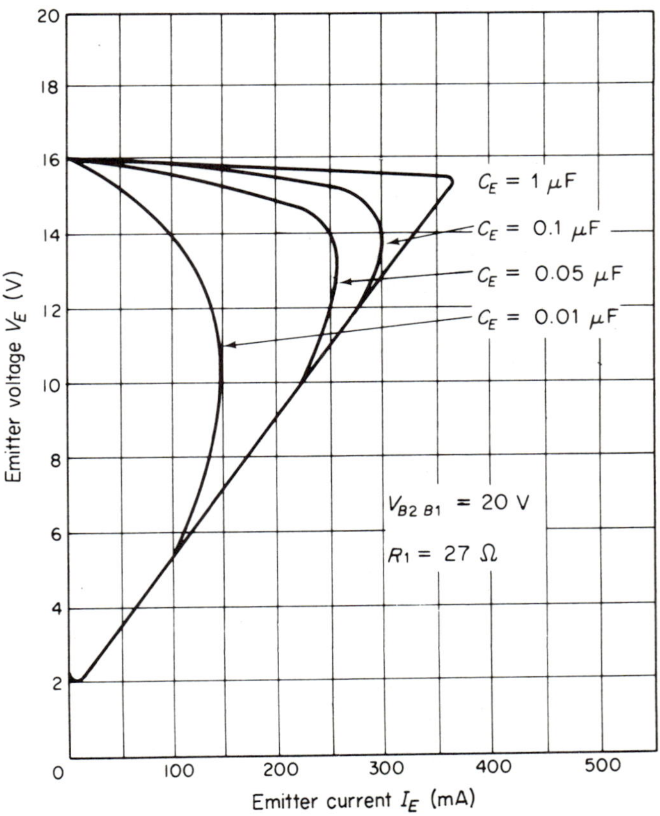

FIGURE 4-44 Dynamic operating path versus emitter capacitance (courtesy Motorola).

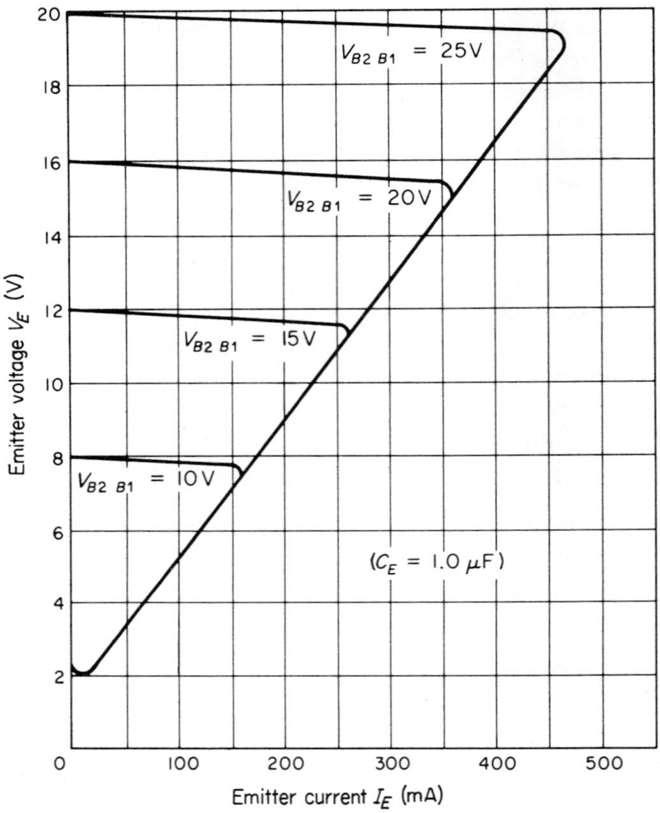

FIGURE 4-45 Dynamic operating path versus interbase voltage (courtesy Motorola).

is because the value of R_1 is generally selected on the basis of required minimum output pulse voltage or on minimum fixed-level voltage. The factors affecting the value of R_1 are discussed in later paragraphs of this section.

The curves of Fig. 4-44 and 4-45 bring up some important tradeoffs. With all other factors equal, the maximum available output voltage is set by the maximum emitter current I_E. As shown in Fig. 4-44, I_E can be increased when C_E is increased. Likewise, as shown in Fig. 4-45, I_E can be increased when interbase voltage (V_{B2B1}) is increased. If either C_E or interbase voltage is increased, the period of oscillation is increased and the frequency is decreased. *Thus, a higher output voltage from the UJT oscillator must be traded for lower operating frequency.*

Keep in mind that the V_{B2B1} voltage shown in Fig. 4-44 and 4-45 is not necessarily the V_1 (source or supply) voltage. Under static conditions, there can be some current flowing through base 1 and base 2. This relationship

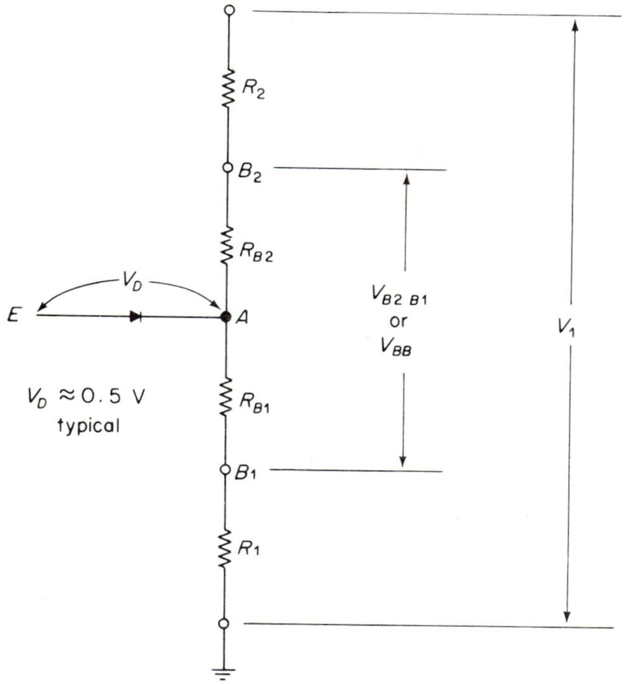

FIGURE 4-46 Electrical equivalent circuit for the UJT with the external compensating resistor R_2 and output resistor R_1. The equivalent circuit is valid in the cutoff region only (courtesy Motorola).

is shown in Fig. 4-46. The amount of current is set by the interbase resistance (r_{BB}), by the value of R_2, and (to a lesser extent) by R_1 (since R_1 is generally quite small in relation to R_2 and r_{BB}). For example, assume that r_{BB} is 7 kΩ, V_1 is 20 V, and R_2 is 1 kΩ; approximately 2.5 mA will flow (7 kΩ + 1 kΩ = 8 kΩ, 20 V/8 kΩ = 2.5 mA). This will produce an approximate drop of 2.5 V across R_2, making the interbase voltage about 17.5 V. (Note that interbase resistance is shown as r_{BB} on some datasheets and R_{BB} on others.)

When an inductive load, like a relay coil, for example, is substituted for R_1, the dynamic operating path will be somewhat different. Figure 4-47(a) shows a relaxation oscillator having a pulse transformer instead of R_1, and Fig. 4-47(b) shows the resulting dynamic operating path. An important difference here is that the emitter no longer ceases to conduct when valley voltage (typically about 2 V) is reached but continues down to less than 0.5 V before turning off. Figure 4-47(c) shows how the turn-off voltage is dependent on the bias voltage.

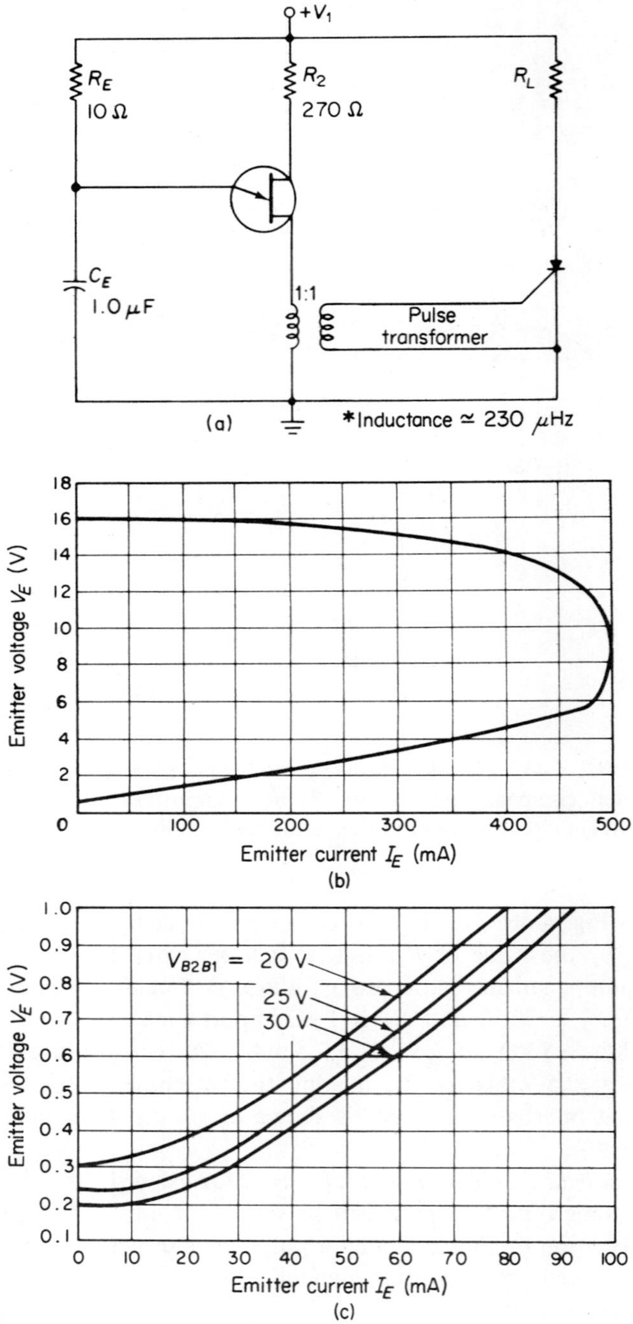

FIGURE 4-47 UJT relaxation oscillator with inductive load (courtesy Motorola).

SELECTING BASE 2 RESISTOR R_2

The primary function of R_2 (in Fig. 4-41 or 4-43) is to provide temperature compensation. Practically all UJT characteristics are temperature dependent, some more than others. The interbase resistance and emitter reverse current increase; whereas the peak and valley voltage (and current), the intrinsic standoff ratio, and the junction diode drop decrease with increasing temperature. Generally, oscillator frequency is the main factor affected by these temperature variations. Output voltage can also be affected, however, and is of some concern in many applications.

The peak-point voltage is given by this equation:

$$V_P = V_D + \eta V_{B2B1} \tag{4-16}$$

Since both V_D and η decrease with temperature, V_P will also decrease. This is, of course, a very undesirable condition in many applications, particularly in timers and oscillator circuits. The change in V_P can be compensated for by adding a resistor R_2 in series with the base 2 terminal.

If R_2 is selected properly, V_P can be made to vary less than 1% over a 50°C temperature variation. The value of R_2 can be selected by using either the equation or the curves of Fig. 4-48.

The curves of Fig. 4-48 show frequency variation as a function of temperature for a typical UJT. Temperature curves for several values of R_2 ranging from 250 Ω to 3 kΩ are shown, and an R_2 of approximately 1.5 kΩ can be seen to compensate very well from −5 to +85°C. A smaller resistance should be used for operation below −5°C.

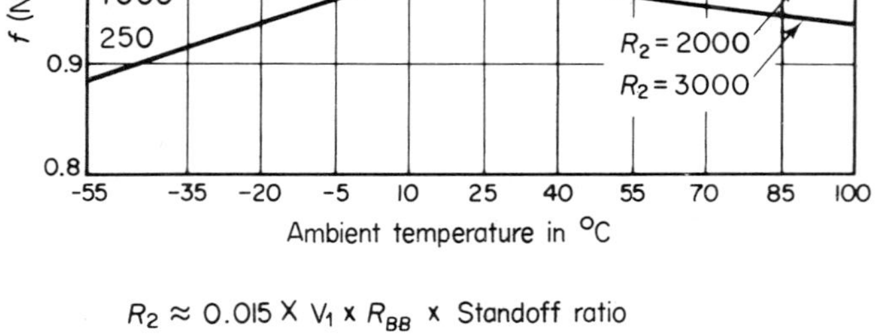

FIGURE 4-48 Frequency versus temperature for a UJT relaxation oscillator. Frequency is normalized to 25°C, and R_2 is a variable parameter (courtesy Motorola).

The equation of Fig. 4-48 can be used for a somewhat more accurate "optimum" value of R_2. The equation takes into account source voltage, interbase resistance, and standoff ratio. Keep in mind that both the curves and equation of Fig. 4-48 are for trial values. Also keep in mind that an increase in R_2 value will decrease interbase voltage. As discussed, this increases frequency but decreases output voltage (all other factors being equal).

SELECTING BASE 1 RESISTOR R_1

The primary function of R_1 (Fig. 4-41 or 4-43) is to provide an output load for the oscillator. In some applications, R_1 is included to provide a path for the interbase current. Such a path is necessary in some circuits to prevent current from flowing through the device being driven by the UJT oscillator.

Typically, R_1 is less than 100 Ω but could be as high as 2 or 3 kΩ in some applications.

When R_1 is selected on the basis of a current path for an external device, a maximum voltage drop is usually specified. For example, if the UJT oscillator is to trigger an SCR (from an output pulse across R_1) and the fixed SCR voltage must not exceed 50 mV (between firing cycles), the drop across R_1 must not exceed 50 mV. Assuming a 2.5-mA interbase current (from the previous example) and the 50-mV maximum voltage drop, the value of R_1 is 20 Ω.

When R_1 is selected on the basis of output voltage, which is usually the case, a minimum output voltage is usually specified. For example, if the UJT oscillator is to trigger an SCR or other thyristor device and the SCR requires a minimum of 3 V for the trigger, the peak drop across R_1 must be 3 V.

If curves such as shown in Fig. 4-43 through 4-45 are available, it is possible to calculate an approximate value of R_1, using the peak emitter current I_E and the desired output voltage. For example, if the output is 3 V and the peak I_E is 300 mA, the value of R_1 is 10 Ω; however, the shape of the dynamic operating curve will change with a change in R_1 (and with changes in C_E).

For this reason, the curves of Fig. 4-49 are provided. In Fig. 4-49, the peak voltage across R_1 is shown as a function of C_E for various values of R_1. These are minimum peak output values. In practice, the peak output across R_1 will usually be 25 to 50% higher.

The values shown in Fig. 4-49 are for a supply voltage of 20 V. The peak voltage amplitude at other values of supply voltage V_1 can be obtained by multiplying the values from Fig. 4-49 by the factor shown. For example, if the supply voltage is 15 V, R_1 is 5 Ω, and C_E is 1 μF, the peak-voltage

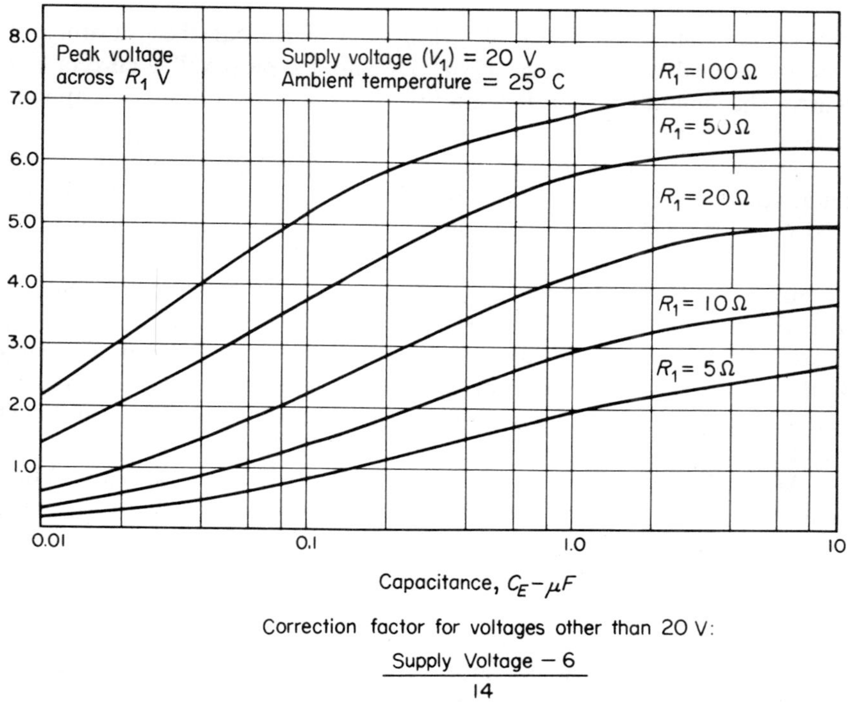

FIGURE 4-49 Peak output voltage across R_1 in UJT trigger circuit—minimum values (courtesy General Electric).

output is about 1.28 V. This is found as follows: The curve of Fig. 4-49 shows a minimum output of 2 V for an R_1 of 5 Ω and a C_E of 1 μF. Using the correction factor for 15 V of approximately 0.64 ($15 - \frac{6}{14}$), the approximate minimum output voltage is $0.64 \times 2\text{ V} = 1.28\text{ V}$.

SELECTING EMITTER CAPACITOR C_E

The value of capacitor C_E is determined by the desired period (or frequency) of operation. The equation for C_E shown in Fig. 4-41 is *very approximate* since it does not take into account the on and off times of the UJT or other characteristics of the circuit such as source voltage (V_1), emitter-to-base 1 voltage drop (V_D), valley voltage (V_V), standoff ratio, or interbase voltage (V_{B2B1} or V_{BB}). However, the period equation of Fig. 4-41 for C_E is accurate enough for a first-trial value.

4-7.2 Design example

Assume that the circuit of Fig. 4-41 is to provide an output of 1.5 V (minimum) across R_1 at a frequency of approximately 5.6 kHz. The available source voltage V_1 is 20 V. A test of the UJT (or datasheet information) shows a typical valley voltage V_V of 2 V and a typical valley current I_V of 6 mA. The standoff ratio is 0.86 and the interbase resistance r_{BB} is approximately 8 kΩ.

Design starts with R_E. Equation 4-15 shows that the absolute minimum value, or $R_{E(\min)}$, is 3 kΩ ($20 - 2 = 18$; $18/0.006 = 3$ kΩ). The first trial value of R_E should be three times $R_{E(\min)}$ or 9 kΩ. Use about 10 kΩ in this case.

In a practical experimental circuit, all the components should be assembled, using a 20- or 25-kΩ potentiometer for R_E. The potentiometer is set to 10 kΩ and C_E is adjusted (or selected) for the desired frequency. With the oscillator operating at the correct frequency (or producing the correct period), R_E is adjusted to twice the selected value (or about 20 kΩ) and to one-half the selected value (or about 5 kΩ).

When R_E is adjusted between the two extremes, the oscillator should continue to oscillate, even though the frequency will shift. If so, the value of R_E (10 kΩ) is such that the UJT is well within the negative resistance region, and the oscillator should continue to operate in spite of changes in supply voltage, temperature, etc.

On the other hand, if the oscillator stops with small changes in R_E value, the UJT is too near the saturation region. Use an R_E value of at least one order higher. Keep in mind that an increase in R_E value will require a corresponding decrease in C_E to produce the same frequency. If R_E is increased without a change in C_E, the period will increase and the frequency will decrease.

With the value of R_E established at 10 kΩ, the value of C_E is selected to provide the correct frequency. The desired frequency is approximately 5.6 kHz, which requires a period of about 175 μs. With a period of 175 μs and R_E of 10 kΩ, and a standoff ratio of 0.87, the equation of Fig. 4-41 shows that C_E should be approximately 0.01 μF.

With a value of 0.01 μF for C_E, and a required minimum of 1.5-V peak output across R_1, Fig. 4-49 shows a value of 50 Ω for R_1.

The value of R_2 is found by using the equation of Fig. 4-48. With a source voltage of 20 V, a standoff ratio of 0.87, and an interbase resistance of about 8 kΩ, the value of R_2 is $0.015 \times 20 \times 0.87 \times 8000 = 2088$ Ω (use 2 kΩ for a practical circuit).

In a practical experimental circuit, the UJT is subjected to variations in temperature and any frequency shift is noted. (A simple test is to hold a soldering tool near, but not touching, the oscillating UJT.) There will be some frequency variation with changes in temperature no matter what value of R_2 is selected; however, the optimum value for R_2 is that which produces the least amount of frequency change with temperature variation.

In practical terms, the circuit of Fig. 4-41 could operate with R_2 values of about 200 to 3000 Ω. If the circuit is to operate over a specific temperature range, test the experimental circuit within that range to find the best value for R_2. Keep in mind that a change in R_2 value will also change the interbase voltage V_{B2B1}, which, in turn, will affect the frequency and output voltage.

4-7.3 Alternate design using UJT datasheets

Section 4-7.2 describes the procedure for selecting UJT oscillator component trial values "from scratch." It is usually not necessary to design a UJT oscillator "from scratch" if the UJT datasheet is available. Typically, the UJT datasheet will show the UJT connected in a working circuit (or "test" circuit) as a relaxation oscillator. All or part of the values are given, together with the rated output voltage. These values can be used as a starting point for design.

For example, if a different frequency is required, use all the component values shown on the datasheet, except for the value of C_E. Change C_E as necessary to produce the desired oscillation frequency.

If a different output voltage is required, change the value of R_1 from that shown on the datasheet. (Use the curves of Fig. 4-49 to find a new trial value for R_1.)

The UJT datasheet will also include absolute maximum ratings (such as power dissipation and temperature range) and typical electrical characteristics (such as standoff ratio and interbase resistance). All of these characteristics should be studied carefully before starting design, as is the case with two-junction transistors and FETs. For example, if the interbase voltage should not exceed 30 V, the source should not exceed 30 V since the full source voltage could be placed across the bases (if R_1 and R_2 are eliminated or are very low value).

When the UJT oscillator is to be used as a trigger source for an SCR or other thyristor, the UJT datasheet will often include a special set of curves. The curves show values of R_1 to operate with specific SCRs (by number or type). This type of data makes design of UJT trigger oscillators quite simple; however, the curves generally show only R_1 values for SCRs of one manufacturer (the manufacturer publishing the datasheet). For that reason, use of curves may be somewhat limited.

One datasheet characteristic often overlooked in design of UJT oscillators is *peak emitter current* or *maximum emitter current* (or whatever similar term is used). When the output voltage across R_1 is connected to a load, most of the load current is drawn through the emitter during the pulse period. To be on the safe side, make sure that the maximum possible load current is at least 10% below the peak emitter-current rating of the UJT.

4-7.4 Synchronization of UJT oscillators

In many applications, it is necessary to synchronize the UJT oscillator with some external signal (such as a clock or master oscillator in a computer or other digital system). For best results, the UJT oscillator should be designed to operate at a frequency slightly below that of the synchronizing signal. This will ensure that the UJT will not trigger between synchronizing pulses.

A UJT relaxation oscillator can be synchronized by means of either positive pulses at the emitter or negative pulses at base 2. This will raise the emitter voltage or lower the peak voltage, respectively. Either way, the UJT oscillator will be locked in with the synchronizing signal. In a typical UJT oscillator, a signal of about 1 V is required at the emitter or about 1.5 V at base 2 to produce synchronization. These voltages are assumed to have an interbase voltage of 25 V (or less) and a synchronizing signal pulse width of 1 μs or wider. A narrower synchronizing pulse width may require a higher voltage value.

The effect of pulse width of the required trigger amplitude is shown in Fig. 4-50. For pulse widths greater than 1 μs, the required trigger amplitude approaches the dc conditions. For pulse widths of less than 1 μs, the required pulse amplitude is inversely proportional to the pulse width.

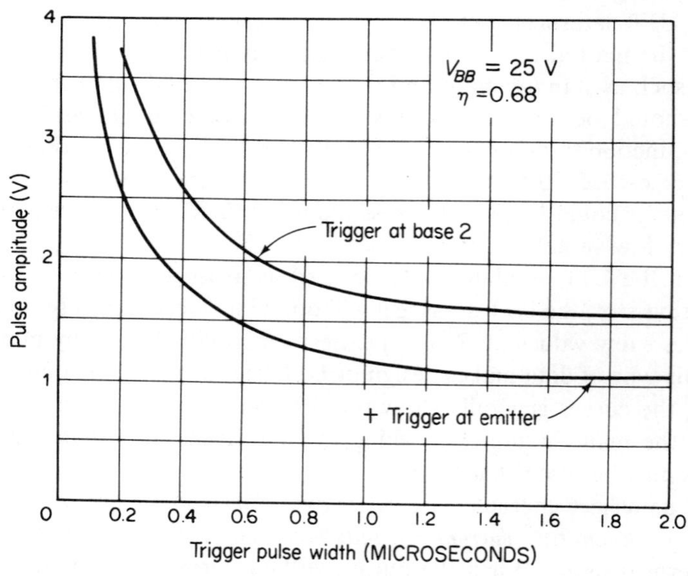

FIGURE 4-50 Minimum trigger amplitude as a function of trigger pulse width for turn on of UJT relaxation oscillator (courtesy General Electric).

Synchronization of UJT oscillators is discussed more fully in Sec. 4-8 of this chapter.

4-7.5 Protecting UJT oscillator circuits

If a base 2 resistor R_2 of the proper value is included in a UJT oscillator circuit and the maximum datasheet rating of peak emitter current and maximum interbase voltage are observed, there should be no damage to the UJT. In general, most UJT circuits assume a peak voltage of 30 V, a C_E of 10 μF or less, and a peak emitter current of 2 A. If any of these factors are exceeded, it is recommended that a resistor be connected in series with C_E. This will limit emitter current. As a trial value, the series resistance should be at least 1 Ω/μF of C_E.

4-7.6 Alternate output connections for basic UJT oscillator

As shown in Fig. 4-41, output pulses can be taken from base 2 as well as from base 1. The base 2 output will be slightly less than base 1 (in amplitude) and will be negative. The design considerations for a base 2 output are the same as for base 1 output.

It is also possible to take an output from the emitter shown in Fig. 4-41; however, the emitter output is essentially a sawtooth waveform.

4-8. UNIJUNCTION TRIGGER CIRCUITS FOR GATED THYRISTORS

UJTs provide a simple, convenient means for obtaining a thyristor trigger pulse that is synchronized to the ac line at a controlled phase angle. This phase control system, using UJTs and thyristors, is one of the most common means of controlling the flow of power to electric motors, lamps, and heaters.

With ac voltage applied to such circuits, the thyristor (SCR, Triac, etc.) remains in its off state for the first portion of each half cycle of the power-line voltage. Then, at a time (phase angle) determined by the control circuit, the thyristor switches on for the remainder of the half cycle. By controlling the phase angle at which the thyristor is switched on, the *relative power* in the load may be controlled.

UJT trigger circuits are based on the simple relaxation oscillator circuit of Fig. 4-51. Note that this circuit is essentially the same as the basic UJT oscillator described in Sec. 4-7, with two exceptions. First, the source voltage V_S is "controllable" in that it can be turned on and off by switch S_1. Second, the base 1 output is applied to the gate of a thyristor, which, in turn, is in series with a load.

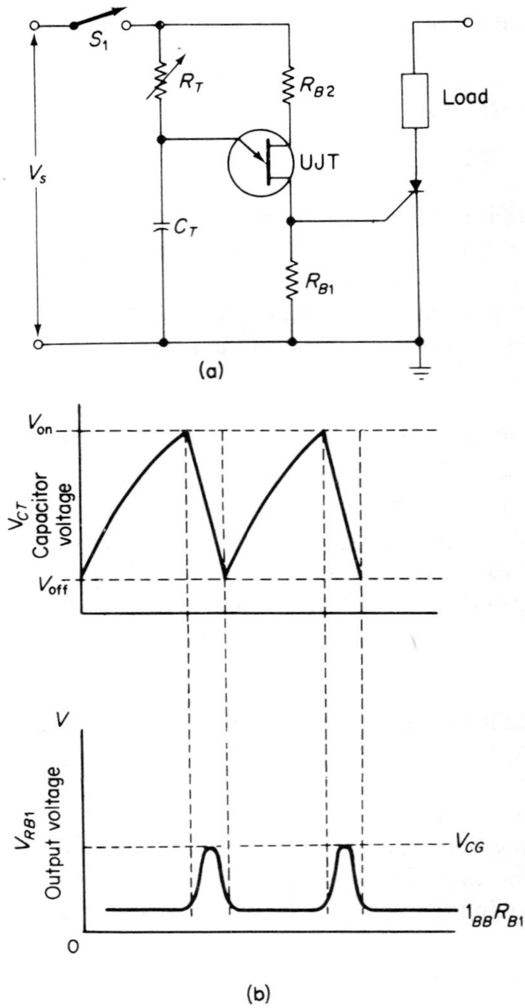

FIGURE 4-51 Basic UJT trigger circuit (courtesy Motorola).

The R_T and C_T in Fig. 4-51 form the timing network that determines the time between the application of voltage to the control circuit (represented by the closing of S_1) and initiation of the pulse to the thyristor.

In the circuit of Fig. 4-51, with V_S pure dc, the UJT oscillator is free-running. R_T and C_T determine the frequency of oscillation. The peak of the output pulse voltage is clipped by the forward conduction voltage of the gate-to-cathode diode in the thyristor.

Operation of the circuit may best be understood by referring to the capacitor voltage waveforms shown in Fig. 4-51(b), along with the base 1 output pulse waveform. When power is applied, C_T charges, at the rate determined by its own capacitance and the value of R_T, until capacitor voltage reaches the peak-point voltage of the UJT. At that time, the UJT switches into conduction in the normal manner, discharging C_T through R_{B1} and the gate of the thyristor.

With V_S pure dc, the cycle repeats immediately. In many cases, however, V_S is derived from the anode voltage of the thyristor so that the timing cycle cannot start again until the thyristor is blocking forward voltage and once again provides V_S.

4-8.1 Design considerations

During the time in which the capacitor is being charged, current flows through the interbase resistance r_{BB} of the UJT. R_{B1} is included in the circuit to provide a path for this current so that the current does not flow through the thyristor gate. This can cause an undesirable turn-on of the thyristor.

The value of R_{B1} is selected so that a maximum voltage developed across R_{B1} is less than 0.2 V. For a typical UJT with r_{BB} of 4 to 9 kΩ and a typical operating voltage of 20 V, the value of R_{B1} is

$$R_{B1} = \frac{0.2 r_{BB(\text{min})}}{V_S} = \frac{0.2 \times 4 \text{ k}\Omega}{20} = 40\ \Omega$$

As discussed in Sec. 4-7, some UJT datasheets specify a recommended value of base 1 resistance for a specific type of thyristor. This recommended value should *always be used as the first trial value.*

Base 2 resistance R_{B2} is necessary only if some degree of temperature compensation is required. The value of R_{B2} is found by using the equation or curves discussed in Sec. 4-7. In the case of a simple power control circuit, particularly in the feedback systems described later in this section, the base 2 resistance may be left out. In such circuits, the base 2 of the UJT is connected directly to the positive side of V_S.

It is often necessary to synchronize the timing of the output pulses to power-line voltage zero-crossing points. One simple method of accomplishing synchronization is shown in Fig. 4-52. Zener diode D_1 clips the rectified supply voltage, resulting in a V_S as shown in Fig. 4-52(b). Since V_{BB} and, thus, the peak-point voltage of the UJT drops to zero each time the line voltage crosses zero, C_T is discharged at the end of every half cycle and begins each half cycle in the discharged state.

With the circuit of Fig. 4-52, even if the UJT has not triggered during one half cycle, the capacitor begins the next half cycle discharged so that the phase angle at which the pulse occurs is directly controlled for each cycle (by the values of C_T and R_T). The Zener diode also provides voltage stabiliza-

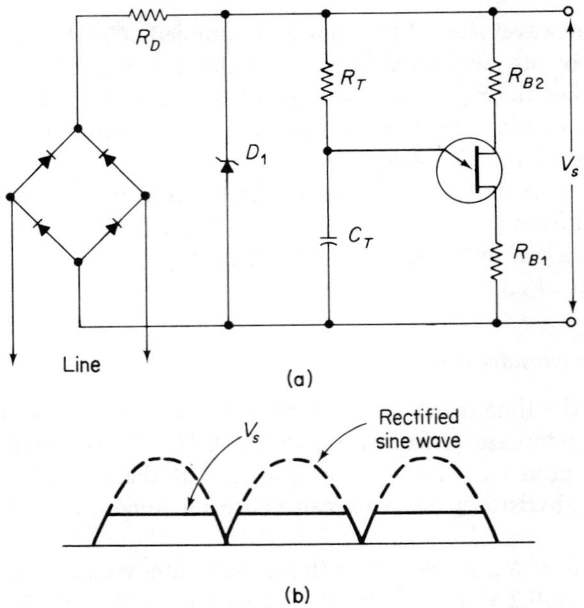

FIGURE 4-52 Basic control circuit for synchronizing a UJT trigger (courtesy Motorola).

tion for the timing circuit, giving the same pulse phase angle regardless of normal line voltage fluctuations.

4-8.2 Design example of basic half-wave trigger circuit

The most elementary application of the UJT trigger is a half-wave circuit shown in Fig. 4-53. The values of R_T and C_T set the frequency of operation. Note that R_T is made variable so that the phase angle (and thus the power applied to the load through the thyristor) can be adjusted. The values of R_T and C_T are suitable for typical 60-Hz line power.

The value of R_{B1} is obtained from the datasheet (for the type of thyristor used), as is the value of R_{B2} (as an optimum value for the anticipated temperature range).

The value of D_1 is determined by the maximum allowable interbase voltage for the UJT. The Zener voltage of D_1 should be slightly less than the maximum allowable UJT interbase voltage. The actual interbase voltage will then be well within the allowable range (due to the additional drop across R_{B1} and R_{B2}).

The value of R_D is selected to limit the current through D_1 so that the diode dissipation capability is not exceeded. Since D_1 is conducting in the regulating mode only during positive half cycles, the allowable positive

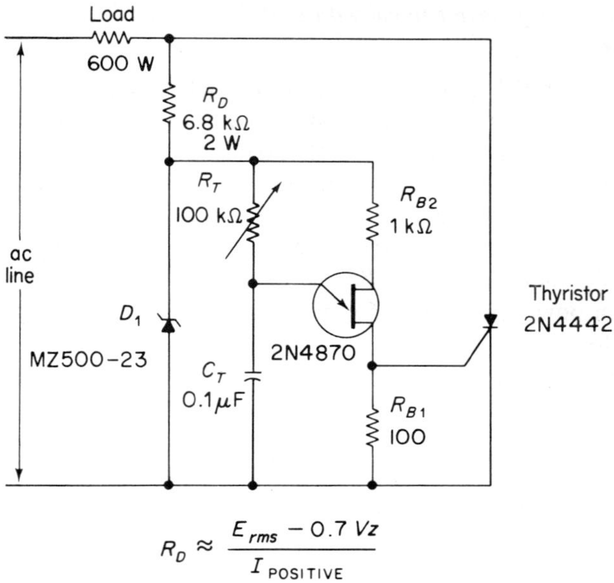

FIGURE 4-53 Half-wave UJT control circuit with typical values for a 600-W resistive load and 60-Hz line (courtesy Motorola).

current can be calculated when the allowable diode dissipation is divided by one-half the Zener voltage.

For example, if D_1 has a 150-mW rating and the Zener voltage is 24 V, the allowable positive current is 12.5 mA ($24 \times 0.5 = 12$; $0.150/12 = 0.0125$). If the Zener datasheet shows a specific current, usually listed as I_{ZT} or some similar term, use that figure for I_{positive}.

Once the positive half-cycle current is determined, the value of R_D is found by subtracting 0.7 times the Zener voltage from the rms line voltage and dividing the result by the positive current as shown by the equation in Fig. 4-53.

For example, assume that the Zener voltage is 24 V, line voltage is 115 V, and I_{positive} is 15 mA. R_D is then $115 - (0.7 \times 24)/0.015 = 6.8$ kΩ (nearest standard value).

Although only the positive half cycles need be considered in calculating the resistance value, the power rating of R_D must be calculated on the basis of full-wave conduction. (D_1 is conducting on the negative half cycle acting as a shunt rectifier, as well as providing V_S on the positive half cycle.)

Using the example of 6.8 kΩ and 15 mA, the minimum power rating of R_D is (I^2R) $15^{-3} \times 15^{-3} = 225 \times 10^{-6}$; $6800 \times 225 \times 10^{-6} = 1.53$ W; use a 2-W value for R_D to provide safe power dissipation.

4-8.3 Providing half-wave control with switching for full-wave operation

In the circuit of Fig. 4-53, the thyristor is acting both as a power control device and a rectifier, providing variable power to the load during the positive half-cycle and no power to the load during the negative half-cycle. The circuit of Fig. 4-53 is designed to be a two-terminal control that can be inserted in place of a switch.

If full-wave power is desired at the upper extremes of this control, a switch can be added that will short-circuit the SCR when R_T is turned to its maximum power position. The switch may be placed in parallel with the SCR if the load is resistive. If the load is inductive, the load must be transferred from the SCR to the direct line, as shown in Fig. 4-54.

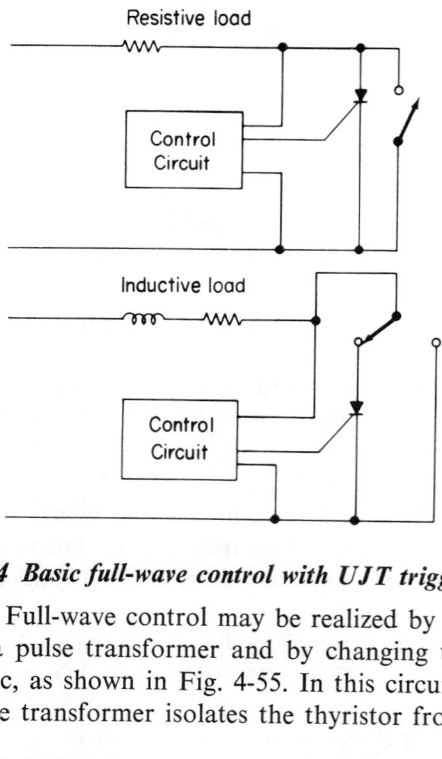

FIGURE 4-54 Half-wave controls with switching for full-wave operation (courtesy Motorola).

4-8.4 Basic full-wave control with UJT trigger

Full-wave control may be realized by the addition of a bridge rectifier or a pulse transformer and by changing the thyristor from an SCR to a Triac, as shown in Fig. 4-55. In this circuit, R_{B1} is not necessary since the pulse transformer isolates the thyristor from the steady-state UJT current.

4-8.5 Full-wave control with constant UJT trigger output

Occasionally a circuit is required that will provide constant output voltage regardless of line-voltage changes. This can be accomplished by adding a potentiometer in series with R_D, as shown in Fig. 4-56. The value of this potentiometer P_1 should be 10% (or less) of the R_D value.

In operation, P_1 is adjusted to provide reasonably constant output over the desired range of line voltage. As line voltage increases, so does the

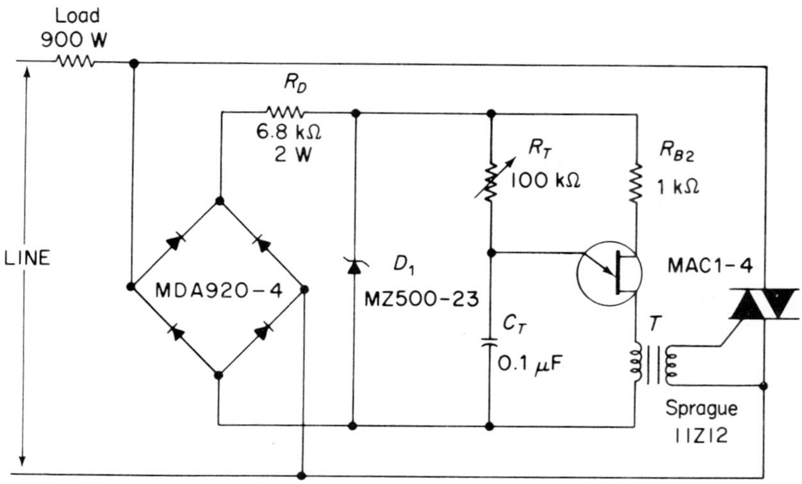

FIGURE 4-55 Simple full-wave UJT trigger circuit with typical values for 900-W load and 60-Hz line (courtesy Motorola).

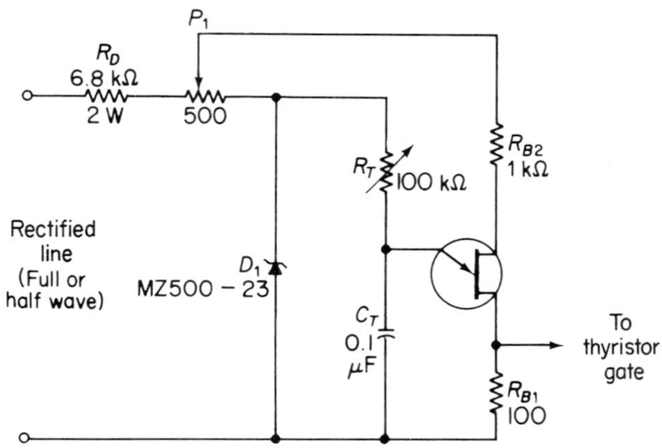

FIGURE 4-56 UJT trigger circuit with line voltage compensation (courtesy Motorola).

voltage on the wiper of P_1, increasing interbase voltage (and the peak voltage of the UJT). The increased peak-voltage point results in C_T charging to a higher voltage and thus taking more time to trigger. The additional delay reduces the thyristor conduction angle and maintains the average voltage at a reasonably constant value.

4-8.6 UJT trigger circuits with feedback

The circuits described thus far have been manual controls; that is, the power output is controlled by a potentiometer R_T turned by hand. Simple feedback circuits may be constructed by replacing R_T with heat- or light-

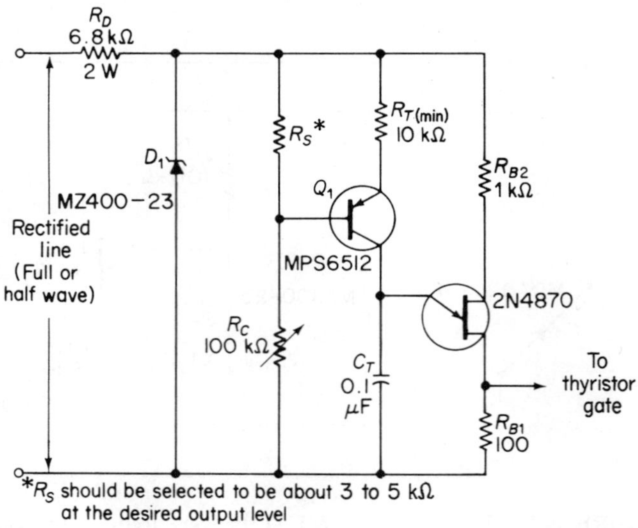

FIGURE 4-57 UJT trigger with feedback control (courtesy Motorola).

dependent sensing resistors; however, such circuits have no means of adjusting the operating levels. By adding a transistor to the basic UJT trigger circuit, it is possible to provide both manual and automatic (feedback) control.

Figure 4-57 shows a feedback control using a sensing resistor R_S for feedback. The sensing resistor may respond to any one of many stimuli, such as heat, light, pressure, moisture, or magnetic field. R_S is the sensing resistor and R_C is the manual control resistor that establishes the desired operating point. Transistor Q_1 is connected as an emitter follower such that an increase in the resistance of R_S decreases the voltage on the base of Q_1, causing more current to flow.

Current through Q_1 causes voltage to charge C_T, triggering the UJT at some phase angle. As R_S becomes larger, more current flows into the capacitor; the voltage builds up faster, causing the UJT to trigger at a smaller phase angle; and more power is applied to the load. Thus, this circuit is for a sensing resistor that decreases in response to excessive power in the load. If the sensing resistor increases with load power, then R_S and R_C should be interchanged.

4-9. TYPICAL PUT CIRCUITS

The following paragraphs describe a few typical PUT circuits. The circuits described by no means represent all the many applications for PUTs. Instead, the circuits given here are chosen because of their universal nature.

In general, a PUT can be used almost anywhere a UJT is required. As discussed in the beginning of this chapter, however, PUTs are best suited for timing applications rather than for phase control. For that reason, we shall concentrate on timing and timing-control circuits.

4-9.1 Voltage-controlled ramp generator

The PUT provides a simple approach to a voltage-controlled ramp generator (VCRG) as shown in Fig. 4-58. The current source formed by Q_1 in conjunction with capacitor C_1 set the duration time of the ramp. As the positive dc voltage at the gate is changed, the peak-point firing voltage of the PUT is changed. In turn, the ramp-time duration is changed. For example, if the peak-point firing voltage is increased, the duration is increased and vice versa.

Figure 4-59 shows a plot of voltage versus ramp duration time for a 0.0047- and a 0.01- μF timing capacitor. As shown, it is possible to have a change in frequency (or period) with either a change in C_1 value or a change in gate voltage.

For example, it is possible to have a 3-ms duration ramp with a gate voltage of 8 V and a C_1 of 0.01 μF. The same 3-ms ramp can be obtained with a C_1 value of 0.0047 μF and a gate voltage of 14 V. From a practical design standpoint, the gate voltage must be kept within the limits specified by the datasheet (5 to 20 V in this case); however, the C_1 value can be changed over a wide range to achieve a given ramp duration. A 10-to-1, or even a 100-to-1 time change can be achieved by a corresponding change in C_1 value (using the remaining values shown in Fig. 4-58).

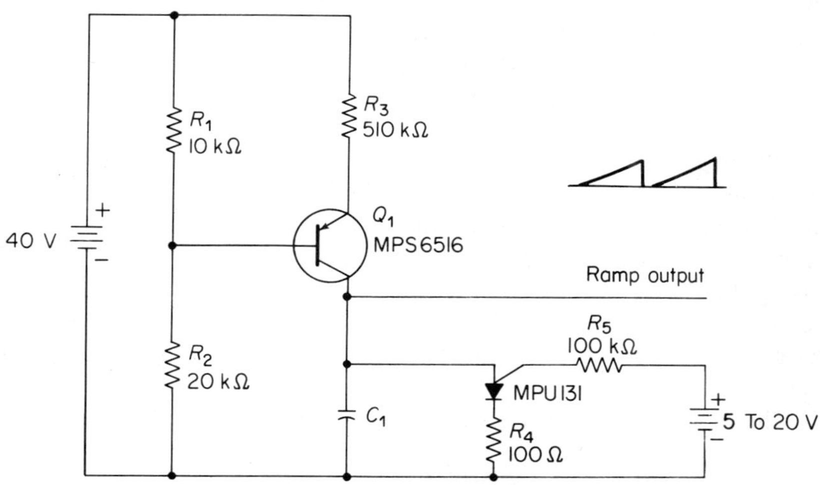

FIGURE 4-58 Voltage-controlled ramp generator (VCRG) using a PUT (courtesy Motorola).

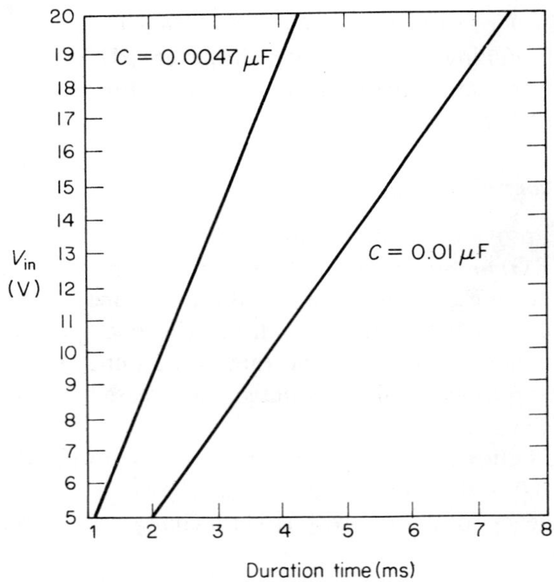

FIGURE 4-59 Voltage versus ramp duration time of VCRG (courtesy Motorola).

4-9.2 Low-frequency divider

The circuit of Fig. 4-60 is a frequency divider, with the ratio of capacitors C_1 and C_2 determining the division. Operation of the circuit is as follows:

Assume that $C_1 = C_2$ (for a division of two, as indicated by the table on Fig. 4-60). Also assume that C_1 and C_2 are discharged and that a pulse is applied to the base of Q_1. Transistor Q_1 turns on and off for each positive pulse.

When Q_1 turns off, both C_1 and C_2 charge to 10 V through R_3. On the next pulse to the base of Q_1, C_1 is again discharged (at the end of the pulse) but C_2 remains charged to 10 V (due to the arrangement of D_1 and D_2). On this second pulse, C_2 charges to the peak-point firing voltage of the PUT, causing the PUT to fire. This discharges C_2 and allows C_1 to charge to the line voltage. As soon as C_2 discharges and C_1 charges, the PUT turns off. The next cycle begins with another pulse on the base of Q_1, which again discharges C_1.

The input and output frequency can be approximated by the equation

$$f_{in} \approx \frac{C_1 + C_2}{C_1} f_{out}$$

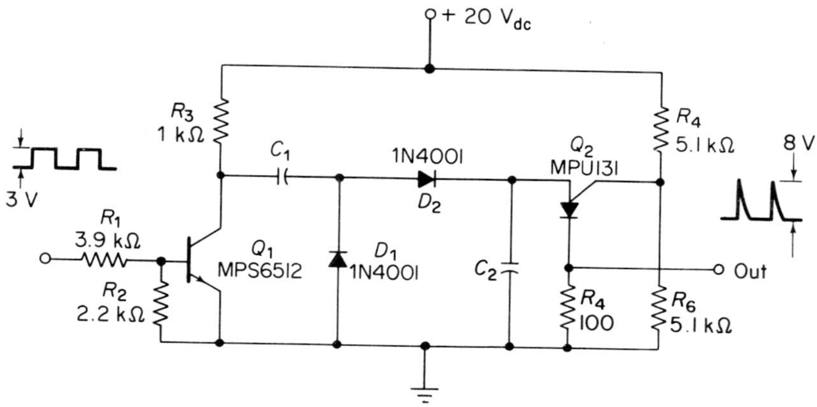

C_1(μF)	C_2(μF)	Division
0.01	0.01	2
0.01	0.02	3
0.01	0.03	4
0.01	0.04	5
0.01	0.05	6
0.01	0.06	7
0.01	0.07	8
0.01	0.08	9
0.01	0.09	10
0.01	0.10	11

FIGURE 4-60 Low-frequency divider using PUT (courtesy Motorola).

For a 10-kHz input frequency with an amplitude of 3 V, the table of Fig. 4-60 shows the values for C_1 and C_2 needed to divide by 2 to 11.

This division range can be changed by using the programmable aspect of the PUT and changing the voltage on the gate by changing the ratio $R_6/(R_6 + R_5)$. Decreasing the ratio with a given C_1 and C_2 decreases the division range; increasing the ratio increases the division range.

The circuit shown in Fig. 4-60 is relatively insensitive to the amplitude, pulse width, and rise and fall times of the incoming pulses.

4-9.3 Long duration timer

A long duration timer circuit that can provide a time delay of up to 20 min is shown in Fig. 4-61. The circuit is a standard relaxation oscillator with a JFET current source in which resistor R_1 is used to provide reverse bias on the gate-to-source of the JFET. This turns the JFET off and increases the charging time of C_1. Capacitor C_1 should be a low-leakage capacitor such as a Mylar type.

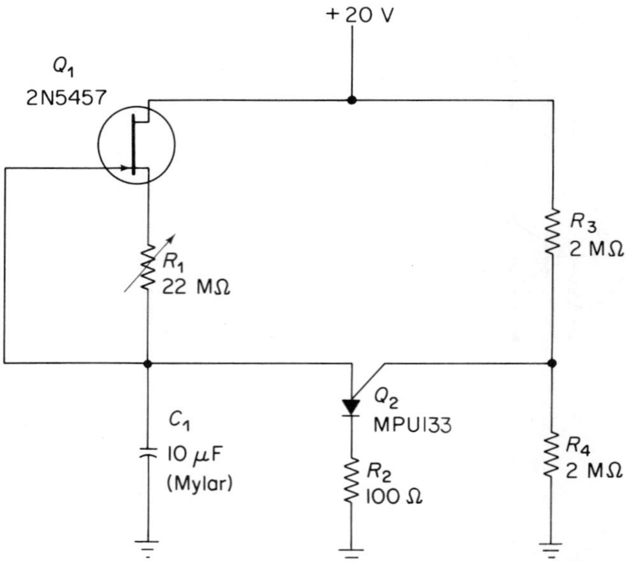

FIGURE 4-61 A 20-minute, long duration timer using a PUT (courtesy Motorola).

The source resistor of the current source can be accomplished using the following equation:

$$V_{GS} = V_P\left(1 - \sqrt{\frac{I_O}{I_{DSS}}}\right)$$

$$\therefore \quad R_1 = \frac{V_{GS}}{I_O}$$

where I_0 is the current out of the current source
V_P is the pinch-off voltage
V_{GS} is the voltage gate-to-source
I_{DSS} is the current, drain-to-source, with the gate shorted to the source

The time needed to charge C_1 to the peak-point firing voltage of Q_2 can be approximated by the following equation:

$$T = \frac{C\,\Delta V}{I}$$

where T is time in seconds
C is capacitance in μF
V is the change in voltage across capacitor C_1
I is the constant current used to charge C_1.

Maximum time delay of the circuit is limited by the peak-point firing current I_P needed to fire Q_2. For charging currents below I_P there is not enough current available from the current source to fire Q_2, causing the circuit to lock up. Thus, a PUT works better than a UJT in a long duration timer because it has a lower peak firing current. Also, because of the programmable aspect of the PUT, I_P can be made very small by making R_G (the equivalent parallel resistance of R_3 and R_4) large (1 MΩ), as shown in Fig. 4-61.

4-9.4 Phase control

Figure 4-62 shows a circuit using a PUT for phase control of an SCR. The relaxation oscillator formed by Q_2 provides conduction control of Q_1 from 1 to 7.8 ms or 21.6 to 168.5 deg. This constitutes control over 97% of the power available to the load.

Only one SCR is needed to provide phase control of both the positive and the negative portions of the sine wave by putting the SCR across the bridge composed of diodes D_1 through D_4.

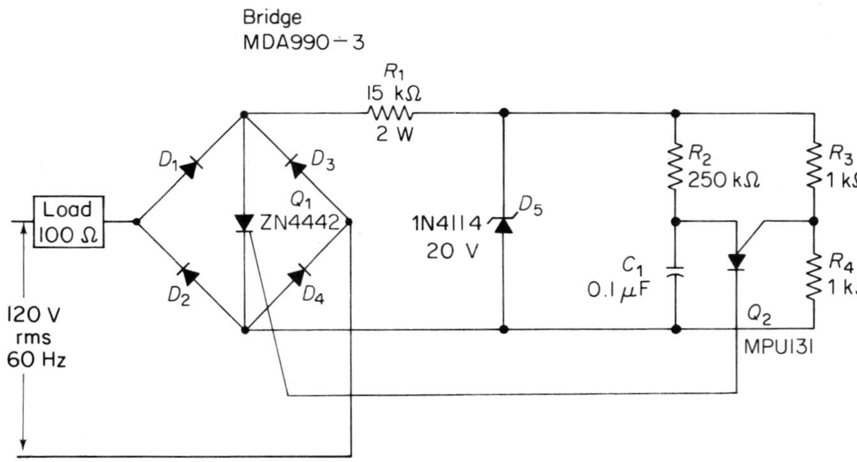

FIGURE 4-62 SCR phase control using a PUT (courtesy Motorola).

4-10. TESTING UNIJUNCTION TRANSISTORS

As is the case for two-junction transistors and FETs, there are a number of commercial test sets for UJTs. Likewise, it is possible to adapt commercial test sets (designed for two-junction transistors) for use with UJTs. The following describes circuits and procedures for testing of all important UJT characteristics.

4-10.1 Emitter and interbase curve tracer

The circuit of Fig. 4-63 can be used to display either the emitter characteristic curves or the interbase characteristic curves on a standard oscilloscope. When the switches are set to display the interbase curves, the meter will indicate the emitter current. When the switches are set to display the emitter curves, the meter will indicate the interbase voltage.

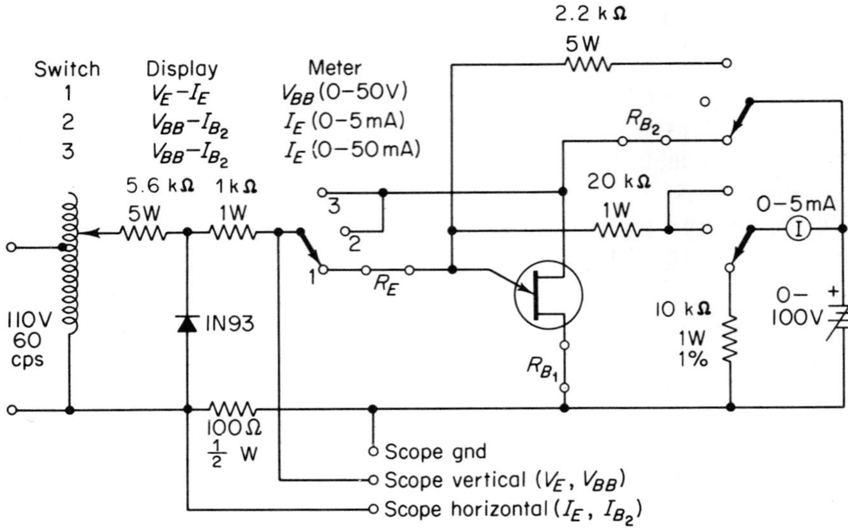

FIGURE 4-63 Emitter and interbase curve tracer (courtesy General Electric).

It is important to set the variac and the dc supply to zero before changing the switches or inserting the UJT in the circuit so as to avoid accidental burnout.

If desired, external resistors may be inserted in series with the emitter, base 1, or base 2 to determine the characteristics of the UJT in a particular circuit.

4-10.2 Commercial test set conversion

Commercial test equipment designed for use with two-junction transistors, such as the Tektronix Transistor Curve Tracer, can also be used to display the characteristic curves of the UJT. Connections for conversion of the commercial tracers to UJT use are shown in Fig. 4-64. Note that for test of UJT interbase characteristics, the test set terminal normally used for a two-junction transistor base is connected to the UJT emitter. For test of emitter curves, the UJT emitter is connected to the collector terminal of the test set. In display of the emitter characteristic curves, the interbase voltage should

Test	Emitter curves	Interbase curves
Circuit	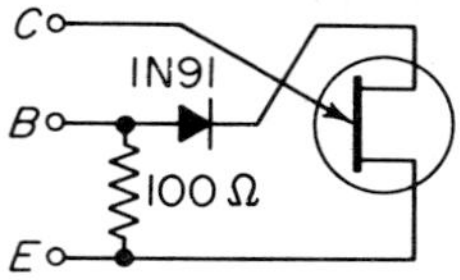	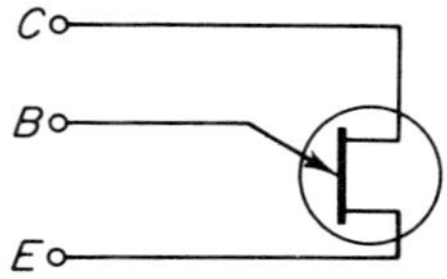
Collector sweep polarity	+	+
Base step polarity	+	+
Collector peak volts range	200 V	20 V
Collector limiting resistor	5K	500 Ω
Base current steps selector	20 mA/step	10 mA/step
Number of current steps	5	5
Vertical current range	2 mA/div I_E	2 mA/div I_{B_2}
Horizontal voltage range	1V/div V_E	2 V/div V_{BB}

FIGURE 4-64 Use of Tektronix 575 transistor curve tracer for display of UJT characteristic curves (courtesy Tektronix).

not exceed 12 V because of the voltage limitation of the base current step generator.

4-10.3 General-purpose UJT test set

A general-purpose test set circuit for the UJT is shown in Fig. 4-65. This circuit can be used for testing of six major UJT characteristics as follows:

Test	Parameter	Range	Conditions
1	Peak-to-peak emitter voltage	0–10 V	$V_{B2B1} = 10$ V; $C = 0.1\ \mu$F
2	Intrinsic standoff ratio	0–1.0	$V_{B2B1} = 10$ V
3	Interbase resistance	3 kΩ–∞	Power < 15 mW
4	Emitter saturation voltage	0–10 V	$V_{B2B1} = 10$ V $I_E = 50$ mA
5	Emitter voltage at 1 mA	0–10 V	$V_{B2B1} = 10$ V $I_E = 1$ mA
6	Emitter leakage current	0–100 μA	$V_{EB2} = 10$ V or less

The oscillator test portion of the circuit (Fig. 4-65) indicates if a UJT can oscillate in a relaxation oscillator circuit. Failure to pass this test usually means total failure of a UJT. The UJT is connected as an oscillator in the Test 1 position, and the peak-to-peak emitter voltage is indicated on the meter. In the Test 1 position, power is applied across base 1 and base 2

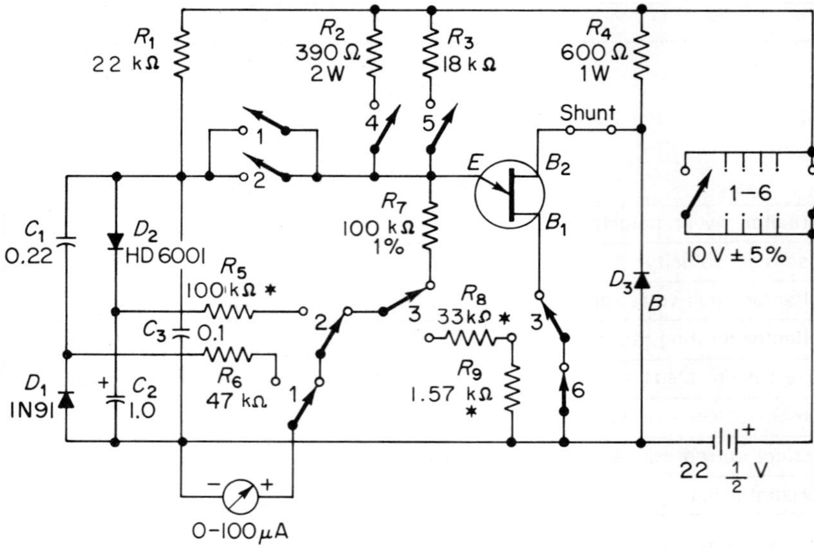

Note: (1) All switches shown in the inactive position.
(2) All components ± 5% except as indicated.
(3) * denotes approximate values, must be calibrated.

FIGURE 4-65 General Electric UJT test set schematic (courtesy General Electric).

(through R_4), and the emitter is connected to the junction of R_1 and C_3. The meter is connected to the junction of C_1 and D_1, through R_6, and indicates the charge across C_1. If the UJT fails to oscillate, C_1 does not charge and there is no meter indication.

The intrinsic standoff ratio test (Test 2) also makes use of a relaxation oscillator and a peak voltage detector. To calibrate the circuit for this test, a temporary connection should be made from the common point of R_4–D_3 to the common point of D_2–C_2, and R_5 is adjusted to give a full-scale deflection on the meter with the button for Test 2 depressed. During test, the temporary connection is removed, the Test 2 switches are closed, and the meter is checked to indicate standoff ratio. For example, if the full-scale deflection is 100 μA and the meter indicates 80 μA during test, the standoff ratio is 0.8.

For measurement of interbase resistance, resistors R_8 and R_9 must be calibrated to give the correct meter deflection. The R_8 and R_9 values shown in Fig. 4-65 are for a typical UJT.

As an additional test, the interbase modulated current $I_{B2(\text{mod})}$ can be made by connecting an external meter in the base 2 shunt position and depressing the button for Test 4.

4-10.4 Intrinsic standoff ratio test

The general-purpose test circuit of Fig. 4-65 is usually satisfactory for measurement of the intrinsic standoff ratio. If greater accuracy is required or if the equivalent emitter-diode voltage V_D is to be measured, however, the circuit of Fig. 4-66 can be used.

In the circuit of Fig. 4-66, the interbase voltage is swept by the 10-Hz oscillator, while the UJT oscillates at about 2 kHz in the simple relaxation oscillator. The interbase voltage is applied to the horizontal axis of the oscilloscope. The emitter voltage is applied to one input of the vertical differential amplifier and the potentiometer voltage is applied to the other input.

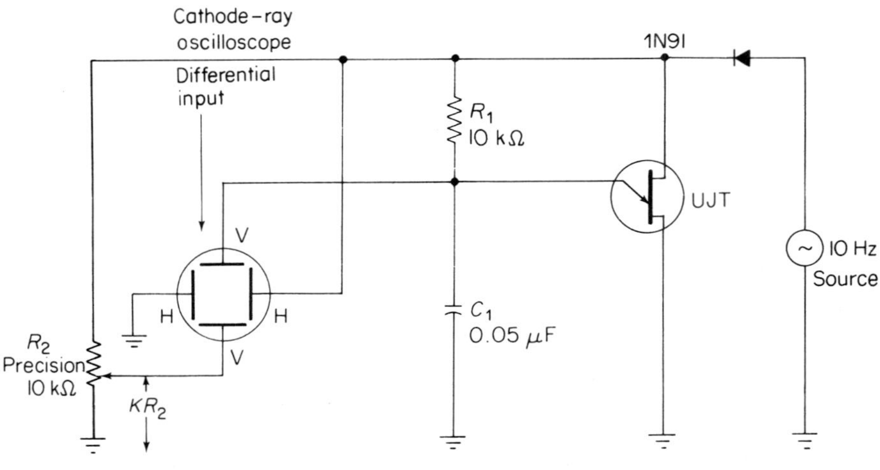

FIGURE 4-66 Circuit for measurement of intrinsic standoff ratio and equivalent emitter diode voltage V_D (courtesy General Electric).

The oscilloscope pattern consists of a plot of V_{B2B1} on the horizontal axis against

$$V_P - KV_{B2B1} = V_D + (-K)V_{B2B1}$$

on the vertical axis. K is the fractional setting of the potentiometer and V_P is the upper envelope of the emitter voltage.

If K is set equal to the intrinsic standoff ratio, the upper envelope will be horizontal and the displacement from the zero axis on the oscilloscope will be equal to V_D.

If a precision potentiometer is used for R_2, the intrinsic standoff ratio can be measured with an accuracy of better than 0.05% and V_D can be measured within 20 mV.

4-10.5 Peak-point current test

The peak-point current I_P can be measured with the circuit of Fig. 4-67. This circuit measures the *minimum emitter current* required for oscillation in a relaxation oscillator circuit. This minimum emitter current is a good approximation to the peak-point emitter current.

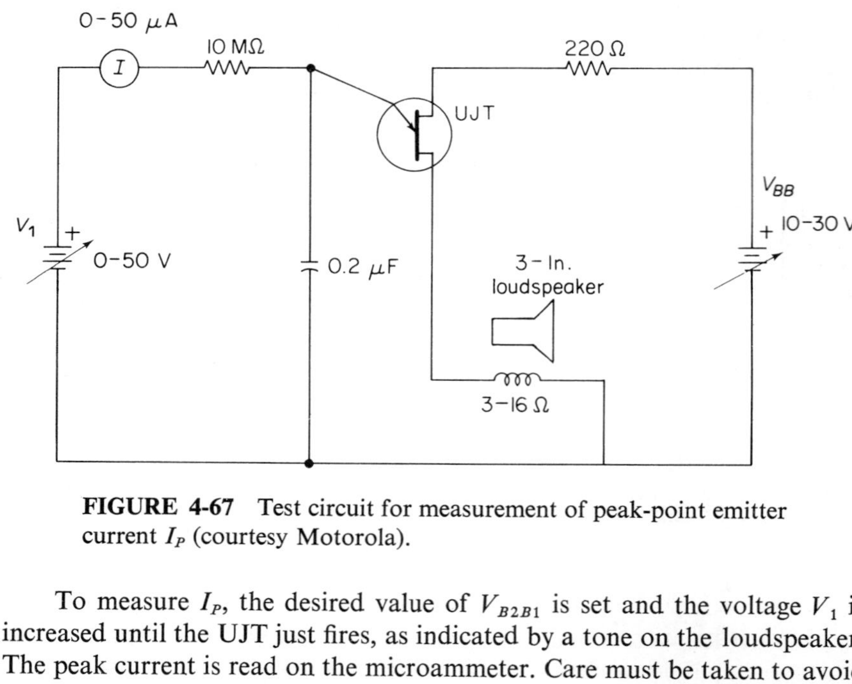

FIGURE 4-67 Test circuit for measurement of peak-point emitter current I_P (courtesy Motorola).

To measure I_P, the desired value of V_{B2B1} is set and the voltage V_1 is increased until the UJT just fires, as indicated by a tone on the loudspeaker. The peak current is read on the microammeter. Care must be taken to avoid any ripple on the V_{B2B1} supply since this can reduce the apparent peak-point current considerably.

4-10.6 Valley-point current and voltage test

The valley point corresponds to the point on a particular UJT emitter-characteristic curve where the emitter voltage is at minimum or where the dynamic resistance is zero. Because of the slight change of emitter voltage near the valley point, it is difficult to locate the position of the valley point by inspection of the emitter-characteristic curve. The circuit of Fig. 4-68 can be used to measure V_V and I_V by the null method.

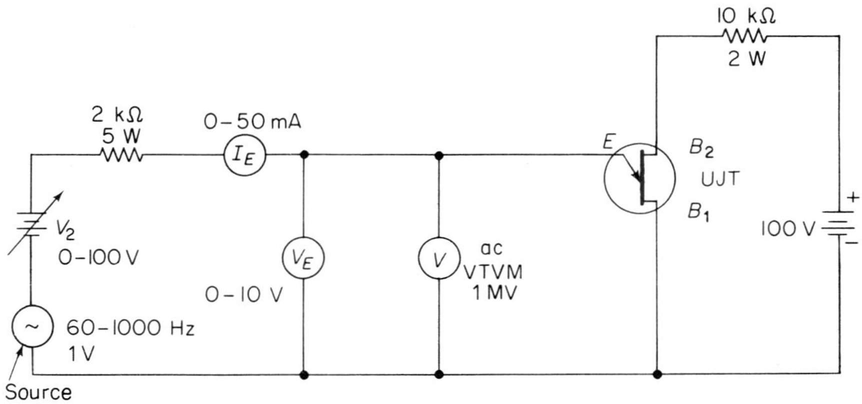

FIGURE 4-68 Test circuit for measurement of valley voltage and valley current (courtesy General Electric).

With the circuit of Fig. 4-68, the supply voltage V_2 is adjusted to obtain a null on the electronic voltmeter. The valley voltage and valley current can then be measured on the V_E and I_E meters, respectively.

The circuit of Fig. 4-68 shows typical bias conditions used for specification of I_V and V_V. Other values of interbase voltage and base 2 series resistance can be used as desired for special tests.

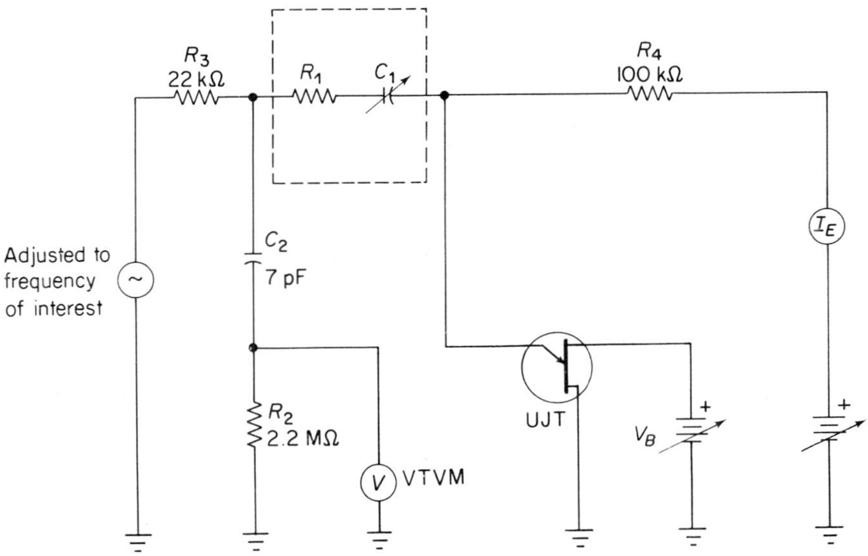

FIGURE 4-69 Circuit for measurement of emitter input impedance of UJT (courtesy General Electric).

Curve	Bias	
(a)	$V = 25$ V, $I_E = 4.2$ mA	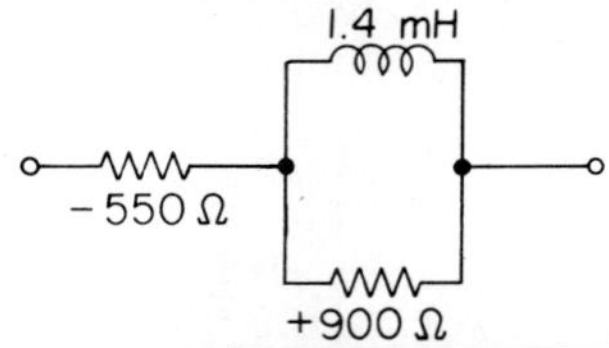 1.4 mH; −550 Ω; +900 Ω
(b)	$V = 25$ V, $I_E = 1.0$ mA	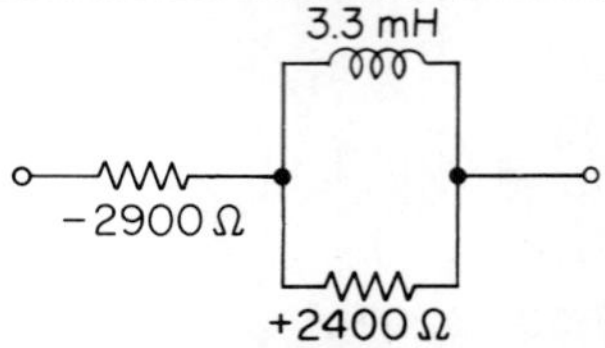 3.3 mH; −2900 Ω; +2400 Ω
(c)	$V = 25$ V, $I_E = 0.5$ mA	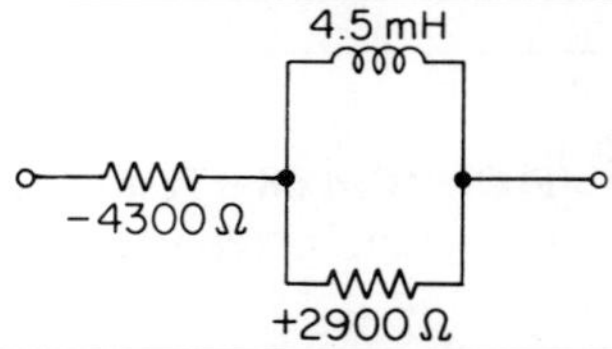 4.5 mH; −4300 Ω; +2900 Ω

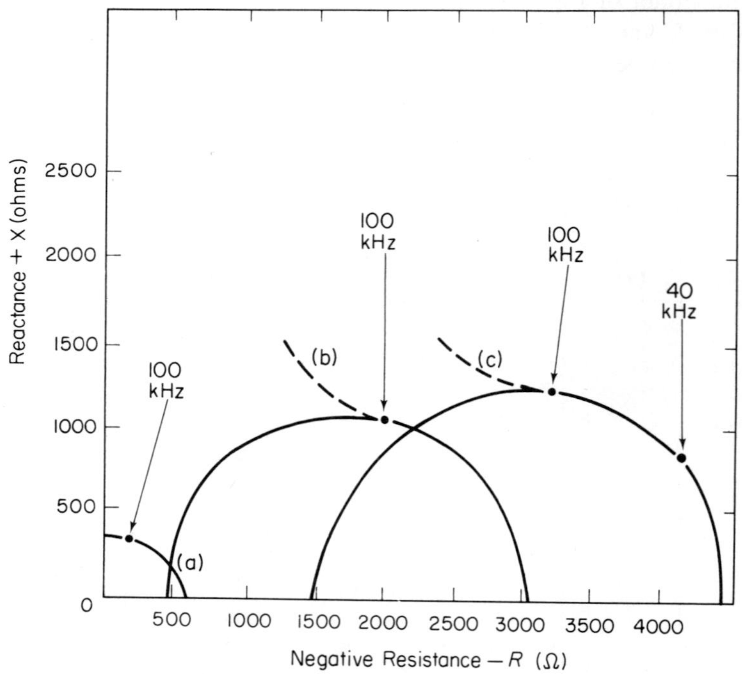

FIGURE 4-70 Emitter impedance diagram of UJT at several bias points (courtesy General Electric).

4-10.7 *Frequency response*

The small-signal frequency-response characteristic of most value is the *emitter input impedance*, which can be measured with the circuit shown in Fig. 4-69. In this circuit, the UJT is biased in the negative resistance region, and a null is obtained on the electronic voltmeter by adjusting R_1 and C_1. At null, the emitter input-impedance is approximately equal to

$$Z_E = -R_1 + \frac{j}{6.28C_1}$$

The capacitor C_2 is used to reduce the effective capacitance between the emitter of the UJT and ground so as to prevent oscillation in the relaxation mode.

An impedance diagram for a typical UJT, showing several bias conditions, is illustrated in Fig. 4-70. Up to 100 kHz, the simple three-element *RL* network of Fig. 4-70 provides an accurate representation of the input impedance.

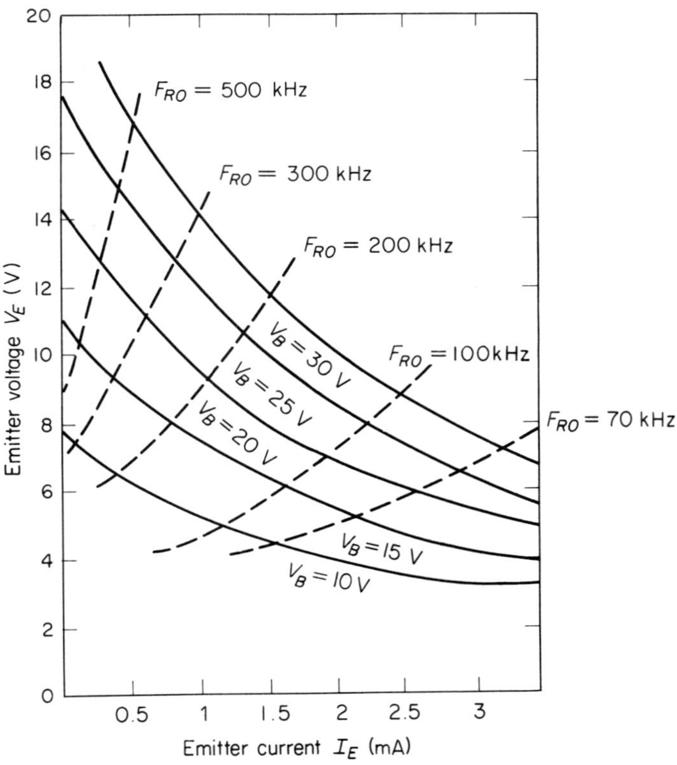

FIGURE 4-71 UJT emitter characteristic curves with contours of constant resistance cutoff frequency (courtesy General Electric).

The frequency at which the resistive component of the emitter input impedance is equal to zero is called the *resistance cutoff frequency* F_{RO}. This is a frequency figure of merit for a UJT since it represents the maximum frequency at which the UJT is regenerative at a given bias point.

The circuit of Fig. 4-69 can be used to measure F_{RO} if R_1 is set equal to zero and a null obtained on the electronic voltmeter by adjusting the frequency and C_1. Since F_{RO} varies with bias conditions, measured values are best presented by means of a plot such as that of Fig. 4-71, which gives *contours of constant* F_{RO} superimposed on the normal emitter-characteristic curves. F_{RO} is proportional to the magnitude of the negative resistance at a given bias point. Thus, a single measurement of F_{RO} will establish the values of F_{RO} over the entire negative resistance portion of the emitter characteristics.

The value of F_{RO}, for a given emitter-characteristic curve, increases as the peak point is approached. The limiting value at the peak point corresponds to the *maximum frequency of oscillation*, or f_{max}. This is the maximum frequency at which the UJT can oscillate in a relaxation oscillator circuit.

For the lower range of V_{B2B1}, where the heating caused by the interbase power dissipation is not excessive, f_{max} is directly proportional to V_{B2B1}.

5

Audio-Frequency Considerations for Transistors

Both two-junction transistors and FETs are used as *AF* amplifiers. UJTs are not generally found in *AF* amplifier work. One exception is the UJT regenerative amplifier discussed in Sec. 5-4. FETs are usually limited to voltage amplifier applications. Two-junction transistors can be used in either voltage-amplifier or power-amplifier circuits.

Most of the classic vacuum-tube *AF*-amplifier circuits can be duplicated with transistors. Since the number of different amplifier circuits is almost unlimited, it is impossible to cover all aspects of *AF* amplifiers in this chapter. *Handbook of Modern Solid-State Amplifiers* by John D. Lenk, Prentice-Hall, Inc., Englewood Cliffs, N.J., 1974, provides a detailed discussion of many amplifier types.

All the design considerations for transistors discussed in Chapters 1 through 4 apply to *AF* amplifiers. Of particular importance are how to interpret datasheets, determining parameters at different frequencies, temperature-related design problems, and basic bias schemes.

5-1 HOW TRANSISTOR AMPLIFIER CHARACTERISTICS AFFECT FREQUENCY

Every electronic component has some impedance and is thus frequency sensitive; that is, the component will not attenuate (or pass) signals of all frequencies equally. Even a simple length of wire has impedance. Wire, being a conductor, has some resistance. If alternating current is passed through the wire, there is some inductive reactance. If the wire is near another conductor (or metal chassis), there is some capacitance between the two conductors and thus some capacitive reactance. The reactance and resistance combine to produce impedance, which, in turn, varies with frequency.

Were it not for reactance, a transistor (by itself) should be capable of operating at any frequency from zero (direct current) on up; that is, the top-frequency limit would be set only by the transit time of electrons across the transistor junctions. There are limitations placed on the operating frequency of any amplifier, however, by the transistor characteristics. Similarly, the other components (capacitors, resistors, inductors, etc.) used in the amplifier circuit also limit the operating frequency. In this section, we shall see how each of the components affects operating frequency of amplifiers.

Of course, many of the impedances presented by components are of little practical concern. On the other hand, certain of the impedances have a very pronounced effect on amplifier design and operation. There are four major components used in *AF*-amplifier designs: transistors, resistors, capacitors, and inductances (both coils and transformers). Let us examine how the impedances and reactances of these components affect *AF*-amplifier operation.

Transistor frequency limitations. All transistors have some capacitance between the elements (emitter-base junction, collector-base junction, gate-drain, etc.). If any of the elements is common or ground, the remaining elements have some capacitance to ground. For example, in a common-emitter amplifier, there is some capacitance from base to ground (across the input) and collector to ground (across the output). Likewise, there is capacitance from collector to base (which forms a feedback path from output to input). This is shown in Fig. 5-1.

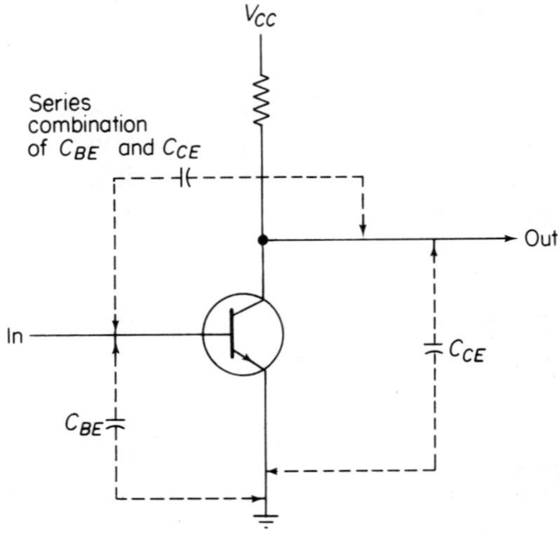

FIGURE 5-1 Typical transistor element capacitances.

Capacitive reactance decreases with an increase in frequency and vice versa. A capacitance in series with a conductor presents less attenuation to the signal as frequency increases. A capacitance across a conductor (for example, in parallel or shunt from the conductor to ground) acts as a short to signals of increasing frequency. Consider a common-emitter amplifier where transistor capacitances are across the input and output. As frequency increases, the capactive reactance drops, producing a short across the input and output, and increases attenuation of the signal. At some frequency, the attenuation equals the transistor amplification so there is no gain. At higher frequencies, the attenuation exceeds amplification and there is a loss, even though the transistor may still continue to operate.

From a practical standpoint, the input and output capacitances of transistors have little effect at audio frequencies. Most modern transistors will operate well beyond the *AF* range and will generally produce equal (or flat) *frequency response*; that is, all signals up to about 20 kHz (or possibly higher) are amplified by the same amount. With most transistors, however, as signal frequencies increase into the *RF* range, amplification begins to drop. (*RF* amplifiers are discussed in Chapter 6.)

All transistors have some inductance in their leads. This produces inductive reactance in series with the transistor elements. Inductive reactance increases with frequency. In the *AF* range the inductive reactance is of little concern. At radio frequencies, however, the inductive reactance can produce considerable attenuation.

Resistor frequency limitations. At audio frequencies, resistors offer relatively few problems since resistors attenuate signals equally. Only at very high frequencies, where the resistor leads and body could produce some kind of reactance, is there any particular concern about frequency limits imposed by resistors; however, resistors do produce *voltage drops*. These voltage drops can be a problem when considering interstage coupling methods (described in Sec. 5-2) and when used in conjunction with coupling capacitors.

Capacitance frequency limitations. Capacitors have three major uses in transistor *AF* amplifiers: bypass, decoupling, and coupling.

Bypass capacitors are used to provide a signal path around high resistances. For example, if the power supply of an *AF* amplifier does not have a filter capacitor, or a battery is used, the collector-emitter current must pass through a high resistance. This can impede the ac component of the signal. A bypass capacitor provides a signal path as shown in Fig. 5-2.

When several stages of amplification are connected, they all join at one point, the common power supply. In multistage amplifiers there is the possibility of one stage feeding back through the power supply to a previous stage, thus causing interference with the signal. To avoid this feedback, one or more of these stages may be *decoupled* from the power supply. Figure 5-2 also shows a typical decoupling network. Resistor *R* is placed in series between

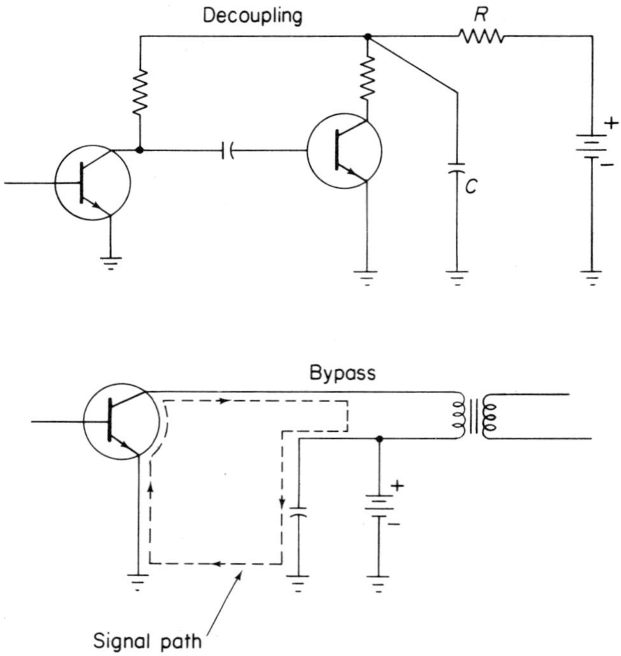

FIGURE 5-2 Effect of decoupling and bypass capacitors in transistor circuits.

the load resistors of the stages and the power supply. Hence, it offers a high-resistance path for the ac signal component to the power supply. Capacitor *C*, on the other hand, offers a low shunt reactance to this component and thus decouples (bypasses) the component to ground.

Actually, the functions of bypass and decoupling capacitors are the same and the terms are interchangeable. In either case, the main concern is that the *reactance be low at the lowest frequency involved.* For example, assume that the lowest frequency is 100 Hz and that the minimum required reactance is 100 Ω. This requires a capacitance value of about 10 μF. If the required frequency is decreased to 10 Hz, the capacitance value must be raised to 160 μF to keep the reactance below 100 Ω.

Coupling capacitors are used at the input and output of each stage to block direct current. For example, if a coupling capacitor is not used between two transistor stages, the collector of the first stage is connected directly to the base of the following stage and both elements are at the same bias voltage. While it is possible to operate transistors in this way, direct coupling does increase problems. For one thing, the amplifier cannot tell the difference between a change of signal level and a change in power supply voltage.

The values of coupling capacitors are dependent on the *low-frequency limit* at which the amplifier is to operate and on the resistances with which the capacitors operate. As frequency increases, capactive reactance decreases and coupling capacitors become (in effect) a short to the signal. Therefore, the high-frequency limit need not be considered in *AF* circuits.

Figure 5-3 shows how a high-pass filter is formed by coupling capacitors. Capacitor C_1 forms a high-pass *RC* filter with R_B. Capacitor C_2 forms another high-pass filter with the input resistance of the following stage (or the load). The input voltage is applied across the capacitor and resistor in series. The output voltage is taken across the resistance. The relation of input voltage to output voltage is

$$\text{Output voltage} = \text{input voltage} \times \frac{R}{Z}$$

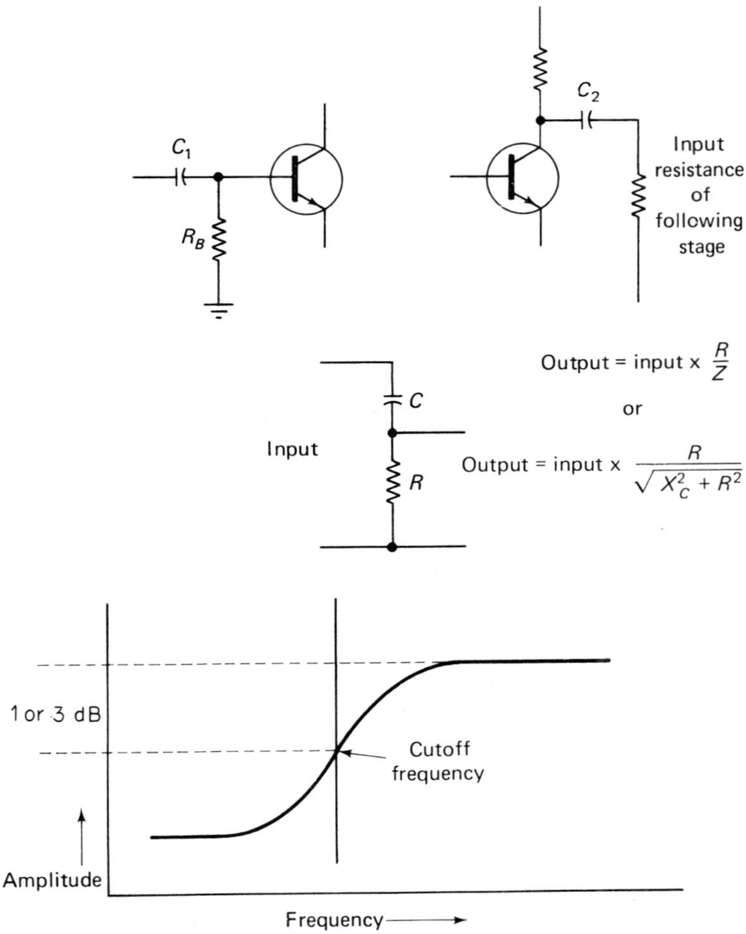

FIGURE 5-3 Formation of high-pass (low-cut) RC filter by coupling capacitors and related resistances. (Lenk, *Handbook of Modern Solid-State Amplifiers*, Prentice-Hall.)

where R is the dc resistance value and Z is the impedance obtained by the vector combination of series capacitive reactance and dc resistance.

When the reactance drops to approximately one-half the resistance, the output will drop to about 90% of the input (or approximately a 1-dB loss). Using the 1-dB loss as the low-frequency cutoff point, the value of C_1 or C_2 can be found by

$$\text{Capacitance} = \frac{1}{3.2FR}$$

where capacitance is in farads, F is the low-frequency limit in hertz, and R is resistance in ohms.

If a 3-dB loss can be tolerated at the low-frequency cutoff point, the value of C_1 or C_2 can be found by

$$\text{Capacitance} = \frac{1}{6.2FR}$$

Inductance frequency limits. Both coils and transformers are used in audio amplifiers. As discussed in Sec. 5-2, coils are sometimes used in place of the collector resistor as a load. This permits the collector to be operated at a higher voltage. Likewise, transformers are used for coupling between stages. This provides impedance matching, as discussed in Sec. 5-2.

The inductive reactance of coils and transformers increases with frequency. At the high end of the audio range, the attenuation produced by this increased reactance is usually sufficient to impair operation of the amplifier. At the low end of the *AF* range, the reactance of a typical transformer drops to a few ohms. This low impedance acts as a short across the line and attenuates the signal. Thus, coils and transformers tend to attenuate signal at both the high and low ends of the *AF* range.

Stray impedances. As discussed, any conductor (wiring, terminals, etc.) can have resistance, reactance, and impedance. Thus, care must be used in the routing of wires and placement of terminals to minimize the effects of this *stray* impedance. Similarly, the effects of stray impedance can alter the characteristics of components. A classic example of this is stray capacitance, which is added to the input and output capacitances of transistors. The effects of stray impedances are usually not critical at audio frequencies; however, the effects of stray impedance on amplifiers operating at radio frequencies (Chapter 6) can be of considerable importance.

5-2. TRANSISTOR AMPLIFIER COUPLING TECHNIQUES

All amplifiers require some form of coupling. Even a single-stage audio amplifier must be coupled to the input and output devices. If more than one stage is involved, there must be *interstage coupling*. Amplifiers are often classified as to coupling method. For example, the four basic coupling methods are *capacitor* (*or capacitance*) *coupling*, *inductive coupling*, *direct coupling*, and *transformer coupling*.

All four coupling methods require resistance and could be called *resistance-coupled* amplifiers. The term *resistance coupled however*, is generally used to indicate that the amplifier does not have inductances or transformers between stages and that the input and/or output impedance is formed by a resistance. Capacitor coupling is often called resistance-capacitance (or *RC*) coupling.

In this section, we shall see how the different methods affect operation of practical transistor *AF* amplifiers. Figure 5-4 shows the four coupling methods.

With direct coupling, Fig. 5-4(a), the collector of one transistor is connected directly to the base of the following transistor. The outstanding char-

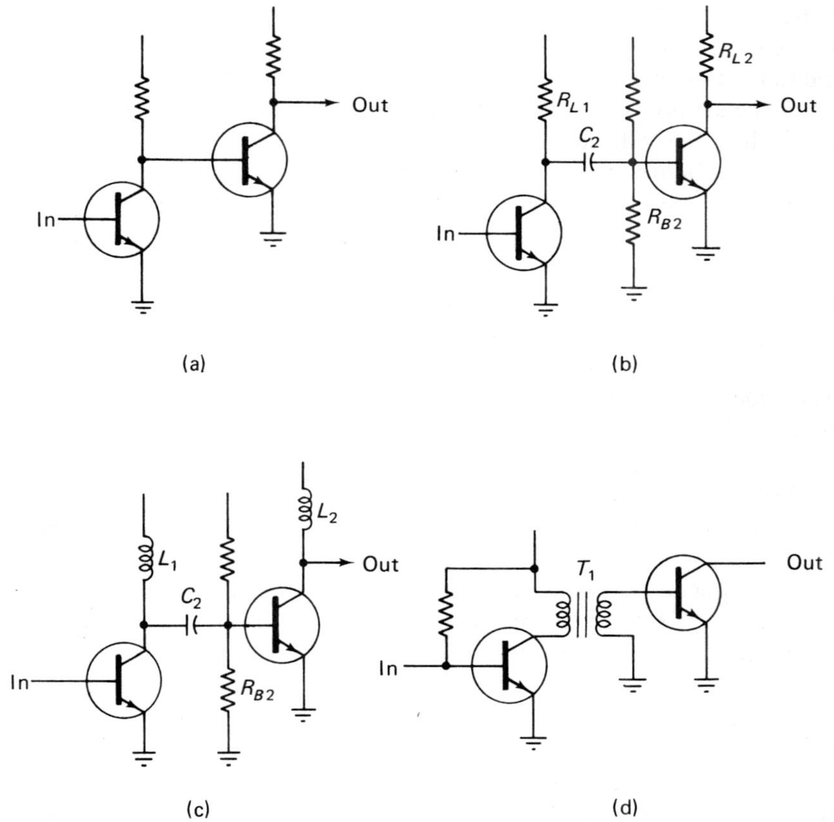

FIGURE 5-4 Four basic types of coupling used in audio amplifiers. (Lenk, *Handbook of Modern Solid-State Amplifiers*, Prentice-Hall.)

acteristic of a direct-coupled amplifier is the ability to amplify direct current and low-frequency signals.

With capacitor coupling, RC *coupling*, Fig. 5-4(b), the coupling is accomplished by means of the load resistor R_{L1} of stage 1, the base resistor R_{B2} of stage 2, and the coupling capacitor C_2. The original signal is acted upon by stage 1 and appears in amplified form as the voltage drop across R_{L1}. The dc component of the amplified signal is blocked by C_2, which passes the ac component to the input section of stage 2 for further amplification. If necessary, more stages may be coupled to the output of stage 2 for still further amplification.

The main advantage of capacitor or *RC* coupling is that the amplifier will amplify uniformly over nearly the entire audio range since resistor values are independent of frequency changes. As discussed in Sec. 5-1, however, *RC*-coupled amplifiers do have a low-frequency limit imposed by reactance of the capacitor (which increases as frequency decreases). Reactance-capacitance coupling is also small, light, and inexpensive and produces no magnetic field to interfere with the signal. One disadvantage of the *RC* coupling method is that the supply voltage is dropped (usually to one-half) by the load resistance. Thus, the collectors must operate at reduced voltage.

With inductive or impedance coupling, Fig. 5-4(c), the load resistors R_{L1} and R_{L2} are replaced by inductors L_1 and L_2. The advantage of impedance coupling over resistance coupling is that the ohmic resistance of the load inductor is less than that of the load resistor. Thus, for a power supply of a given voltage, there is a higher collector voltage.

Impedance coupling also suffers from some disadvantages. Impedance coupling is larger, heavier, and costlier than resistance coupling. To prevent the magnetic field of the inductor from affecting the signal, the inductor turns are wound upon a closed iron core and usually are shielded further. The main disadvantage of impedance coupling is frequency discrimination.

With impedance coupling at very low frequencies, the gain is low due to the capacitive reactance of the coupling capacitor, just as in the *RC*-coupled amplifier. The gain increases with frequency, leveling off at the middle frequencies of the audio range (however, the frequency spread of this level portion is not so great as for the *RC* amplifier).

With impedance coupling at very high frequencies, the gain drops because of the increased reactance. Impedance coupling is rarely, if ever, used at frequencies above the audio range.

With transformer coupling, Fig. 5-4(d), the transformer T_1 serves several purposes. As the fluctuating collector current of the first stage flows through the primary winding of T_1, the current induces an alternating voltage with similar waveform in the secondary of T_1. This voltage forms the input signal to the second stage. Since the secondary of T_1 conveys the ac component of the signal directly to the base of the second stage, there is no need for a coupling capacitor. Also, since the secondary winding furnishes a base return path, there is need for a base resistance.

Compared to the *RC*-coupled amplifier, the transformer-coupled amplifier has essentially the same advantages and disadvantages as the impedance-coupled amplifier. The transistor collectors can be operated at higher voltages. The impedances are set by the transformer primary and secondary windings. Transformers are frequency sensitive (impedance changes with frequency). Therefore, the frequency range of the transformer-coupled amplifiers is limited.

The inductances and transformers used in *AF* work are generally of the iron-core type. If air-core transformers are used at audio frequencies, the inductive reactance (and the impedance) will be so small as to be ineffective. At frequencies above the audio range (or at the high end), the reactance of iron-core inductances and transformers is so large that signals cannot pass (or are greatly attenuated). Thus, air-core transformers and inductances are used for higher-frequency amplifiers, as discussed in Chapter 6.

Coupling transformers also provide for impedance matching between stages. Because the transistor is a current-operated device, impedance matching between the output of one stage to the input of the next is desirable for maximum transfer of power. This can be accomplished by making the primary and secondary transformer windings of different impedance. Typically, the input impedance of a transistor stage is less than the output impedance. Thus, the secondary impedance of an interstage transformer is typically lower than the primary impedance. When two common-emitter stages are impedance matched, the overall gain is greater than when identical stages are resistance coupled.

Transformer coupling is also effective when the final amplifier output must be fed to a low-impedance load. For example, the impedance of a typical loudspeaker is in the order of 4 to 16 Ω; whereas the output impedance of a transistor stage is several hundred (or thousand) ohms. A transformer at the output of an audio amplifier can offset the obviously undesired effects of such a mismatch.

5-2.1 Effects of coupling on audio-amplifier frequency response

A simplified *frequency-response graph* or curve is illustrated in Fig. 5-5, which shows the effects of coupling methods on amplifier frequency response. The response is measured by the gain of the amplifier at various frequencies in the audio range.

Note that the gain falls off at the very low frequencies. In an *RC*-coupled amplifier, this drop in gain (generally referred to as *roll-off*) at the low end is due to the capacitive reactance of the coupling capacitor. Since the coupling

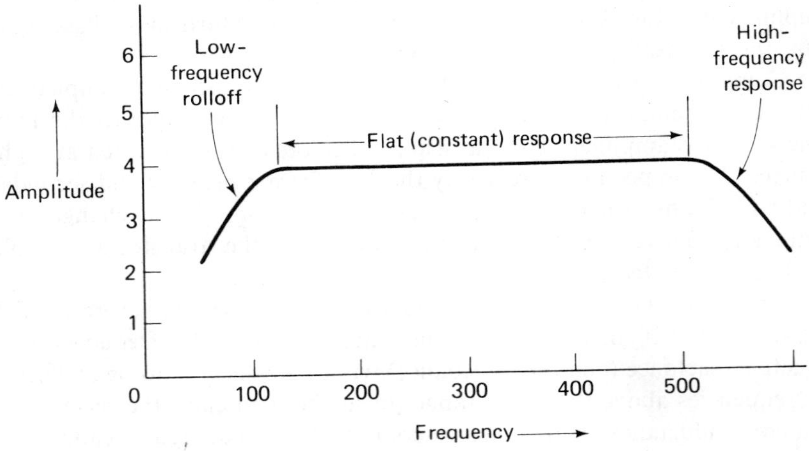

FIGURE 5-5 Simplified frequency-response graph or curve. (Lenk, *Handbook of Modern Solid-State Amplifiers*, Prentice-Hall.)

capacitor is between the output of the first stage and the input to the second stage, the signal is attenuated by the voltage drop across the capacitor. Hence, the lower the frequency, the larger the capacitive reactance, and the smaller is the signal input to the second stage. In impedance-coupled or transformer-coupled amplifiers, the low-frequency roll-off is caused by the very low inductive reactance, which acts as a short across the signal path. In effect, the low reactance bypasses some of the signal to ground.

As shown in Fig. 5-5, the gain also falls off at the higher frequencies. In *RC*-coupled amplifiers, this high-frequency roll-off is due to the output capacitance of the first stage, the input capacitance of the second stage, and the stray capacitance furnished by the coupling network. These capacitances act to bypass some of the signal to ground. The higher the frequency, the smaller the capacitive reactance becomes, and the greater the amount of signal so bypassed. Hence, the overall gain falls. At frequencies between these two extremes, the gain remains fairly constant. In impedance-coupled or transformer-coupled amplifiers, the high-frequency roll-off is due to the large inductive reactance that attenuates the signal.

In sum, then, resistance coupling produces the lowest gain; transformer coupling, the highest. As a guideline, three stages of *RC*-coupled amplification will produce approximately the same gain as two stages of *comparable* transformer-coupled amplification. *RC*-coupling produces the least frequency distortion of the signal. Transformer coupling has the added advantage of providing an impedance match at the amplifier input and output.

5-3. TRANSISTOR AMPLIFIER CLASSIFICATIONS BASED ON OPERATING POINT

Amplifiers are often classified as to *operating point*, or the amount of current flow under no-signal conditions. The following is a brief summary of the four basic operating-point classifications.

In all four classifications, the base-collector junction is always reverse biased at the operating point, as well as under all signal conditions. Thus, no base-collector current flows (with the possible exception of reverse leakage current I_{CBO}). The base-emitter junction is biased such that base-emitter current will flow under certain conditions and possibly under all conditions. When base-emitter current flows, emitter-collector current also flows.

5-3.1 Class A amplifier

In the class A amplifier, the base-emitter bias and the input voltages are such that the transistor operates only over the *linear portion* of the *characteristic curve*. Such a curve, representing the relationship between base voltage (input) and collector current (output), is shown in Fig. 5-6. At no point of the input signal cycle does the base become so positive or negative as to cause the transistor to operate at the nonlinear portion of the curve. The transistor collector current is never cut off nor does the transistor ever reach saturation.

The main advantage of the class A amplifier is the relative lack of distortion. The output waveform follows that of the input waveform, except in amplified form. With any class of amplifier, however, there is some distortion.

The disadvantages of class A amplifiers are their relative inefficiency (low-power output for a high-power input dissipated by the transistor) and their inability to handle large signal voltage swings. Rarely is a class A amplifier over about 35% efficient. Thus, if the power input to a class A amplifier is 1 W (generally, the maximum power dissipation capability of a single transistor), the output will be less than 0.3 W.

The peak-to-peak output signal voltage swing of a class A amplifier is limited to something less than the total supply voltage. Since the output voltage must swing both positive and negative, the peak output is less than one-half the supply voltage.

For example, assume that the supply voltage is 20 V and the amplifier is biased so that the *Q*-point collector voltage is one-half the supply, or 10 V. (Such a *Q* point is generally typical for a class A amplifier.) Under these conditions, the output voltage swing cannot exceed ± 10 V. If distortion must be kept to a minimum, the output will usually be on the order of ± 5 V so

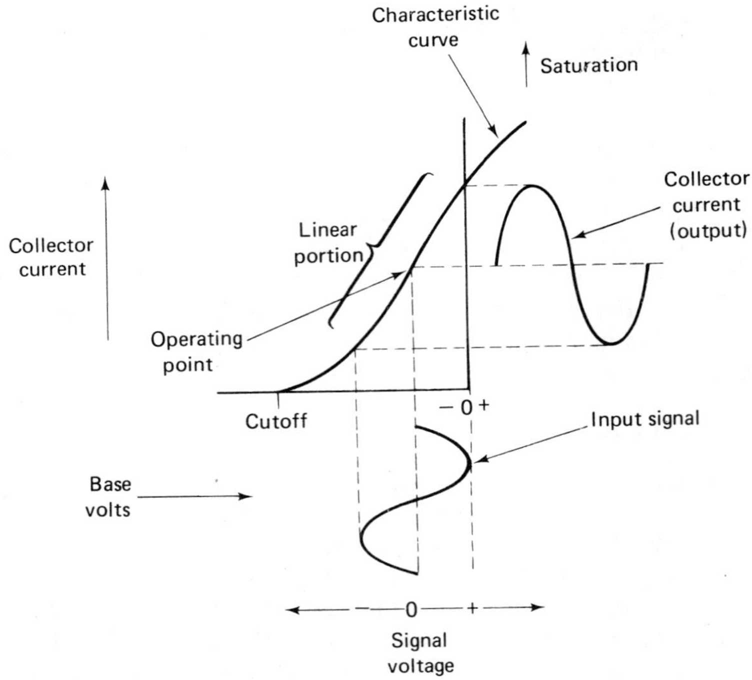

FIGURE 5-6 Typical class A amplifier characteristic curve. (Lenk, *Handbook of Modern Solid-State Amplifiers*, Prentice-Hall.)

as to keep the transistor on the linear portion of the characteristic curve. (In most cases, the curve becomes nonlinear near the cutoff and saturation points.) This can be determined, however, only from an actual test of the amplifier circuit.

The input voltage swing of a class A amplifier is limited by the output voltage swing capability and the voltage amplification factor. For example, if the output is limited to ± 10 V and the voltage amplification factor is 100, the input is limited to ± 0.1 V (100 mV).

Because of these limitations, class A amplifiers are generally used as voltage amplifiers rather than power amplifiers. Typically, a class A amplifier stage is used ahead of a power amplifier stage.

5-3.2 Class B amplifier

If the base-emitter bias is changed so that the operating point coincides with the transistor cutoff point, we obtain class B amplification. For an *NPN* transistor this means making the base more negative than for class A operation. (For *PNP* transistors, class B is obtained by making the base more positive than for class A.) Either way, the base-emitter reverse bias is increased for class B operation.

As shown in Fig. 5-7, when the input signal voltage is zero, there is no flow of collector current. During the positive half cycle of the signal voltage (Fig. 5-7 is for an *NPN* transistor), the collector current rises to its peak and then falls back to zero in step with the variations of that half cycle. During the negative half cycle of the signal voltage, there is no collector current since the base-emitter reverse bias is at all times greater than the cutoff voltage of the transistor. Hence, collector current flows only during half the input signal cycle.

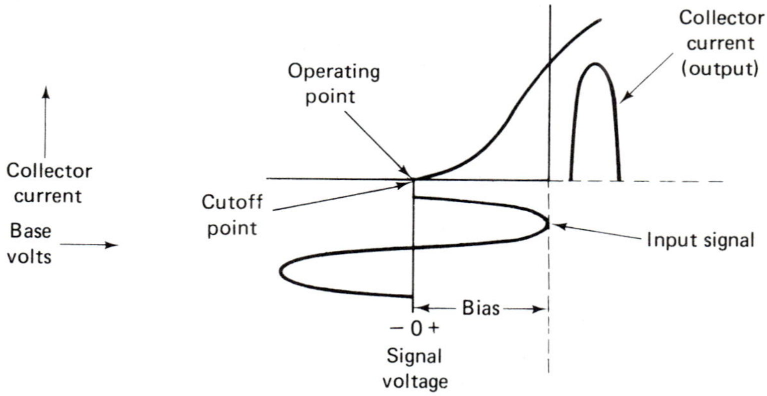

FIGURE 5-7 Typical class B amplifier characteristic curve. (Lenk, *Handbook of Modern Solid-State Amplifiers*, Prentice-Hall.)

If a single transistor is operated class B, there will be considerable distortion. This is because the waveform of the resulting collector current resembles that of the positive half cycle of the input signal and, consequently, does not resemble the complete waveform of the input. It is possible to use two transistors, one for each half cycle of the input signal, and by combining the outputs of these transistors in *push-pull* to reconstruct an output whose waveform resembles the full waveform of the input.

The peak output voltage swing of a class B amplifier is slightly less than the supply voltage. Since the output appears only on half cycles, it is possible to operate class B amplifiers at a higher current (or power) rating than class A, all other factors being equal. For example, if a transistor is capable of 0.3-W dissipation (without damage) as class A, the same transistor can be operated at 0.6 W, class B, since the transistor is conducting collector current only half the time. (This is a theoretical example. In practice, there are factors that limit class B power dissipation to something less than twice that of class A.)

The peak output of a class B amplifier is equivalent to the peak-to-peak output of a class A amplifier. Thus, if two transistors are connected in push-pull and operated as class B, the *output voltage can be twice* that of class A.

Because of these voltage and power factors, class B amplifiers are generally used as power amplifiers rather than as voltage amplifiers. Typically, two push-pull transistors are operated in class B, preceded by a single class A amplifier stage. The class A stage provides voltage amplification; whereas the class B stage produces the necessary power amplification.

5-3.3 Class AB amplifier

Class B is the most efficient operating mode for audio amplifiers since it draws the least amount of current; that is, the transistors are cut off at the *Q* point and draw collector current only in the presence of an input signal. Class B operation can, however, result in a form of distortion known as *crossover distortion*. The effects of crossover distortion can be seen by comparing the input and output waveforms in Fig. 5-8.

In true class B operation, the transistor *remains cut off* at very low signal inputs (because transistors have very low current gain at cutoff) and turns on abruptly with a large signal. For example, a silicon transistor does not have appreciable collector-current flow until the base-emitter junction is forward biased by about 0.5 V. Assuming that the input signal starts at 0 V, there will be little or no collector-current flow (and thus no change in the output voltage) during the time that the input signal is going from 0 to 0.5 V. When the input reaches 0.5 V, the collector current increases rapidly and follows the input signal in a linear fashion.

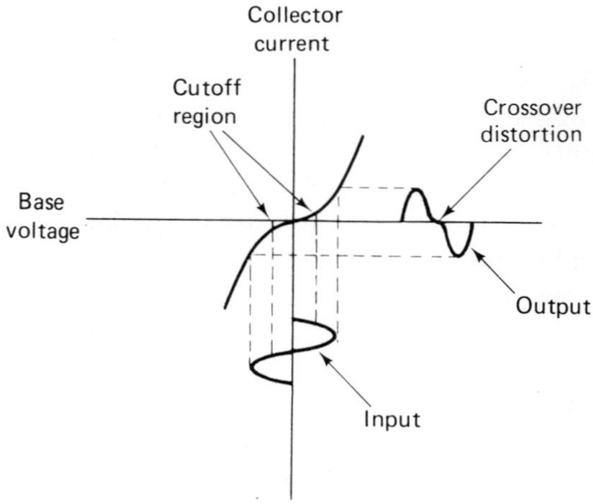

FIGURE 5-8 Example of crossover distortion. (Lenk, *Handbook of Modern Solid-State Amplifiers*, Prentice-Hall.)

Crossover distortion can be minimized by operating the stage as class AB (or somewhere between B and AB); that is, the transistors are forward biased just enough for a *small amount* of collector current to flow at the Q point. For a typical silicon transistor, the forward bias is just below 0.5 V for class AB. Therefore, some collector current is flowing at the lowest signal levels, and there is no abrupt change in current gain. Class AB is less efficient than class B since more current must be used. Generally class AB is only used in push-pull circuits.

5-3.4 Class C amplifier

If the transistor is reverse biased considerably below the cutoff point, we obtain a class C amplifier. As shown in Fig. 5-9, during the positive half cycle of the input signal, the signal voltage starts from zero, rises to the positive peak value, and falls back to zero. (Figure 5-9 is for an *NPN* transistor.) A *portion of the input signal* causes the base-emitter junction to be forward biased. As a result, there is a flow of collector current for a portion of one-half the input cycle. The negative half cycle of the input signal lies well below the cutoff point of the transistor.

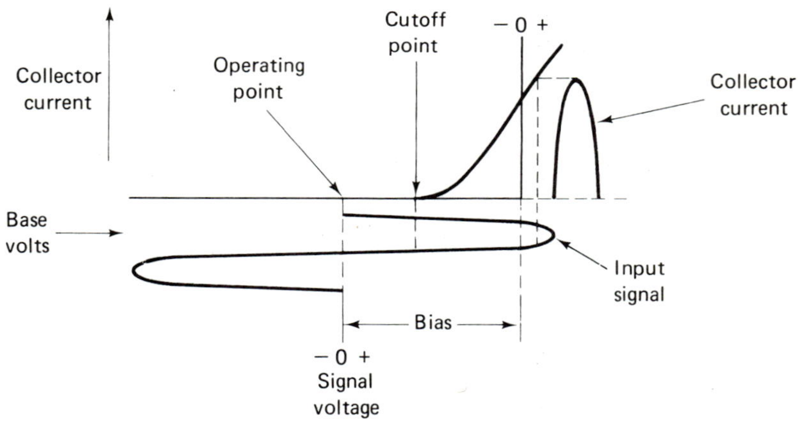

FIGURE 5-9 Typical class C amplifier characteristic curve. (Lenk, *Handbook of Modern Solid-State Amplifiers*, Prentice-Hall.)

Collector current flows only during that portion of the positive half cycle of the input signal between the cutoff point and the peak. The resulting collector current is a pulse, the duration of which is considerably less than a half cycle of the input signal.

Obviously, the waveform of the output signal cannot resemble that of the input signal nor can this resemblance be restored by the push-pull method mentioned in the discussion of class B and AB amplifiers. For this reason, class C is limited to those applications where distortion is of no concern.

Generally, class C operation is limited to use in *RF* amplifiers and is not found in *AF* amplifiers.

5-4. UJT REGENERATIVE AMPLIFIER

The basic UJT relaxation oscillator described in Chapter 4 can be adapted to form a regenerative pulse amplifier. Such an amplifier is shown in Fig. 5-10. Note that the circuit is essentially the basic relaxation oscillator with two resistors added. Resistor R_L provides a load for a trigger input or a pulse output. Resistor R_3 forms a voltage divider with resistor R_E. The voltage at the junction of R_3 and R_E is set so that the UJT will not fire except in the presence of a trigger; that is, R_3 and R_E have a ratio such that the emitter voltage does not exceed the peak-point (firing) voltage for the quiescent state.

As shown in Fig. 5-10, inputs can be applied to any of three points (base 1, base 2, or emitter). Likewise, outputs can be taken from any of these points. In operation, a trigger pulse is applied to one of the inputs, and a correspond-

$$R_L \approx Z \text{ of trigger or } Z \text{ of output load}$$

$$R_3 \approx \frac{R_E}{V_1 - V_{bias}} \times V_{bias}$$

$$R_E \gg \frac{V_1 - V_V}{I_V} \qquad V_{bias} = V_P - 0.5 \text{ trigger}$$

$$C_E \quad \frac{\text{period}}{R_E \times \ln \frac{1}{1 - \text{standoff ratio}}}$$

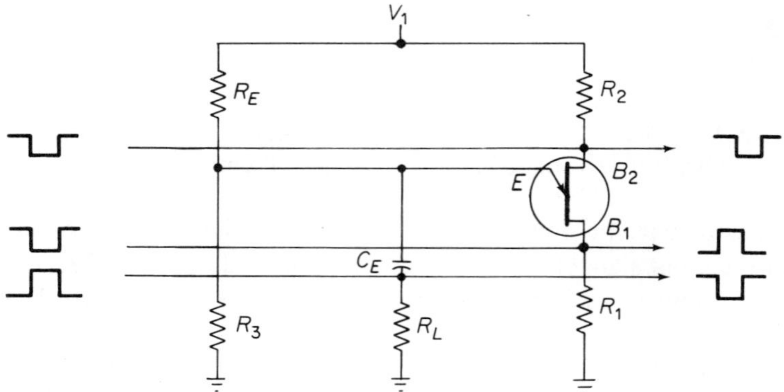

FIGURE 5-10 Basic UJT regenerative amplifier (courtesy General Electric).

ing trigger output (in greatly amplified form) is taken from another of the points. For example, a trigger of 0.1 V can be applied to the emitter, with a 7-V output taken from base 1.

5-4.1 Design considerations

The values of R_1, R_2, and C_E are selected on the same basis as discussed for the UJT relaxation oscillator in Chapter 4. The time constant for R_E and C_E should be approximately the same as the period of the trigger.

The value of R_E is found in the same way essentially as for the basic oscillator but with a minor difference. The maximum value of R_E need not be of great concern in the regenerative amplifier. The UJT is held in the off condition until a trigger is applied by the voltage drop across R_3 (or by the voltage division at the junction of R_E and R_3). Thus, no calculation need be made for $R_{E(\text{max})}$. The minimum value for R_E is found in the same way as for the basic oscillator; that is, the minimum R_E that can be used with the regenerative amplifier is set by

$$R_E > \frac{V_1 - V_V}{I_V} = R_{E(\text{min})}$$

In order to assure turn-off after the trigger pulse is removed, however, the value of R_E should be two to three times larger than $R_{E(\text{min})}$.

The value of R_L is selected to match the load impedance of the input trigger source, or the output, depending on application.

The value of R_3 is set by the desired bias voltage. The ratio of R_E and R_3 sets the fixed bias that must be overcome by the input trigger. V_{bias} is set at V_P (peak voltage of the UJT), less one-half the input trigger.

5-5. BASIC TWO-JUNCTION TRANSISTOR AMPLIFIER STAGE

Figure 5-11 is the working schematic of a basic, single-stage audio amplifier using a two-junction transistor. Note that the basic audio circuit is similar to the bias circuits discussed in Chapter 2, except that input and output coupling capacitors C_1 and C_2 are added to prevent dc flow to and from external circuits. A bypass capacitor C_3 is shown connected across the emitter resistor R_E. C_3 is required only under certain conditions, as discussed in later paragraphs.

Input to the amplifier is applied between base and ground across R_B. Output is taken across the collector and ground. The input signal adds to, or subtracts from, the bias across R_B. Variations in bias voltage cause corresponding variations in base current, collector current, and the drop across

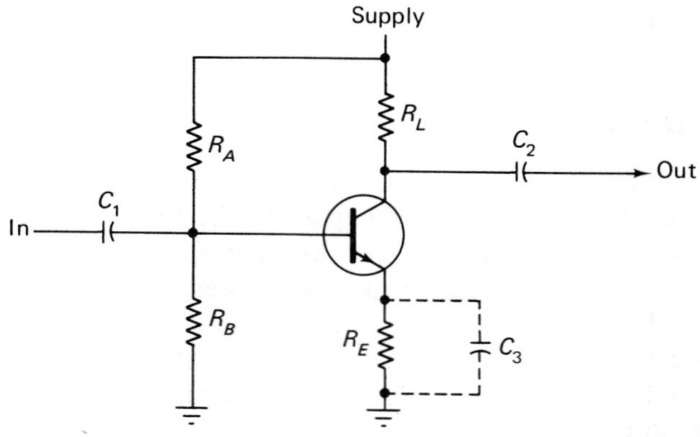

$Z_{in} \approx R_B$

$Z_{out} \approx R_L$

$I_{gain} \approx \frac{R_B}{R_E}$

$V_{gain} \approx \frac{R_L}{R_E}$

$R_L > 5R_E$

$R_L \approx 10R_E$

$R_B \approx 10R_E$

$R_B < 20R_E$

$R_E \approx 100 - 1000\ \Omega$

$V_{collector} \approx 0.5 \times \text{supply (set by } R_A)$

$I_{gain} \approx \text{stability} \approx \frac{R_B}{R_E}$

$S \approx 20$ for high gain

$S \approx 10$ for stability

$S \approx 5$ for power gain

$C_1 \approx \frac{1}{3.2\,FR}$ for 1 dB

$C_2 \approx \frac{1}{6.2\,FR}$ for 3 dB

FIGURE 5-11 Basic single-stage, audio amplifier using a two-junction transistor. (Lenk, *Handbook of Modern Solid-State Amplifiers*, Prentice-Hall.)

collector resistor R_L. Therefore, the collector voltage (or circuit output) follows the input signal waveform except that the output is inverted in phase. (If the input swings positive, the output swings negative and vice versa.)

Variations in collector current also cause variations in emitter current. This results in a change of voltage drop across the emitter resistor R_E and a change in the base-emitter bias relationship. As discussed in Chapter 2, the change in bias that results from the voltage drop across R_E tends to cancel the initial bias change caused by the input signal and serves as a form of negative feedback to increase stability (and limit gain). This form of emitter feedback (current feedback) is known as *stage feedback* or *local feedback* since only one stage is involved. As discussed in later sections, *overall feedback* or *loop feedback* is sometimes used where several stages are involved.

5-5.1 Circuit analysis

The outstanding characteristic of the circuit in Fig. 5-11 is that circuit characteristics (gain, stability, impedance) are determined (primarily) by circuit values rather than transistor characteristics (beta). The circuit is shown with an *NPN* transistor. The power supply polarity must be reversed if a *PNP* transistor is used.

The maximum peak-to-peak output voltage is set by the source voltage. For class A operation, the collector is operated at approximately one-half the source voltage. This permits the maximum positive and negative swing of output voltage. Generally, the absolute maximum peak-to-peak output can be between 90 and 95% of the source. For example, if the source is 20 V, the collector will operate at 10 V (*Q* point) and swing from about 1 to 19 V. There is less distortion, however, if the output is one-half to one-third of the source. In any circuit, the maximum collector voltage rating of the transistor cannot be exceeded.

The input and/or output impedances are set by the values of R_B and R_L, as shown by the equations of Fig. 5-11. Maximum power transfer occurs when R_B and R_L match the impedances of the previous stage and following stage, respectively.

Stability versus gain tradeoff. As shown by the equations, maximum voltage gain occurs when R_L is made larger in relation to R_E. Similarly, current gain increases when R_B is made larger in relation to R_E. Circuit stability is greatest, however, when R_B and R_L are made smaller in relation to R_E; that is, the circuit gain will remain more constant in the presence of temperature, supply voltage, or input signal changes when R_E in made larger in relation to R_B and R_L. Thus, there is a tradeoff between gain and stability.

In a practical amplifier, the circuit of Fig. 5-11 should be limited to a maximum current gain of 10 and a maximum voltage gain of 20. Higher gains are possible but the stability is generally poor. Of course, even though gain is set by circuit values, the minimum ac beta of the transistor must be higher than the desired gain. For example, if the circuit values are chosen for a gain of 20, the minimum beta must be 20 across the entire frequency range.

The *Q* point is affected by the values of all four resistors in Fig. 5-11; however, the value of R_A is the *final determining factor for* Q *point.* That is, in a practical amplifier, the remaining resistor values are selected for the desired gain, impedance, and stability of the circuit; then the value of R_A is selected (or adjusted) to give a desired operating point.

The values of C_1 and C_2 are dependent on the *low-frequency limit* at which the amplifier is to operate. C_1 forms a high-pass (or low-cut) *RC* filter with R_B. Capacitor C_2 forms another filter with the input resistance of the following stage (or the load). For a given resistance value, a lower frequency requires a larger capacitor value. Of course, if the resistance can be made larger (with the same desired frequency), the capacitor value can be reduced. Since two-junction transistors are low-impedance, current-operated devices, the coupling capacitors in two-junction amplifiers are generally large; that is, the coupling capacitors are generally large in relation to those of FET (and vacuum-tube) amplifiers.

5-5.2 Emitter bypass for transistor amplifier stages

Figure 5-11 shows (in phantom) a bypass capacitor C_3 across emitter resistor R_E. This arrangement permits R_E to be removed from the circuit as far as the signal is concerned but leaves R_E in the circuit (in regard to direct current). With R_E removed from the signal path, the voltage gain is approximately R_L divided by the dynamic resistance of the transistor, and the current gain is approximately equal to ac beta of the transistor. Thus, the use of an emitter-bypass capacitor permits the highly temperature stable dc circuit to remain intact while providing a high signal gain.

An emitter-bypass capacitor also creates some problems. Transistor input impedance changes with frequency, and from transistor to transistor, as does beta. Thus, current and voltage gains can only be approximated. When the emitter resistance is bypassed, the circuit input impedance is approximately beta times transistor input impedance, making circuit input impedance subject to variation and unpredictable.

The emitter bypass is generally used where maximum gain must be obtained from a single stage of amplification, and a stable gain is of little concern. The value of the emitter-bypass capacitor should be such that the *reactance is less than the transistor input impedance* at the lowest frequency of operation. This will effectively short the emitter (signal path) around R_E.

The value of C_3 can be found by

$$\text{Capacitance} = \frac{1}{6.2FR}$$

where capacitance is in farads

F is the low-frequency limit in hertz

R is the maximum input impedance of the transistor in ohms.

5-5.3 Basic audio amplifier with partially bypassed emitter

Figure 5-12 shows a basic single-stage two-junction transistor used as an audio amplifier with a partially bypassed emitter resistor. This circuit is a compromise between the basic amplifier without bypass and the fully bypassed emitter. The dc characteristics of both the unbypassed and partially bypassed circuits are essentially the same. The circuit values (except C_3 and R_C) are calculated in the same way for both circuits. The voltage and current gains for a partially bypassed amplifier are greater than an unbypassed circuit, however, but less than for the fully bypassed circuit.

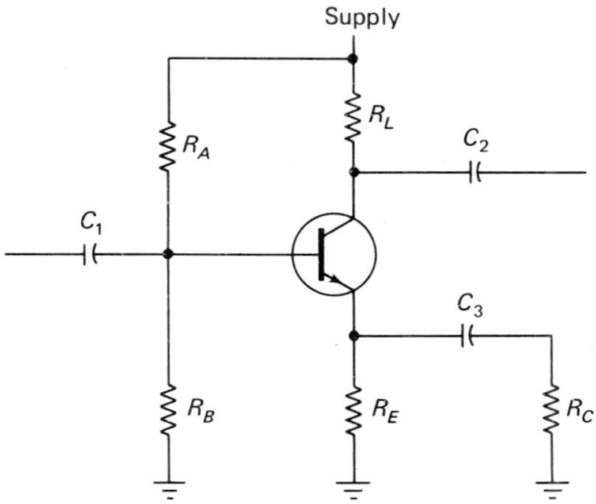

$$I_{gain} \approx \frac{R_B}{R_C} \qquad R_L < 50R_C$$

$$V_{gain} \approx \frac{R_L}{R_C} \qquad R_B < 50R_C$$

$$C_3 \approx \frac{1}{6.2FR_C}$$

FIGURE 5-12 Basic, single-stage, two-junction transistor used as an audio amplifier with partially bypassed emitter resistor. (Lenk, *Handbook of Modern Solid-State Amplifiers*, Prentice-Hall.)

The value of R_C is chosen on the basis of voltage gain, even though current gain will be increased when voltage gain increases. R_C should be substantially smaller than R_E. Otherwise, there will be no advantage to the partially bypassed circuit. A smaller value for R_C will require a larger value for C_3, however, since the C_3 value is dependent on the R_C value and the desired low-frequency cutoff point.

With the circuit of Fig. 5-12, current gain and voltage gain are dependent on the ratios of R_B/R_C and R_L/R_C, respectively. Thus, the value of R_E has little or no effect on circuit gain.

5-6. BASIC FET AMPLIFIER STAGE

Figure 5-13 shows a basic single-stage FET amplifier. The basic amplifier circuit is similar to the basic common-source circuit described in Chapter 3, except that input and output coupling capacitors C_1 and C_2 are added to prevent dc flow to and from external circuits. Bypass capacitor C_3, connected across R_S, is required only under certain conditions. Also note that

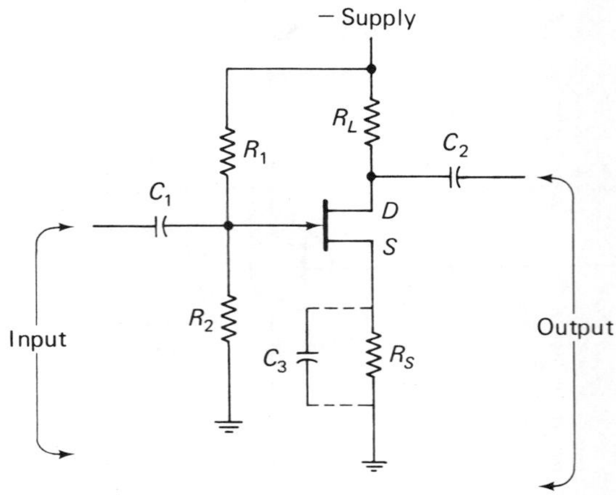

$$A_V \approx \frac{R_L}{\frac{1}{Y_{fs}} + R_S} \approx \frac{R_L}{R_S}$$

Drain voltage = 0.5 x supply

$$Z_{in} \approx R_1 \,||\, R_2 \approx R_2 \qquad C_1 \approx \frac{1}{3.2\,FR_2} \text{ for 1 dB}$$

$$Z_{out} \approx R_L \qquad C_1 \approx \frac{1}{6.2\,FR_2} \text{ for 3 dB}$$

FIGURE 5-13 Basic common-source FET amplifier stage. (Lenk, *Handbook of Modern Solid-State Amplifiers*, Prentice-Hall.)

the circuit of Fig. 5-13 is the FET version of the Fig. 5-11 circuit. Thus, many of the circuit characteristics are similar.

One major difference is that voltage gain in the FET circuit (Fig. 5-13) is set by factors of R_L, R_S, and y_{fs}. Another difference is that the values of coupling capacitors for FET circuits are much smaller than for two-junction transistors. This is because FETs are high-impedance, voltage-operated devices.

Sufficient feedback. The design of an FET amplifier stage can be checked by noting if there is sufficient feedback. Such a condition occurs when the calculated gain is at least 75% of the R_L/R_S ratio. If so, there is sufficient feedback to be of practical value. As an example, assume that R_L is 16.9 kΩ, R_S is 2.2 kΩ, any y_{fs} is 2000 μmhos. The ratio of R_L/R_S is slightly over 7.6, with the gain slightly over 6.

$$\frac{R_L}{R_S} = \frac{16.9}{2.2} = 7.6+; \qquad \text{gain} \frac{16.9\text{ k}\Omega}{(\frac{1}{2000}\ \mu\text{mho}) + 2.2\text{ k}\Omega} = \frac{16.9\text{ k}\Omega}{2.7\text{ k}\Omega} = 6+$$

Since 75% of 7.6 is 5.7, the gain of 6 is greater and there is sufficient feedback. Under these conditions the design should be stable.

5-6.1 Source bypass for FET amplifier stage

Figure 5-13 shows (in phantom) a bypass capacitor C_3 across source resistor R_S. This arrangement permits R_S to be removed from the circuit as far as signal is concerned but leaves R_S in the circuit (in regards to direct current). With R_S removed from the signal path, the voltage gain is approximately equal to $y_{fs} \times R_L$. Thus, the use of a bypass capacitor permits a temperature-stable dc circuit to remain intact, while providing a high signal gain.

5-6.2 Other single-stage FET amplifiers

Figures 5-14 through 5-18 show the circuit diagrams and design equations for some typical FET amplifier configurations. The following notes apply to these circuits.

The value of R_C in Fig. 5-14 is chosen on the basis of desired voltage gain. R_C should be substantially smaller than R_S. Otherwise, there will be no advantage to the partially bypassed design. As shown by the equations, voltage gain is approximately equal to R_L/R_C. This holds true unless both y_{fs} and R_C are very low (where $1/y_{fs}$ is about equal to R_C). In such a case,

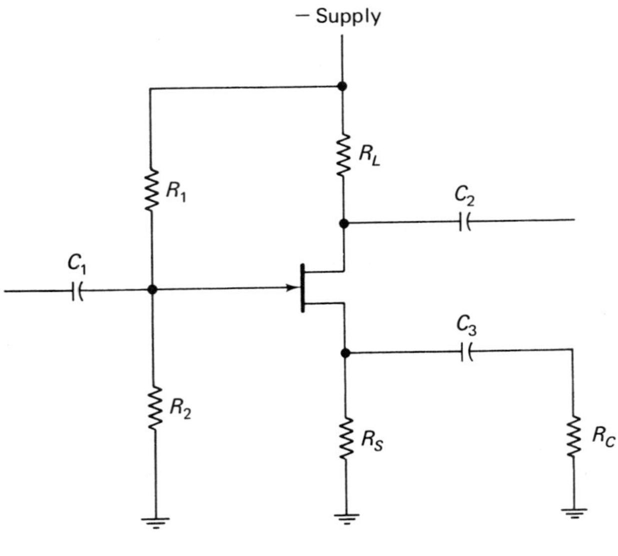

$$A_V \approx \frac{R_L}{R_C} \qquad C_3 = \frac{1}{6.2\,FR_C}$$

FIGURE 5-14 Basic FET audio amplifier with partially bypassed source resistor. (Lenk, *Handbook of Modern Solid-State Amplifiers*, Prentice-Hall.)

a more accurate gain approximation is

$$\frac{R_L}{(1/y_{fs}) + R_C}$$

The circuit of Fig. 5-15 is used primarily where high input impedance and low output impedance (with no phase inversion) are required but no gain is needed. The source follower is the FET equivalent of the two-junction transistor emitter follower and the vacuum-tube cathode follower.

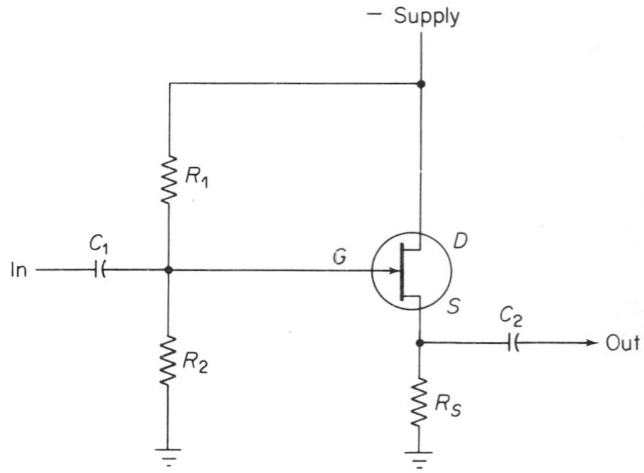

$$C_1 = \frac{1}{6.2\,FR_2} \qquad A_V \approx \frac{R_S}{\frac{1}{Y_{FS}} + R_S} \approx 0.6 \times \text{Input}$$

$$Z_{in} \approx R_1 \,||\, R_2 \approx R_2$$

$$Z_{out} \approx R_S \,||\, \frac{1}{Y_{FS}}$$

FIGURE 5-15 Basic common-drain (source-follower) FET amplifier stage.

The circuit of Fig. 5-16 is used primarily where low input impedance and high output impedance (with no phase inversion) are required. Gain is determined (primarily) by circuit values rather than FET characteristics. The common-gate amplifier is the FET equivalent of the two-junction common-base amplifier and the vacuum-tube common-grid amplifier.

The circuit of Fig. 5-17 has no fixed bias. The major difference in this circuit and the FET amplifiers with fixed bias is that the amount of I_D at the Q point is set entirely by the value of R_S. It may not be possible to achieve a

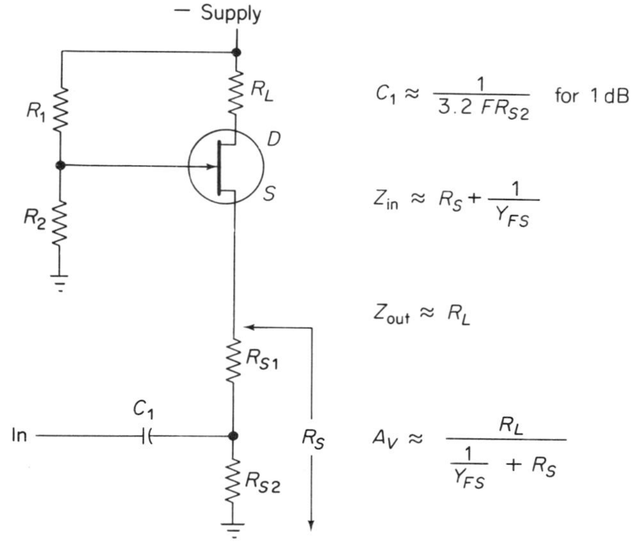

FIGURE 5-16 Basic common-gate FET amplifier stage.

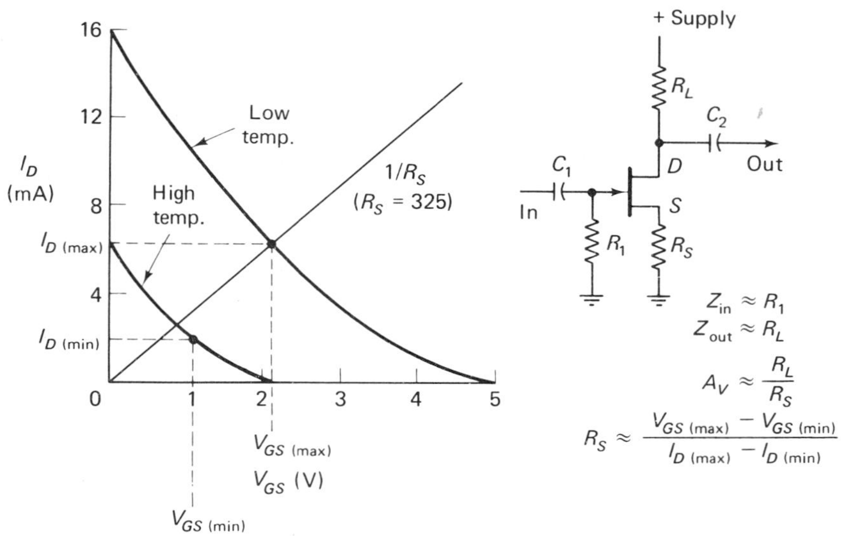

FIGURE 5-17 Basic FET amplifier without fixed bias. (Lenk, *Handbook of Modern Solid-State Amplifiers*, Prentice-Hall.)

desired I_D with a practical value of R_S. Thus, it may not be possible to operate at the zero-temperature-coefficient point. If this is of less importance than minimizing the number of circuit components (elimination of one resistor), the circuit of Fig. 5-17 can be used in place of the fixed-bias FET amplifier.

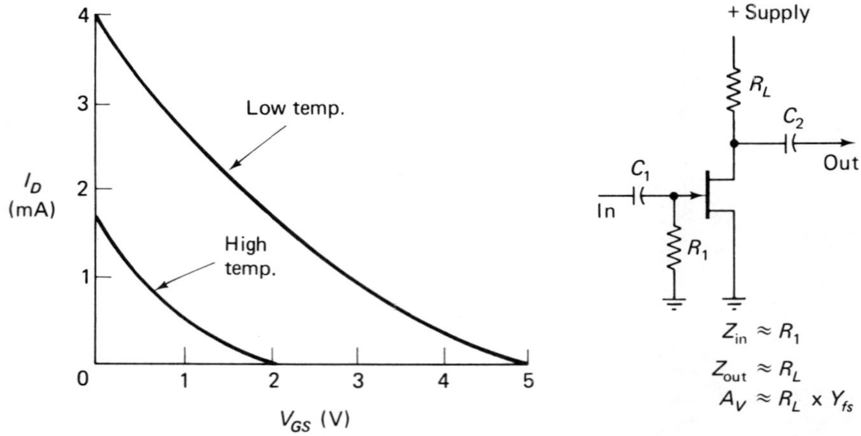

FIGURE 5-18 Basic FET amplifier with zero bias. (Lenk, *Handbook of Modern Solid-State Amplifiers*, Prentice-Hall.)

The circuit of Fig. 5-18 is also without fixed bias and without stabilizing feedback. Any FET will have some value of I_D at zero V_{GS}. If the FET has characteristics similar to those of Fig. 5-18, the I_D will vary between about 1.75 and 4 mA, depending on temperature, and from FET to FET. Therefore, with a zero-bias circuit it is impossible to set the Q point I_D at any particular value. Similarly, drain-voltage Q point is subject to considerable variation. Since there is no source resistor, there is no negative feedback. Thus, there is no means to control this variation in I_D. For these reasons, the zero-bias circuit is used where circuit stability is of no particular concern.

6

Radio-Frequency Considerations for Transistors

Both two-junction transistors and FETs are used as *RF* amplifiers. Generally, FETs are limited to voltage-amplifier applications. Two-junction transistors can be used in either voltage-amplifier or power-amplifier circuits. Practically any of the classic vacuum-tube *RF*-amplifier circuits can be duplicated with transistors. Since the number of different amplifier circuits is almost unlimited, it is impossible to cover all aspects of *RF* amplifiers in this chapter. As discussed in Chapter 5, *Handbook of Modern Solid-State Amplifiers*, by John D. Lenk, Prentice-Hall, Inc., Englewood Cliffs, N.J., 1974, provides a detailed discussion of many amplifier types.

The most needed information regarding solid-state *RF* amplifiers is the design of tuning networks. Such networks not only provide tuning to the desired operating frequency but match transistor characteristics to input and output circuit impedances. For that reason, we shall concentrate on problems associated with these networks.

All the design considerations for transistors discussed in Chapters 1, 2, and 3 apply to *RF* amplifiers. Of particular importance are how to interpret datasheets, determine parameters at different frequencies, temperature-related design problems, and basic bias schemes.

6-1. RESONANT CIRCUITS FOR *RF* AMPLIFIERS

RF-amplifier design is based on the use of resonant circuits (tank circuits) consisting of a capacitor and a coil (inductance) connected in series or parallel, as shown in Fig. 6-1. At the resonant frequency, the inductive and

Capacitive Reactance

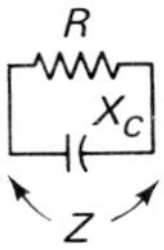

$$Z = \frac{RX_C}{\sqrt{R^2 + X_C^2}} \qquad Q = \frac{R}{X_C}$$

Parallel

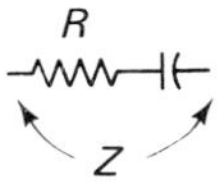

$$Z = \sqrt{R^2 + X_C^2} \qquad Q = \frac{X_C}{R}$$

Series

$$X_C = \frac{159}{F\ (\text{kHz}) \times C\ (\mu\text{F})}$$

$$F\ (\text{kHz}) = \frac{159}{(X_C) \times C\ (\mu\text{F})}$$

$$C\ (\mu\text{F}) = \frac{159}{F\ (\text{kHz}) \times X_C}$$

Inductive Reactance

$$Z = \frac{RX_L}{\sqrt{R^2 + X_L^2}} \qquad Q = \frac{R}{X_L}$$

Parallel

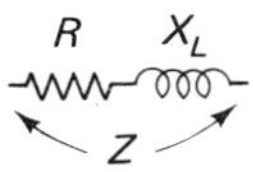

$$Z = \sqrt{R^2 + X_L^2} \qquad Q = \frac{X_L}{R}$$

Series

$$X_L = 6.28 \times F\ (\text{kHz}) \times L\ (\text{mH})$$

$$F = \frac{X_L}{6.28L}$$

$$L = \frac{X_L}{6.28F}$$

Impedance and Resonance

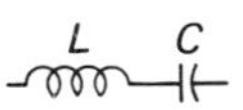

Series (zero impedance)

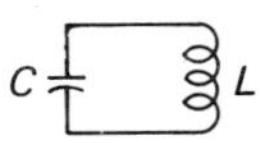

Parallel (infinite impedance*)

$$F\ (\text{kHz}) = \frac{159}{\sqrt{L\ (\mu\text{H}) \times C\ (\mu\text{F})}}$$

$$L\ (\mu\text{H}) = \frac{2.54 \times 10^4}{F\ (\text{kHz})^2 \times C\ (\mu\text{F})}$$

$$C\ (\mu\text{F}) = \frac{2.54 \times 10^4}{F\ (\text{kHz})^2 \times L\ (\mu\text{H})}$$

FIGURE 6-1 Resonant circuit equations. (Lenk, *Handbook of Modern Solid-State Amplifiers*, Prentice-Hall.)

capacitive reactances are equal, and the circuit acts as a high impedance (if it is a parallel circuit) or a low impedance (if it is a series circuit). In either case, any combination of capacitance and inductance has some resonant frequency.

Either (or both) the capacitance or inductance can be variable to permit tuning of the resonant circuit over a given frequency range. When the inductance is variable, tuning is usually done by means of a metal slug (usually powdered iron) inside the coil. The metal slug is screwdriver adjusted to change the inductance (and thus the inductive reactance) as required.

Typical *RF* circuits used in receivers (*AM*, *FM*, communications, etc.) often include two resonant circuits in the form of a transformer (*RF* or *IF* transformer, etc.). Either the capacitance or inductance can be variable.

In the case of *RF* transmitters, it is sometimes necessary to design the coil portion of the resonant circuit. This is because coils of a given inductance and physical size may not be available from commerical sources.

6-1.1 Basic design considerations for resonant circuits

The two most important considerations for *RF* resonant circuits are *resonant frequency* and the *Q* (or *quality*) factor.

Resonant frequency. Figure 6-1 contains equations that show the relationship among capacitance, inductance, reactance, and frequency as they relate to resonant circuits. Note that there are two sets of equations. One set includes reactance (inductive and capacitive). The other set omits reactance. The reason for two sets of equations is that some design approaches require the reactance to be calculated for resonant networks. Solid-state *RF* transmitter circuits are a classic example of this.

Quality factor and selectivity. A resonant circuit has a *Q*, or quality, factor. This *Q* factor is directly related to the selectivity of the circuit and is dependent on the ratio of reactance to resistance. If a resonant circuit has pure reactance, the *Q* is high (theoretically infinite); however, this is not practical. For example, any coil will have some dc resistance, as will the leads of a capacitor. Also, as frequency increases, the ac resistance presented by the leads will increase due to skin effect. The sum total of these resistances is usually lumped together and is considered as a resistor in series or parallel with the resonant circuit. The total resistance is usually termed the *effective* resistance and is not to be confused with the reactance.

The resonant circuit *Q* is dependent on the individual *Q* factors of the inductance and capacitance used in the circuit. For example, if both the inductance and capacitance have a high *Q*, the circuit will have a high *Q*, provided that a minimum of resistance is produced when the inductance and capacitance are connected to form a resonant circuit.

Usually, resonant circuit *Q* is measured at points on either side of the

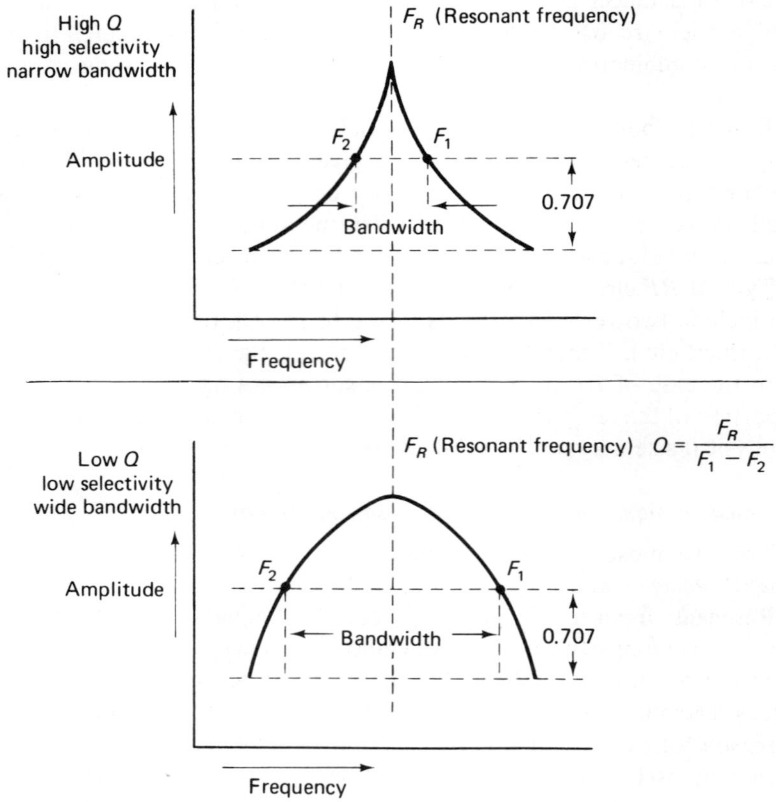

FIGURE 6-2 Relationship of Q to bandwidth and selectivity. (Lenk, *Handbook of Modern Solid-State Amplifiers*, Prentice-Hall.)

resonant frequency where the signal amplitude is down 0.707 of the peak resonant value; see Fig. 6-2. As shown, a resonant circuit with a high Q produces a sharp resonance curve (narrow bandwidth); whereas a low Q produces a broad resonance curve (wide bandwidth). For example, a high-Q resonant circuit will provide good harmonic rejection (tend to pass only the fundamental frequency) and efficiency in comparison with a low-Q circuit, all other factors being equal. Thus, the *selectivity* of a resonant circuit is related directly to Q.

A very high Q (or high selectivity) is not always desired. In some applications, it is necessary to add resistance to a resonant circuit to broaden the response (increase the bandwidth, decrease the selectivity). An example of this is the damping resistor used across peaking coils in a video amplifier.

If a given bandwidth must be maintained, but the resonant frequency is increased, the Q must also increase. For example, if the resonant frequency is 10 MHz with a bandwidth of 2 MHz, the required circuit Q is 5. If the resonant frequency is increased to 50 MHz, with the same 2-MHz bandwidth, the required Q is 25. Also, Q must be decreased for increases in bandwidth if the same resonant frequency is to be maintained.

6-1.2 Basic design considerations for RF *coils*

The equations necessary to calculate the self-inductance of a single-layer air-core coil are given in Fig. 6-3. Such coils are most efficient (that is, maximum inductance for minimum physical size) when the ratio of coil radius to coil length is 1.25 (that is, when the length is 0.8 times the radius).

The equations of Fig. 6-3 are approximations only and do not take into account such factors as uneven sizes of turns, spacing between turns, and the like. From a practical standpoint, use the equations to find the nearest number of turns (for a given inductance), and then spread or compress the turns as necessary to obtain a precise value of inductance (as measured on an inductance bridge).

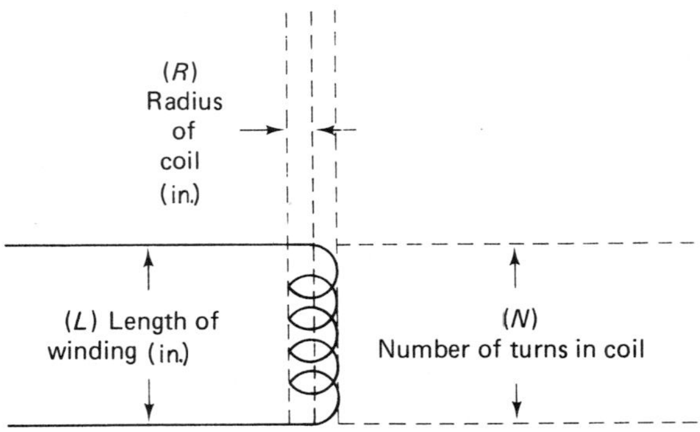

$$\text{Inductance } (\mu\text{H}) \approx \frac{(RN)^2}{9R + 10L}$$

When $L = 0.8 \times R$ or $R/L = 1.25$

$$N \approx \sqrt{\frac{17 \times \text{inductance } (\mu\text{H})}{R}}$$

$$\text{Inductance } (\mu\text{H}) \approx \frac{(RN)^2}{R17}$$

FIGURE 6-3 Calculations for self-inductance of single-layer air-core coil. (Lenk, *Handbook of Modern Solid-State Amplifiers*, Prentice-Hall.)

6-2. BASIC *RF*-AMPLIFIER DESIGN APPROACHES

Transistor *RF* amplifiers can be designed using two-port networks. Basically, the method consists of characterizing the transistor as a linear active two-port network (LAN) with admittances (y parameters) and using the parameters to solve design equations for stability, gain, and input/output admittances. The two-port design approach is best suited for *voltage amplifiers*, using *small-signal* characteristics or parameters. The two-port system is recommended for all FET *RF* amplifiers and for low-power two-junction *RF* amplifiers.

Design based on *large-signal* characteristics (transistor input/output resistance and capacitance) is recommended for *RF* power amplifiers (using high-power two-junction transistors). Both design approaches are discussed throughout the remainder of this chapter.

With either approach, it is difficult, at best, to provide simple, step-by-step procedures for designing *RF* amplifiers to meet all possible circuit conditions. In practice, the procedure often results in considerable trial and error. There are several reasons for this problem of *RF* amplifier design.

First, not all the characteristics are always available in datasheet form. For example, input and output admittances may be given at some low frequency but not at the desired operating frequency.

Often, manufacturers do not agree on terminology. A classic example of this is in y parameters, where one manufacturer uses letter subscripts (y_{fs}) and another uses number subscripts (y_{21}). Of course, this can be solved by conversions.

In some cases, manufacturers will give the required information on datasheets but not in the required form. For example, instead of giving input admittance in mhos, the input capacitance is given in farads. The input admittance is found when the input capacitance is multiplied by $6.28f$ (where f is the frequency of interest). This is based on the assumption that the input admittance is primarily capacitive and thus dependent on frequency. This assumption is not always true for the frequency of interest. Thus, it may be necessary to make actual tests of the transistor, using complex admittance measuring equipment.

The input and output tuning circuits of an *RF* amplifier must perform two functions. Obviously, the circuits (capacitors and coils) must tune the amplifier to the desired frequency. The circuits must also match the input and output impedances of the transistor to the impedances of the source and load. Otherwise, there will be considerable loss of signal.

Finally, there is some feedback between output and input of any transistor amplifier. If the admittance factors are just right, the feedback will be of sufficient amplitude and of proper phase to cause oscillation of the amplifier. The amplifier is considered as *unstable* when this occurs.

Amplifier instability in any form is always undesirable and can be corrected by one of two procedures: feedback (neutralization) or by changes in the input/output tuning networks. Although the neutralization and tuning circuits are relatively simple, the equations for determining stability (or instability) and impedance matching are long and complex. Generally, such equations are best solved by computer-aided design methods.

In an effort to cut through this maze of information and complex equations, we shall discuss all the steps involved in *RF*-amplifier design. Armed with this information, the reader should be able to interpret datasheets or test information and use the information to design tuning networks that will provide stable *RF* amplification at the frequencies of interest. With each step we shall discuss the various alternative procedures and types of information available. Specific design examples are given. These examples summarize the information contained in the various steps.

On the assumption that all readers may not be familiar with two-port networks, we shall start with a summary of the *y*-parameter system.

6-2.1 y *parameters*

Impedance (Z) is a combination of resistance (the real part) and reactance (the imaginary part). Admittance (y) is the reciprocal of impedance and is composed of conductance (the real part) and susceptance (the imaginary part). A y parameter is an expression for admittance in the form

$$y_{is} = g_{is} + jb_{is}$$

where g_{is} is the real (conductive) part of common-source input admittance and b_{is} is the imaginary (susceptive) part of input admittance.

The term $y_{is} = g_{is} + jb_{is}$ expresses the y parameter in *rectangular form.* Some manufacturers describe the y parameter in *polar form.* For example, they will give the *magnitude* of forward transadmittance as $|y_{fs}|$ and the *angle* of forward transadmittance as $\angle y_{fs}$. Quite often, the manufacturers will mix the two systems of vector algebra on their datasheets.

Conversion of vector algebra forms. It is assumed that the readers are already familiar with the basics of vector algebra; however, the following notes summarize the steps necessary to manipulate vector algebra terms. This should be sufficient to perform all calculations involved in design of *RF*-amplifier networks.

Converting from rectangular to polar form:

1. Find the magnitude from the square root of the sum of the squares of the components.

$$\text{Polar magnitude} = \sqrt{g^2 + jb^2}$$

2. Find the angle from the ratio of the component values.

$$\text{Polar angle} = \text{arc tan } \frac{jb}{g}$$

The angle is leading if the jb term is positive and lagging if the jb term is negative.

For example, assume that the y_{fs} is given as $g_{fs} = 75$ and $jb_{fs} = 140$. This is converted to polar form by

$$|y_{fs}| \text{ polar magnitude} = \sqrt{(75)^2 + (140)^2} = 159$$

$$\angle y_{fs} \text{ polar angle} = \text{arc tan } \tfrac{140}{75} = 6.13 \text{ deg}$$

Converting from polar to rectangular form:

1. Find the real (conductive or g) part when polar magnitude is multiplied by cosine of polar angle.
2. Find the imaginary (susceptance or jb) part when polar magnitude is multiplied by sine of the polar angle.

If the angle is positive, the jb component is also positive. When the angle is negative, the jb component is also negative.

For example, assume that the y_{fs} is given as $|y_{fs}| = 125$ and $\angle y_{fs} = -43$ deg. This is converted to rectangular form by

$$125 \cos \theta - j125 \sin \theta = g_{fs}91.7 - jb_{fs}85.2$$

The four basic y parameters. A y-equivalent circuit is shown in Fig. 6-4. This equivalent circuit is for an FET; however, a similar circuit can be drawn for a two-junction transistor when analyzing small-signal characteristics.

Note that y parameters can be expressed with number subscripts or letter subscripts. The number subscripts are universal since they can apply to two-junction transistors, FETs, and even integrated-circuit amplifiers; however, the letter subscripts are most popular on FET datasheets.

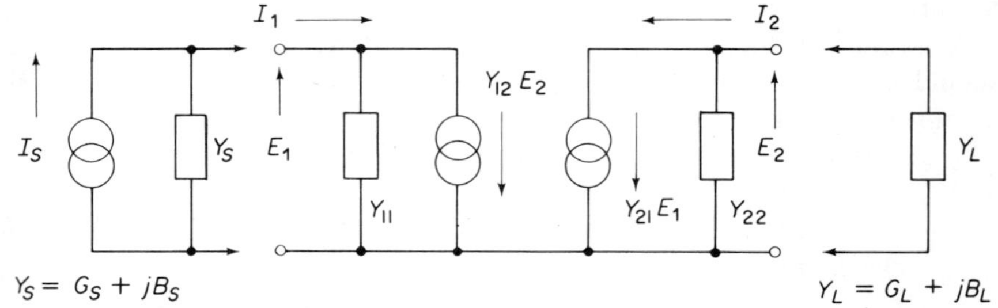

FIGURE 6-4 MOSFET y-equivalent circuit with source and load.

The following notes can be used to standardize y-parameter nomenclature (note that the letter s in the letter subscripts refers to common-source operation of an FET amplifier).

y_{11} is input admittance and can be expressed as y_{is} on an FET datasheet.
y_{12} is reverse transadmittance and can be expressed as y_{rs} on an FET datasheet.
y_{21} is forward transadmittance and can be expressed as y_{fs} on an FET datasheet.
y_{22} is output admittance and can be expressed as y_{os} on an FET datasheet.

Input admittance, with $Y_L = \infty$ (short circuit) is expressed as

$$y_{11} = g_{11} + jb_{11} = \left.\frac{\Delta i_1}{\Delta e_1}\right|_{e_2=0}$$

Some datasheets do not show y_{11} at any frequency. Instead, input capacitance is given. If we assume that the input admittance is entirely (or mostly) capacitive, then the input impedance can be found when input capacitance is multiplied by $6.28F$ (F is frequency in hertz) and the reciprocal is taken. Since admittance is the reciprocal of impedance, admittance is found when input capacitance is multiplied by $6.28F$ (where admittance is capacitive). For example, if the frequency is 100 MHz and the input capacitance is 6 pF, the input admittance is

$$6.28 \times (100 \times 10^6) \times (6 \times 10^{-12}) \approx 3.8 \text{ mmhos}$$

This assumption is accurate only if the real part of y_{11} (or g_{11}) is negligible. Such an assumption is reasonable for an FET but not necessarily for a two-junction transistor. Figure 6-5 shows input admittance curves for a

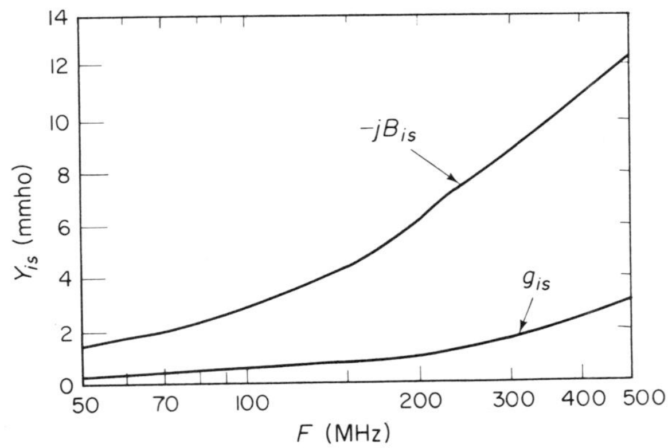

FIGURE 6-5 Input admittance, $Y_{is} = g_{is} + jb_{is}$ of MFE3007 (courtesy Motorola).

typical FET. Note the imaginary part (jb_{11}) is the more significant factor across the entire frequency range. The real or g_{11} part of two-junction transistor input admittance can be quite large in relation to the imaginary jb_{11} part.

Forward transadmittance, with $Y_L = \infty$ (short circuit), is expressed as

$$y_{21} = g_{21} + jb_{21} = \left.\frac{\Delta i_2}{\Delta e_1}\right|_{e_2=0}$$

Some FET datasheets show y_{21} at a low frequency (typically 1 kHz), and then show $Re(y_{fs})$, or the real part of y_{21}, at a high frequency (typically 100 to 200 MHz). Other FET datasheets specify that the g_{21} is the real part of y_{21} and that the values given are for a low frequency. Then some value is given for y_{21} at a high frequency. Often, two-junction transistor datasheets do not give any value for y_{21}. Instead, they show forward transadmittance by means of the hybrid system (Chapter 1) using h_{fe} or h_{21}. No matter what system is used, it is essential that the values of forward transadmittance be considered at the frequency of interest.

These considerations are illustrated in Fig. 6-6, which shows more accurate and complete forward transadmittance curves for a typical FET. Note that the real and imaginary parts actually cross over at about 180 MHz. Also, the real part becomes a negative quantity at about 400 MHz.

Output admittance, with $Y_S = \infty$ (short circuit), is expressed as

$$y_{22} = g_{22} + jb_{22} = \left.\frac{\Delta i_2}{\Delta e_2}\right|_{e_1=0}$$

Figure 6-7 shows output admittance curves for a typical FET. Note that the real part is negligible over the entire frequency range. Such a condition is not true for a typical two-junction transistor.

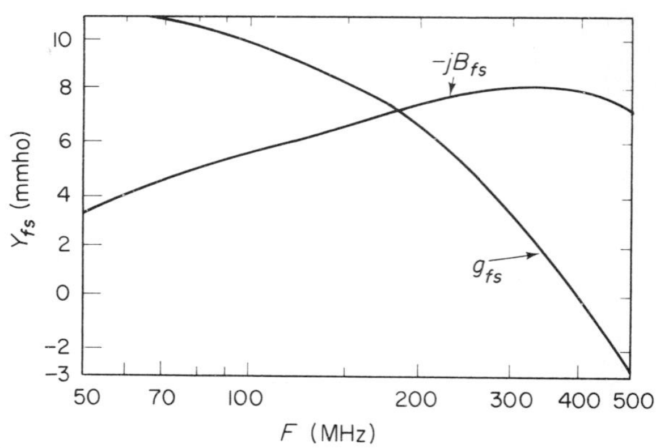

FIGURE 6-6 Forward transfer admittance, $Y_{fs} = g_{fs} + jb_{fs}$ of MFE3007 (courtesy Motorola).

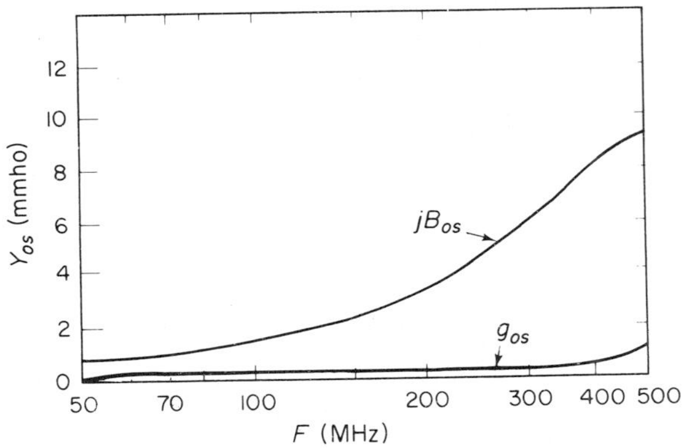

FIGURE 6-7 Output admittance, $Y_{os} = g_{os} + jb_{os}$ of MFE3007 (courtesy Motorola).

Reverse transadmittance, with $Y_S = \infty$ (short circuit), is expressed as

$$y_{12} = g_{12} + jb_{12} = \left.\frac{\Delta i_1}{\Delta e_2}\right|_{e_1=0}$$

Many datasheets do not list y_{12} at any frequency. Instead, reverse transfer capacitance is given. If we assume that the reverse transadmittance (or reverse transfer admittance, as it may be called) is entirely (or mostly) capacitive (jb_{12}), then the reverse transadmittance can be found when reverse transfer capacitance is multiplied by $6.28F$ (F equals frequency in hertz).

This assumption is generally accurate in the case of y_{12} for FETs, as shown in Fig. 6-8. The real part of y_{12} (or g_{12}) is zero across the entire frequency range. Thus, when the term $Re(y_{12})$ appears in an equation (as it

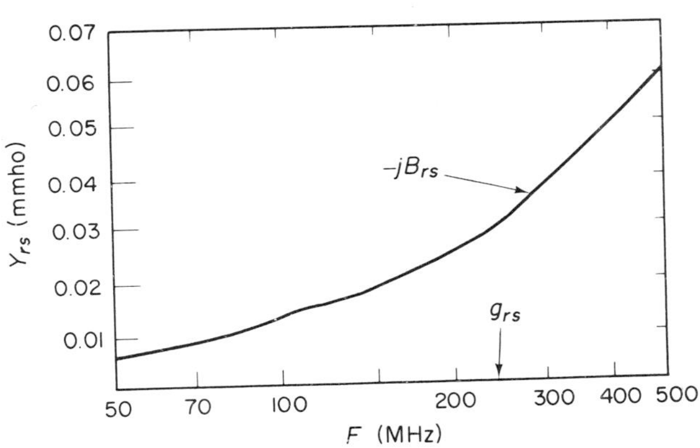

FIGURE 6-8 Reverse transfer admittance, $Y_{rs} = g_{rs} + jb_{rs}$ of MFE3007 (courtesy Motorola).

does frequently in FET *RF* design equations), the term can be considered as zero for all FET applications. This is not necessarily true for two-junction transistors.

***y* parameter measurement.** As seen thus far, *y*-parameter information is not always available or not in convenient form. In practical design, it may be necessary to measure the *y* parameters, using laboratory equipment.

Note that all *y* parameters are based on *ratios* of input/output current to input/output voltage. For example, y_{12} is the ratio of output current to input voltage.

For FETs, y_{21} and y_{22} can be measured by using signal generators, voltmeters, and simple circuits. Likewise, y_{11} and y_{12} can be found by measuring c_{iss} and c_{rss} (input and reverse capacitances) using a simple capacitance meter and then calculating y_{11} and y_{12} based on the frequency of interest. These procedures are described in Chapter 3.

More accurate results are obtained if precision laboratory equipment is used. All four *y* parameters can be measured on a general radio transfer function and immittance bridge. A possible exception is y_{12}, which is typically very small in relation to other FET parameters. In the case of y_{12}, it is often more practical to measure c_{rss} and multiply by $6.28F$.

The main concern in measuring *y* parameters, with either FETs or two-junction transistors, is that the measurements are made under *conditions simulating those of the final circuit.* For example, supply voltages, bias voltages, and operating frequency should be identical (or close) to the final circuit. Otherwise, the tests can be misleading.

Two-junction transistor admittance measurements. Although the datasheets for transistors to be used as power amplifiers usually contain input and output admittance information, it may be helpful to know how this data is obtained.

A typical test-amplifier circuit for two-junction transistors is shown in Fig. 6-9. During test, the transistor is placed in the test circuit designed with *variable components* to provide wide tuning capabilities. This feature is necessary to ensure correct matching while characterizing a transistor at several power levels. Then the circuit is tuned for maximum power gain at each power level for which admittance data is desired.

After the test amplifier has been tuned for maximum power gain, the dc power, signal source, circuit load, and test transistor are disconnected from the circuit. The signal source and output load circuit connections are terminated in their characteristic resistances so that the total circuit impedance is unchanged by this removal. After performing these substitutions, complex admittances are measured at the base and collector circuit connections of the test transistor (points *A* and *B*, respectively, of Fig. 6-9). The transistor input and output admittances are the *conjugates* of the base circuit connection and the collector circuit connection admittances, respectively. For example, if the

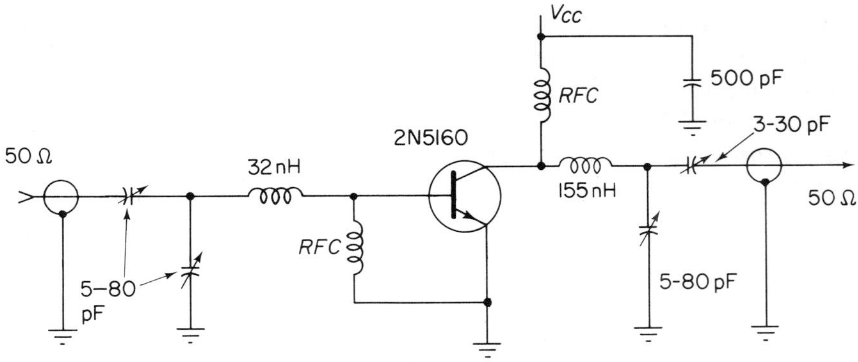

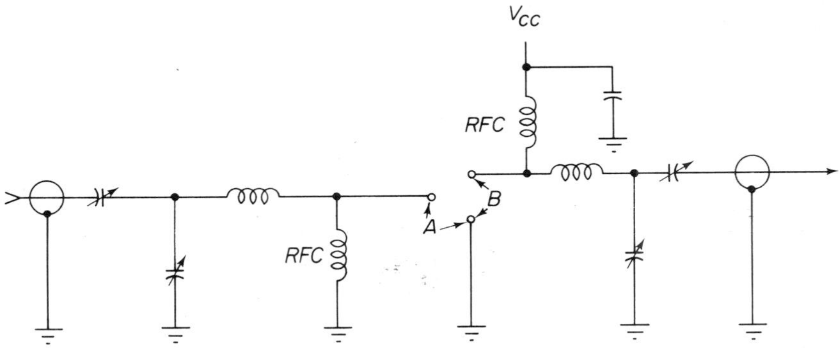

FIGURE 6-9 Typical test-amplifier circuit for two-junction transistors (values are for test at 175 MHz).

base circuit connection admittance is $7 + j3$, the input admittance of the transistor is $7 - j3$.

In some systems of two-junction transistor *RF*-amplifier design, the networks are calculated on the basis of input/output resistance and capacitance instead of admittance. In such cases, the admittances measured in the circuit of Fig. 6-9 must be converted to resistance and capacitance.

Admittances are expressed in mhos (or millimhos, mmhos). To find resistance, the real part of the admittance is divided into 1. To find capacitance, the imaginary part of the admittance is divided into 1 (to find reactance). Then the reactance is used in the equation $C = 1/(6.28FX_C)$ to find actual capacitance.

6-2.2 Stability factors

There are two factors used to determine the potential stability (or instability) of transistors in *RF* amplifiers. One factor is known as the Linvill

C factor; the other is the Stern k factor. Both factors are calculated from equations requiring y-parameter information (to be taken from datasheets or by actual measurement at the frequency of interest).

The main difference between the two factors is that the Linvill C factor assumes the transistor is not connected to a load; whereas the Stern k factor includes the effect of a specific load.

The Linvill C factor is calculated from

$$C = \frac{y_{12}y_{21}}{2g_{11}g_{22} - Re(y_{12}y_{21})}$$

If C is less than 1, the transistor is unconditionally stable; that is, using a conventional (unmodified) circuit, no combination of load and source admittance can be found that will cause oscillation. If C is greater than 1, the transistor is potentially unstable; that is, certain combinations of load and source admittance will cause oscillation.

The Stern k factor is caculated from

$$k = \frac{2(g_{11} + G_S)(g_{22} + G_L)}{y_{12}y_{21} + Re(y_{12}y_{21})}$$

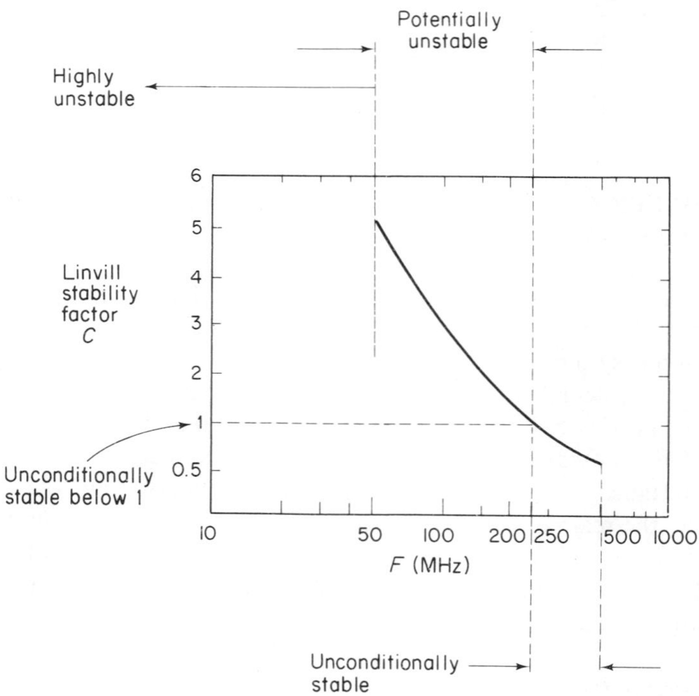

FIGURE 6-10 Linvill stability factor, C, for the MFE3007 between 50 and 500 MHz (courtesy Motorola).

where G_S and G_L are source and load conductances, respectively ($G_S =$ 1/source resistance; $G_L =$ 1/load resistance).

If k is greater than 1, the amplifier circuit is stable (opposite from Linvill). If k is less than 1, the amplifier is unstable. In practical design, it is recommended that a k factor of 3 or 4 be used rather than 1 to provide a margin of safety. This will accommodate parameter and component variations (particularly in regard to bandpass response).

Note that both equations are fairly complex and require considerable time for their solution (unless computer-aided design techniques are used). Some manufacturers provide alternate solutions to the stability and load-matching problems, usually in the form of a datasheet graph. Such a graph is shown in Fig. 6-10, which is a Linvill C factor chart for a typical MOSFET. Note that the MOSFET is unconditionally stable at frequencies above 250 MHz but potentially unstable at frequencies below 250 MHz. At frequencies below about 50 MHz, the MOSFET becomes highly unstable.

6-2.3 Solutions to stability problems

There are two basic solutions to the problem of unstable *RF* amplifiers.

First, the amplifier can be neutralized; that is, part of the output can be fed back (after it is shifted in phase) to the input so as to cancel oscillation. Neutralization permits the amplifier to be matched perfectly to the source and load. Such a match is known as a *conjugate match.* A perfect conjugate match means that transistor input and source, as well as the transistor output and load, are *matched* resistively and that *all reactance* is tuned out. Neutralization requires extra components and creates a problem when frequency is changed.

The other solution is to *introduce some mismatch* into either the source or the load tuning networks. This solution, sometimes known as the Stern solution, requires no extra components but does produce a reduction in gain.

A comparison of the two methods is shown in Fig. 6-11. The higher gain curve represents the unilateralized (or neutralized) operation. The lower gain curve represents the circuit power gain when the Stern k factor is 3.

Assume that the frequency of interest is 100 MHz. If the amplifier is matched directly to the load (perfect conjugate match), without regard to stability or with neutralization to produce stability, the top curve applies and the power gain is about 38 dB. If the amplifier is matched to a load and source where the Stern k factor is 3 (resulting in a mismatch with the actual load and source), the lower curve applies and the power gain is about 29 dB.

The upper curve of Fig. 6-11 is found by the *general power-gain equation:*

$$G_P = \frac{\text{power delivered to load}}{\text{power delivered to input}} = \frac{(y_{21})^2 G_L}{(Y_L + y_{22})^2 Re\left(y_{11} - \dfrac{y_{12}y_{21}}{y_{22} + Y_L}\right)}$$

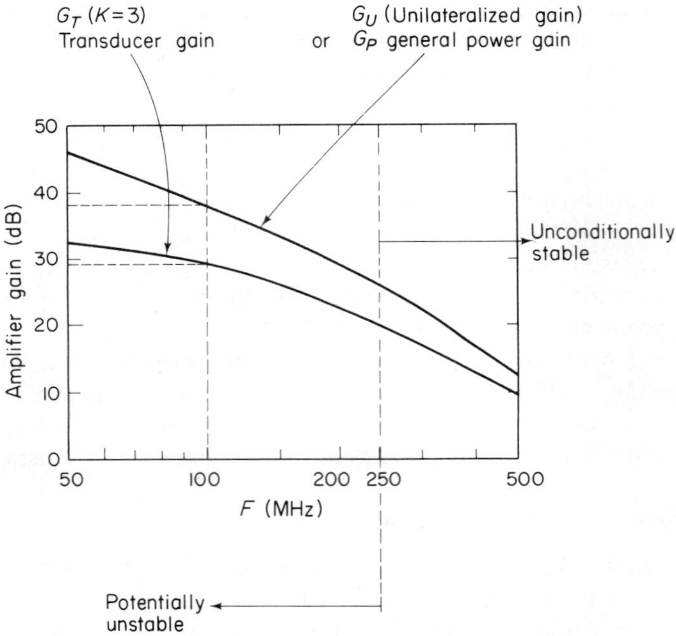

FIGURE 6-11 Amplifier gain characteristics for MFE3007 (courtesy Motorola).

The general power-gain equation applies to circuits with no external feedback and to circuits that have external feedback (neutralization), provided the composite y parameters of both transistor and feedback network are substituted for the transistor y parameters in the equation.

The lower curve of Fig. 6-11 is found by the *transducer-gain expression:*

$$G_T = \frac{\text{power delivered to load}}{\text{maximum power available from source}}$$

$$= \frac{4G_S G_L (y_{21})^2}{[(y_{11} + Y_S)(y_{22} + Y_L) - y_{12}y_{21}]^2}$$

The transducer-gain expression includes input mismatch. The lower curve of Fig. 6-11 assumes that the mismatch is such that a Stern k factor of 3 results; that is, the circuit tuning networks are adjusted for admittances that produce a Stern k factor of 3. The transducer-gain expression considers the input and output networks as part of the source and load.

With either gain expression, the input and output admittances of the transducer are modified by the load and source admittances.

The input admittance of the transistor is given by

$$Y_{\text{in}} = y_{11} - \frac{y_{12}y_{21}}{y_{22} + Y_L}$$

The output admittance of the transistor is given by

$$Y_{\text{out}} = y_{22} - \frac{y_{12}y_{21}}{y_{11} + Y_S}$$

At low frequencies, the second term in the input and output admittance equations is not particularly significant. At *VHF*, the second term makes a very significant contribution to the input and output admittances.

The imaginary parts of Y_S and Y_L (B_S and B_L, respectively) must be known before values can be calculated for power gain, transducer gain, input admittance, and output admittance. Except for some very special cases, exact solutions for B_S and B_L consist of time-consuming complex algebraic manipulations.

As a fairly good simplifying approximation for the equations, let $B_S \approx -b_{11}$ and $B_L \approx -b_{22}$ so that

General power-gain expression:

$$G_P \approx \frac{(y_{21})^2 G_L}{(G_L + g_{22})^2 Re\left(y_{11} - \dfrac{y_{12}y_{21}}{g_{22} + G_L}\right)}$$

Transducer-gain expression:

$$G_T \approx \frac{4G_S G_L (y_{21})^2}{[(g_{11} + G_S)(g_{22} + G_L) - y_{12}y_{21}]^2}$$

Input admittance:

$$Y_{\text{in}} \approx y_{11} - \frac{y_{12}y_{21}}{g_{22} + G_L}$$

Output admittance:

$$Y_{\text{out}} \approx y_{22} - \frac{y_{12}y_{21}}{g_{11} + G_S}$$

Two other gain expressions often found on the datasheets of transistors used in *RF* applications (particularly FET datasheets) are maximum available gain (*MAG*) and maximum usable gain (*MUG*).

MAG is usually applied as the gain in a conjugately matched, neutralized circuit and is expressed as

$$MAG = \frac{(y_{21})^2 R_{\text{in}} R_{\text{out}}}{4}$$

where R_{in} and R_{out} are the input and output resistances, respectively, of the transistor.

An alternate *MAG* expression is

$$MAG = \frac{(y_{21})^2}{4Re(y_{11})Re(y_{22})}$$

where $Re(y_{11})$ is the real part (g_{11}) of the input admittance and $Re(y_{22})$ is the real part (g_{22}) of the output admittance.

MUG is usually applied as the *stable gain* that may be realized in a *practical* (neutralized or unneutralized) *RF* amplifier. In a typical unneutralized FET circuit, *MUG* is expressed as

$$MUG \approx \frac{0.4y_{21}}{6.28Fc_{rss}}$$

where F is frequency in hertz and c_{rss} is reverse transfer capacitance.

MAG and *MUG* are often omitted on datasheets for two-junction power transistors. Instead, gain is listed as h_{fe} at a given frequency. This is supplemented with graphs that show available power output at given frequencies with a given input.

6-2.4 Neutralized solution

There are several methods for neutralization of *RF* amplifiers. The most common method is the *capacitance-bridge* technique shown in Fig. 6-12. Capacitance-bridge neutralization becomes more apparent when the circuit

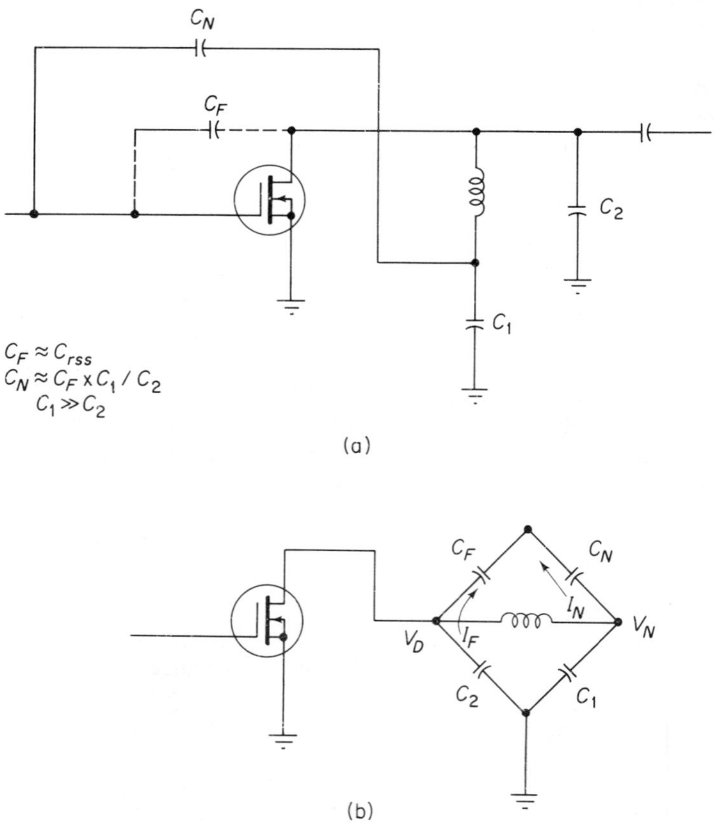

FIGURE 6-12 Capacitance-bridge neutralization circuit.

is redrawn as shown in Fig. 6-12(b). The condition for neutralization is that $I_F = I_N$.

The equations normally used to find the value of the feedback neutralization capacitor are long and complex. For practical work, however, if the value of C_1 is made quite large in relation to C_2 (at least four times), the value of C_N can be found by

$$C_N \approx C_F \frac{C_1}{C_2}$$

where C_F is the reverse capacitance of the transistor.

In simple terms, the value of C_N is approximately equal to the value of reverse capacitance times the ratio of C_1/C_2. For example, if reverse capacitance is 0.2 pF, C_1 is 30 pF and C_2 is 3 pF, the C_1/C_2 ratio is 10, and $C_N \approx 10 \times 0.2$ pF $= 2$ pF.

6-2.5 The Stern solution

A stable design with a potentially unstable transistor is possible without external feedback (neutralization) by proper choice of source and load admittances. This can be seen by inspection of the Stern k factor equation; G_S and G_L can be made large enough to yield a stable circuit, regardless of the degree of potential instability. Using this approach, a circuit stability factor k (typically $k = 3$) is selected, and the Stern k factor equation is used to arrive at values of G_S and G_L that will provide the desired k.

Of course, the actual G of the source and load cannot be changed. Instead, the input and output tuning circuits are designed as if the actual G values were changed. This results in a mismatch and a reduction in power gain, but it does produce the desired degree of stability.

To obtain a particular circuit stability factor, the designer may choose any of the following combinations of matching or mismatching of G_S and G_L to the transistor input and output conductances, respectively.

G_S matched and G_L mismatched.
G_L matched and G_S mismatched.
Both G_S and G_L mismatched.

Often a decision on which combination to use will be dictated by other performance requirements or practical considerations.

Once G_S and G_L have been chosen, the remainder of the design may be completed by using the relationships that apply to the amplifier without feedback. Power gain and input/output admittances may be computed by using the appropriate equations (Sec. 6-2.3).

Simplified Stern approach. Although the procedure above may be adequate in many cases, a more systematic method of source and load admittance determination is desirable for designs that demand maximum power gain per degree of circuit stability. Stern has analyzed this problem and developed equations for computing the best G_S, G_L, B_S, and B_L for a particular circuit stability factor (Stern k factor). Unfortunately, these equations are very complex and quite tedious if they must be done frequently. The complete Stern solution is best solved by computer.

Programs have been written to provide essential information for transistors used as *RF* amplifiers, including the effects of various specific sources and loads. These programs permit the designer to experiment with theoretical "breadboard" circuits in a matter of seconds.

When a Stern solution must be obtained *without the aid of a computer*, it is best to use one of the many shortcuts that have been developed over the years. The following shortcut is by far the simplest and most widely accepted; yet it provides an accuracy close to that of the computer solutions.

1. Let $B_S \approx -b_{11}$ and $B_L \approx b_{22}$, as in the case of the Sec. 6-2.3 equations. This method permits the designer to approximate the exact Stern solution closely for Y_S and Y_L, while avoiding the portion of the computations that is the most complex and time consuming. Further, the circuit can be designed with tuning adjustments for varying B_S and B_L, thereby creating the possibility of finding true B_S and B_L (by experiment) for maximum gain as accurately as if all the Stern equations had been solved.
2. Mismatch G_S to g_{11} and G_L to g_{22} by an *equal ratio*; that is, find a ratio that produces the desired Stern k factor, and then mismatch G_S to g_{11} (and G_L to g_{22}). For example, if the ratio is 4 : 1, make G_S four times the value of g_{11} (and G_L four times the value of g_{22}).

If the mismatch ratio R is defined as

$$R = \frac{G_L}{g_{22}} = \frac{G_S}{g_{11}}$$

then R may be computed for any particular circuit stability k factor using the equation

$$R = \left\{\sqrt{k\left[\frac{y_{21}y_{12} + Re(y_{12}y_{21})}{2g_{11}g_{22}}\right]}\right\} - 1$$

As an example, assume that it is desired to mismatch input and output circuits so that there is a Stern k factor of 4, using a transistor that has the following characteristics:

$$y_{12}y_{21} = 0.545, \qquad g_{11} = 4.95, \qquad g_{22} = 0.05,$$
$$Re(y_{12}y_{21}) = 0.258$$

(all values in mmhos).

$$R = \left\{\sqrt{4\left[\frac{0.545 + 0.258}{2(4.95)(0.05)}\right]}\right\} - 1 \approx 1.56$$

Using the value of 1.56 for R and the equation

$$1.56 = \frac{G_S}{g_{11}} = \frac{G_L}{g_{22}}$$

then

$$G_S = (1.56)(4.95)(10^{-3}) = 7.72 \text{ mmhos}$$

and

$$R_S = \frac{1}{G_S} = 130\ \Omega$$

$$G_L = (1.56)(0.05)(10^{-3}) = 0.078 \text{ mmho}$$

and

$$R_L = \frac{1}{G_L} = 12{,}800\ \Omega$$

The shortcut Stern method may be advantageous if the source and load admittances and power gains for several different values of k are desired. Once the R for a particular k has been determined, the R for any other k may be quickly found from the equation

$$R = \frac{(1 + R_1)^2}{(1 + R_2)^2} = \frac{k_1}{k_2}$$

where R_1 and R_2 are values of R corresponding to k_1 and k_2, respectively.

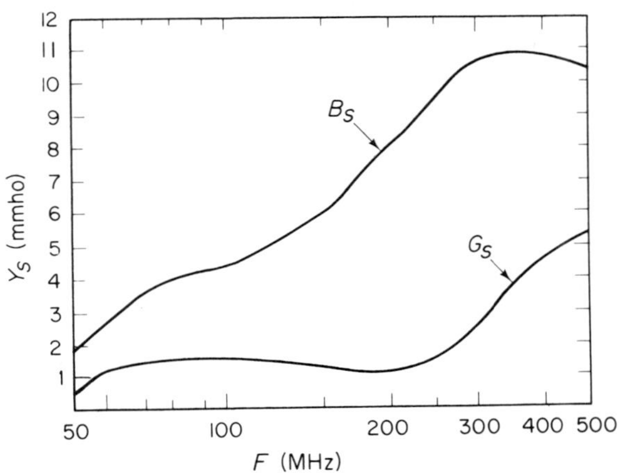

FIGURE 6-13 Best source admittance, $Y_S = G_S + jB_S$ (courtesy Motorola).

The Stern solution with datasheet graphs. It is obvious that the Stern solution, even with the shortcut method, is somewhat complex. For this reason, some manufacturers have produced datasheet graphs that show best source and load admittances for a particular transistor over a wide range of frequencies. Examples of these graphs are shown in Fig. 6-13 and 6-14.

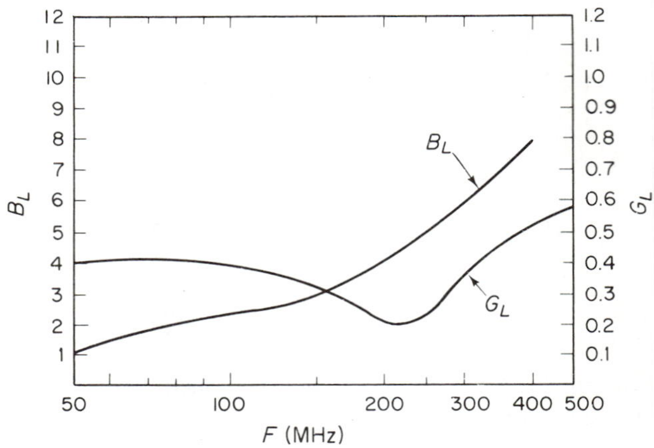

FIGURE 6-14 Best load admittance, $Y_L = G_L + jB_L$ (courtesy Motorola).

Figure 6-13 shows both the real (G_S) and imaginary (B_S) values that will produce maximum gain, but with a stability (Stern k) factor of 3, at frequencies from 50 to 500 MHz. Figure 6-14 shows corresponding information for G_L and B_L. To use these illustrations, simply select the desired frequency, and note where the corresponding G and B curves cross the frequency line. For example, assuming a frequency of 100 MHz, $Y_L = 0.35 - j2.1$ mmhos and $Y_S = 1.3 - j4.4$ mmhos.

If tuning circuits are designed to match these admittances rather than the actual admittances of the source and load, the circuit will be stable. Of course, the gain will be reduced. Use the *transducer-gain expression* G_T of Sec. 6-2.3 to find the resultant power gain.

Index